Mineralogy

Edited by Miloš René

Published in London, United Kingdom

Mineralogy
http://dx.doi.org/10.5772/intechopen.97912
Edited by Miloš René

Contributors
Mustapha Hidouri, Khouloud Kthiri, Mohammed Mehnaoui, Samira Jebahi, Khaled Boughzala, Zizheng Wang, Xin-Sheng Jiang, Bang-Guo Zhou, Sheng-Lin Lu, Sergey I. Kostrovitsky, Robert Cooper Liebermann, Yamei Li, Lidong Dai, Haiying Hu, Wenqing Sun, Yu He, Aziz El Kassimi, Mohammdine El Haddad, Rachid Laamari, Mamoune El Himri, Hicham Yazid, Youness Achour, Miloš René, Muhammad Imran Rashid, Hans Wolfgang Wagner, Takafumi Miyanaga, Reki Nakamura, Yoshiki Oka, Saya Okita, Kazuma Yamauchi, Yushi Suzuki, Dereje Ayalew, Bekele Abebe, Praise Akisanmi

First published in London, United Kingdom, 2022 by IntechOpen
IntechOpen is the global imprint of INTECHOPEN LIMITED, registered in England and Wales, registration number: 11086078, 5 Princes Gate Court, London, SW7 2QJ, United Kingdom

British Library Cataloguing-in-Publication Data
A catalogue record for this book is available from the British Library

Additional hard and PDF copies can be obtained from orders@intechopen.com

Mineralogy
Edited by Miloš René
p. cm.
Print ISBN 978-1-80355-465-5
Online ISBN 978-1-80355-466-2
eBook (PDF) ISBN 978-1-80355-467-9

For EU product safety concerns:
IN TECH d.o.o., Prolaz Marije Krucifikse Kozulić 3, 51000 Rijeka, Croatia,
info@intechopen.com or visit our website at intechopen.com.

Meet the editor

Dr. Miloš René, Ph.D., is a scientist at the Institute of Rock Structure and Mechanics, Czech Academy of Sciences. He obtained his MSc in Mineralogy and his Ph.D. in Economic Geology from Charles University, Prague, Czech Republic, in 1968 and 1981, respectively. He has published about eighty-five scientific papers in peer-reviewed journals in the fields of mineralogy, petrology, geochemistry, and economic geology. His recent scientific activity is concentrated on petrology and geochemistry of granitic rocks, mineralogy of accessory minerals, and mineralogy of uranium deposits, especially from the Central European Variscan belt.

Contents

Preface

Mineralogy is a branch of geosciences that deals with crystal structures, physical properties, and chemical composition of minerals in various igneous, metamorphic, and sedimentary rocks and processes, from high to low temperatures and in various parts of Earth.

Mineralogy is also an important background for some other geosciences and of significance for many other scientific and technical disciplines including environmental science, the mining and materials industry, and biomedical science.

Mineralogy includes thirteen chapters that discuss the methodology of specific mineralogical methods, the composition of minerals from different igneous rocks, and the composition of minerals from different sedimentary rocks.

Chapter 1, "Simultaneous Time-Resolved Photoluminescence and X-Ray Absorption Fine Structure Operando Measurement during Ag Cluster Formation in Ag Zeolite X," by Miyanaga Takafumi et al. presents a detailed study of synthetic zeolites.

Chapter 2, "Some New Progress in the Experimental Measurements on Electrical Property of Main Minerals in the Upper Mantle at High Temperatures and High Pressures," by Dai Lidong et al. is a detailed methodological study of electrical properties of selected silicate minerals occurring in the upper mantle.

Chapter 3, "Mineral Physics," by Robert Lieberman focuses on a detailed methodology of the physical properties of minerals.

Chapter 4, "Testing and Validating Instruments for Feedstocks of Mineral Carbonation" by Rashid Muhammad Imran discusses measurement techniques used for carbon capture and storage in different minerals and rocks.

Chapter 5, "Normative Mineralogy Especially for Shales, Slates, and Phyllites," by Wolfgang H. Wagner provides a new method for normative recalculation of chemical analyses of different low-grade metasedimentary rocks.

Chapter 6, "Minerals as Prebiotic Catalysts for Chemical Evolution towards the Origin of Life," by Yamei Li presents a detailed methodological study of the oldest, prebiotic evolution of the Earth.

Chapter 7, "Ionic Conductivity of Strontium Fluoroapatites Co-doped with Lanthanides" by Khouloud Kthiri et.al discusses the preparation of fluoroapatites co-doped with selected lanthanides.

Chapter 8, "Mg-Ilmenite from Kimberlites, Its Origin" by Sergey I. Kostrovitsky contains detailed chemical analyses of different ilmenites occurring in kimberlite pipes in the Yakutian province (Siberian Craton).

Chapter 9, "Investigation of Accessory Minerals from the Blatná Granodiorite Suite, Bohemian Massif, Czech Republic," by Miloš René discusses selected accessories from significant granitic batholith of the Central European Variscan belt.

Chapter 10, "Mineralogy of Peralkaline Silicic Volcanics: Information from Kone Volcano, Ethiopian Rift Valley" by Ayalew Dereje and Abebe Bekele contains detailed descriptions of selected minerals from volcanic rocks of the East African rift system in Ethiopia.

Chapter 11, "Use of Natural Safiot Clay for the Removal of Chemicals Substances from Aqueous Solutions by Absorption: A Combined Experimental and Theoretical Study," by El Kassimi Aziz et al. discusses the use of selected clay minerals for removing different pollutants from aqueous solutions.

Chapter 12, "Classification of Clay Minerals," by Akisanmi Praise contains a detailed classification and description of clay minerals.

Chapter 13, "Sedimentary Characteristics of the Member 1 of the Middle Devonian Suotoushan Formation and Its Ore-Bearing Potential in Northeastern Yunnan" by Zizheng Wang et al. describes high-purity quartz sandstone deposits from China.

Miloš René
Institute of Rock Structure and Mechanics,
Czech Academy of Sciences,
Prague, Czech Republic

Section 1

Mineralogical Methods

Chapter 1

Simultaneous Time-Resolved Photoluminescence and X-Ray Absorption Fine Structure Operando Measurement during Ag Cluster Formation in Ag Zeolite X

Yushi Suzuki, Takafumi Miyanaga, Kazuma Yamauchi, Saya Okita, Yoshiki Oka and Reki Nakamura

Abstract

We use operando X-ray absorption fine structure (XAFS) to analyze the relation between the properties of photoluminescence (PL) and the structures of Ag clusters and Ag ions. The Ag clusters are generated by evacuation in the cavity of Ag-type zeolite-X. The Ag clusters in the zeolite cavity collapse when exposed to the atmosphere. The results reported herein indicate that the collapsing Ag cluster plays an important role in generating strong PL bands and that Ag clusters might not be a direct species of PL. Results of XAFS analysis show that the Ag cluster formed in the zeolite cavity by evacuation can be tetrahedral with four atoms. By evacuation, 9 or 10 Ag tetrahedral are formed, two of which are expected to be responsible for strong PL. This result suggests that the Ag ion position after cluster collapse plays an important role in PL band generation and that Ag clusters are not direct luminescent species of PL.

Keywords: zeolite, Ag cluster, photoluminescence, XAFS, operando measurement

1. Introduction

Zeolites are crystalline aluminosilicates with large ordered cavities (cages) that exhibit unique properties because of their cages [1–4]. They are used widely in applications such as catalysis, ion exchange, and separation [5–9].

The fact that Ag-exchanged zeolites exhibit luminous properties despite containing no rare earth metal component indicates zeolite as a promising luminescent material for future use. In recent years, many researchers have investigated the luminescence behavior of silver-exchanged zeolites [10–18]. Many reports have described that Ag clusters formed with a zeolite framework are luminescent species. In studies using inorganic material matrices such as silica, light emission from silver clusters formed in the matrix has been confirmed [19, 20]. However, the existence of Ag clusters in PL studies of zeolites has been confirmed only indirectly. In earlier studies, PL and XAFS measurements of zeolites were taken to confirm the relation between PL bands and the local structure of Ag clusters [21, 22]. Results confirmed that Ag clusters were formed during heat treatment processing, however, the Ag cluster decomposed when cooled to room temperature (RT). The XAFS spectra of the unheated sample and the sample cooled after heating were very similar. The Ag clusters formed during heating but collapsed after cooling.

To elucidate the PL mechanism of Ag zeolite, it is fundamentally important to elucidate the relation between the formation and collapse processes of Ag clusters and PL. Accurate PL cannot be observed by thermal excitation. Therefore, it is impossible to observe this process by heat treatment. Ag clusters can be formed not only by heat treatment but also by evacuation. In the case of evacuation, clusters can form at RT. Therefore, as described herein, Ag clusters are formed by evacuation; then the clusters are destroyed by introducing air. In these processes, the relation between cluster formation/collapse and PL is elucidated by the time-resolved operando measurement of PL and XAFS.

For these purposes, we developed an in-situ/operand measurement cell, which can be possible to measure XAFS and PL simultaneously. In this chapter, we discuss the result for time-resolved operando measurement of PL and structure change measured by Quick measurement mode-XAFS.

2. Experimental

2.1 Sample preparation

Fully Ag^+-exchanged zeolite X (hydrated 86Ag-X) powder samples were prepared by immersing the hydrated 86Na-X ($Na_{86}[(AlO_2)_{86}(SiO_2)_{106}]$, purchased from Tosoh Corp.) in 0.1 M $AgNO_3$ solution for 24 h at 25°C. The solution was stirred continuously with a magnetic stirrer during ion exchange. After careful filtration, the 86Ag-X powder was dried under air at RT in the darkroom.

2.2 PL measurements

The PL measurements were taken using a UV–VIS spectrometer (flame; Ocean Insight) and a 313 nm UV light source (UVF-203S; San-Ei Electric) with a bandpass filter (313 nm, Edmund Optics Inc.).

2.3 XAFS measurements

The X-ray absorption spectra of *K*-edge of Ag (25.5 keV) were measured at NW10A of the Photon Factory at KEK using transmission mode. A time-resolved operando measurement of Quick-XAFS was performed during the evacuation process and the air introduction process. In addition, step scan XAFS measurements with high spectral quality were taken in the atmosphere before evacuation, in a vacuum, and after introduction into the atmosphere. The EXAFS oscillation function was extracted from the X-ray absorption spectra and Fourier transformed by XANADU code [23]. To obtain the structural parameters, the EXAFS function was fitted in *k*-space using the nonlinear least-squares method with theoretical parameters calculated using FEFF 8.10 [24]. Experimental details are presented in reports of several earlier studies [25, 26].

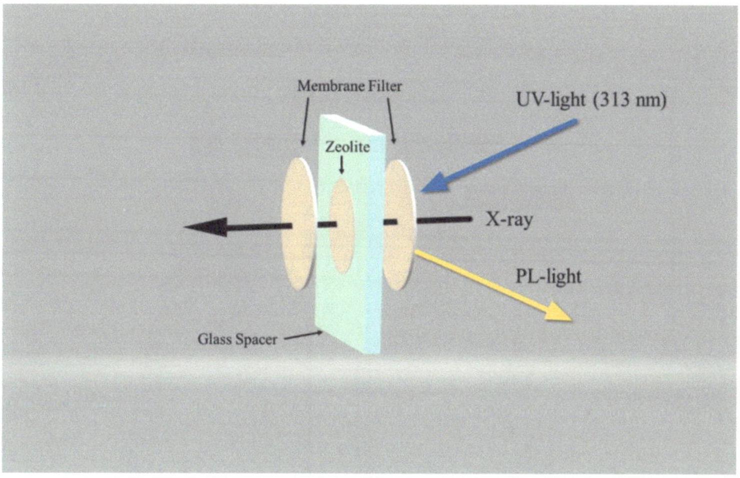

Figure 1.
Schematic drawing of zeolite sample holder for simultaneous time-resolved PL and XAFS operando measurements.

2.4 Simultaneous time-resolved operando measurement of PL and quick-XAFS

The transmission method must be adopted to ensure the spectrum quality by Quick-XAFS. Performing the transmission method with PF-AR NW10A requires the setting of the sample vertically because of the beam-line structure. A cover must have necessary powdery zeolite material installed vertically. Earlier studies have used microscope coverslips [26]. However, the inflow and outflow of air from the front are interrupted in the case of glass. The inflow and outflow of air only from the side face take a long time, especially in the process of introducing air. To date, long-term evacuation and atmospheric exposure have been used to avoid this shortcoming. However, time-resolved measurement has no such time margin. Therefore, a PTFE membrane filter (0.5 μm, T050A025A; Advantec) was used for the cover. With a membrane filter, air can enter and exit from the sample surface at a sufficiently high speed. For the sample, a prepared glass plate with a hole was used as a spacer for filling. The simultaneous time-resolved operando measurement of PL and Quick-XAFS has been realized. **Figure 1** portrays a schematic diagram. The time resolution of PL measurements and Quick-XAFS measurements was set to 1 min.

3. Results of PL and XAFS spectra

First, we discuss the PL measurement results. **Figure 2** shows a time-resolved operando measured PL curve of Ag-zeolite in the evacuation process (a) and the air introduction process (b). In both cases, a PL band having a peak near 556 nm is observed. When the evacuation was started, the PL band intensity decreased sharply and then decreased gradually. Similarly, during the air introduction process, the PL band intensity recovered rapidly in the early stage. Subsequently, it increased gradually.

Figure 3 shows the normalized operando time-resolved Quick-XAFS measurement spectra obtained for the Ag-X zeolite sample with step scan XAFS spectra in

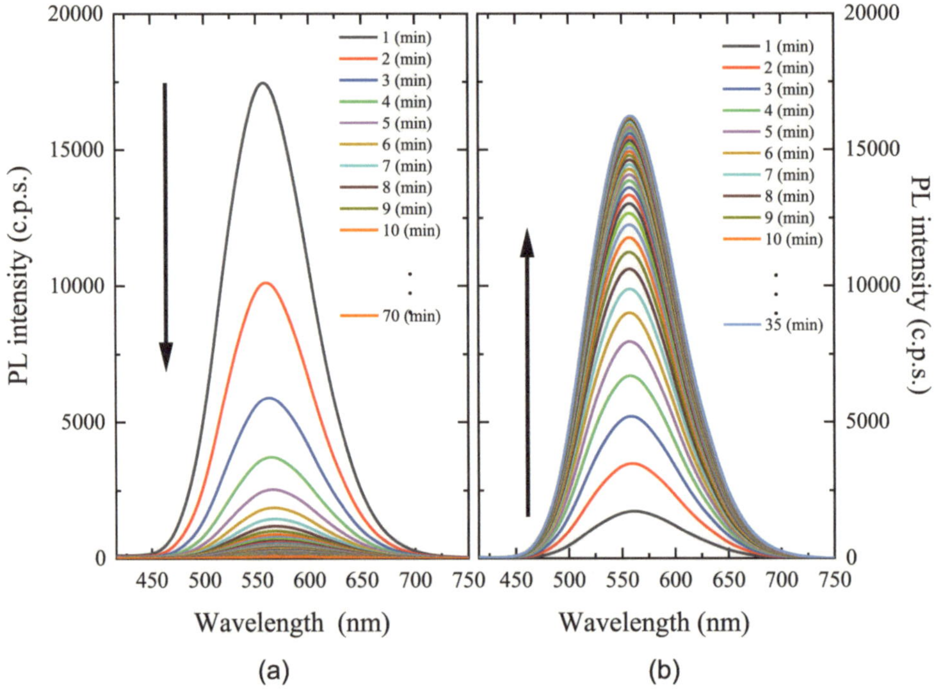

Figure 2.
Evolution of the photoluminescence spectra of 86Ag-X: (a) evacuation process and (b) air introduction process.

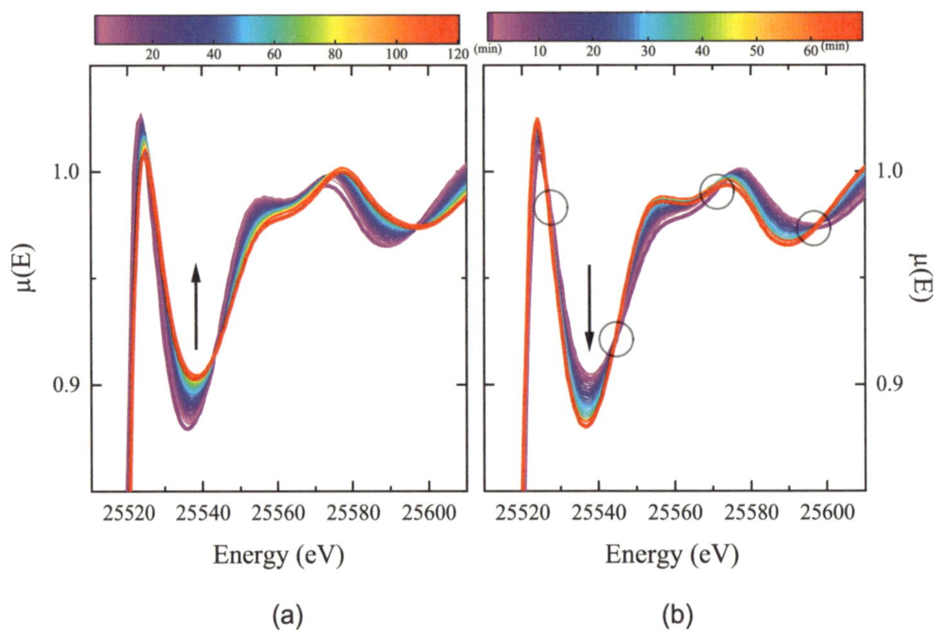

Figure 3.
Evolution of the X-ray absorbance spectra of 86Ag-X: (a) evacuation process and (b) air introduction process.

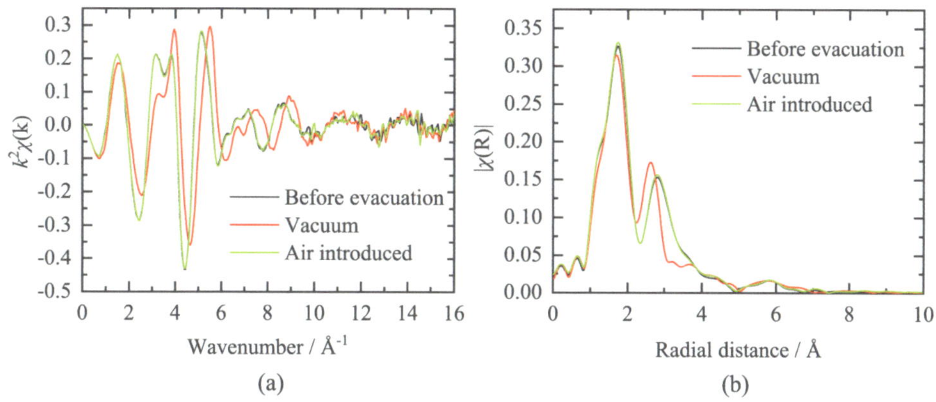

Figure 4.
(a) The k^2 $\chi(k)$ spectra and (b) their Fourier transforms of 86Ag-X.

the evacuation process (a) and the air introduction process (b). In both the evacuation process and the atmosphere introduction process, the XAFS spectrum changes continuously from the initial state to the final state. It is exciting that well-defined isosbestic points (e.g., marked by circles in **Figure 3(B)**) exist, which indicates that no clear intermediate state exists. The transient spectrum is a mixed spectrum of the initial sample and the final sample [27].

Figure 4 shows (a) the $k^2\chi(k)$ spectra and (b) their Fourier transforms for before and after evacuation and air introduction. To elucidate the structural parameters around Ag atom, we applied least-squares fitting (curve-fitting) to XAFS data. These results are discussed later.

4. Discussion

To reproduce XAFS spectrum of the intermediate state, a linear combination of the step scan spectra under atmospheric and vacuum conditions was made. The XAFS spectra simulated from the linear combination of them is shown in **Figure 5**.

Least-squares method was then used to ascertain which intermediate state spectra could be approximated by which linear combined spectra. Based on the results, we were able to plot the proportion of the XAFS spectrum in the atmosphere as a function of the evacuation time as shown in **Figure 6** (as open circle). In this figure, the integrated intensity of the PL band measured simultaneously is also shown (closed circles). Similarly, the results obtained during the process of introducing air are depicted in **Figure 7**.

In **Figure 6**, whereas the XAFS spectrum changes almost linearly from the initial state to the final state, the PL intensity decreases immediately after the start of evacuation. At 8 min after the start of evacuation, the change in XAFS is about one-fifth from the initial state, whereas the PL intensity has decreased to 1/12.5. In other words, the PL spectrum intensity change speed preceded the local structure change speed. No simple correlation was found between the speeds of both changes. This finding suggests that the PL almost disappears when about 20% of the Ag in the Ag-X zeolite changes from the initial state. However, in the process of introducing the atmosphere

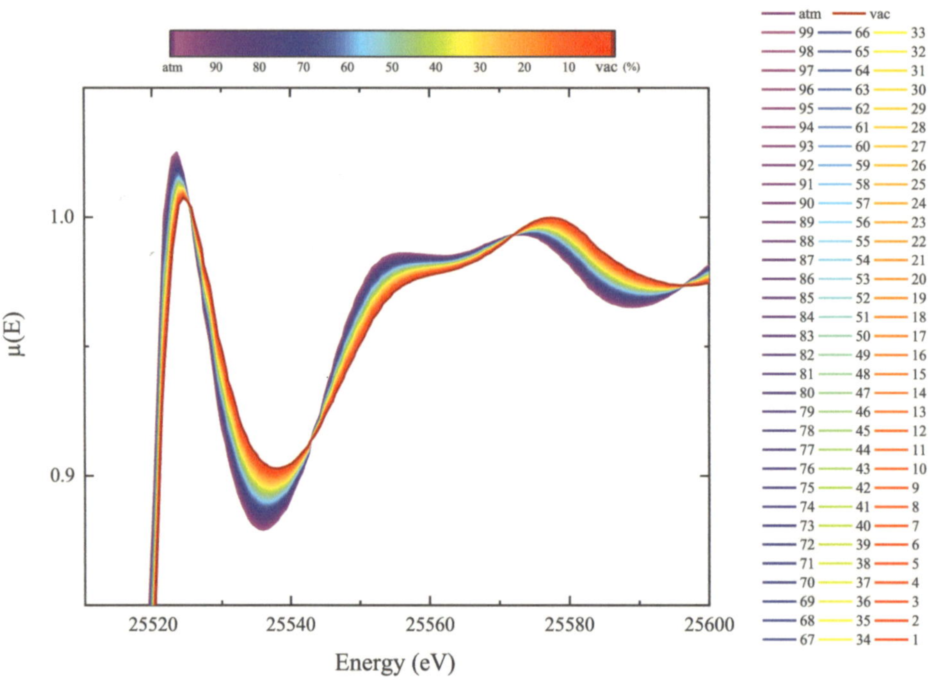

Figure 5.
Linear combination of step scan XAFS spectra under atmospheric and vacuum conditions.

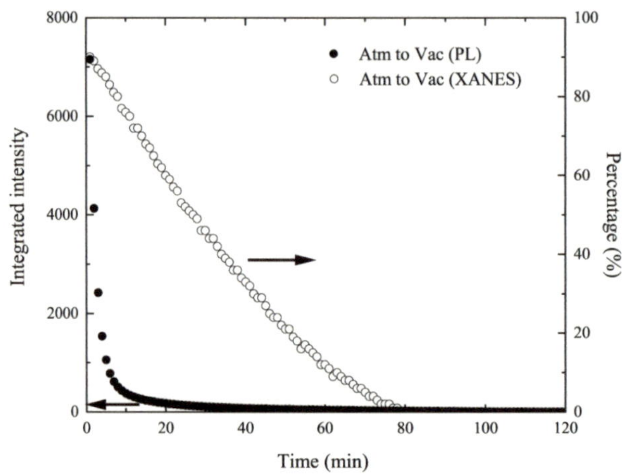

Figure 6.
Evolution of the PL intensity (solid circle) and the proportion of the XAFS spectrum (open circle) in the evacuation process as a function of the evacuation time.

(**Figure 7**), the XAFS changes almost linearly, whereas the change in PL intensity is somewhat complicated. The PL increases rapidly for about the first 10 min after the introduction of the atmosphere. Thereafter, the PL intensity increases slightly until about 35 min. Subsequently, the PL intensity again shows a gradual decrease. The change in the XAFS spectrum ends approximately 40 min after introduction into the atmosphere. Even in the process of introducing the atmosphere, the change in PL intensity and the change in the XAFS spectrum do not match. The XAFS spectrum

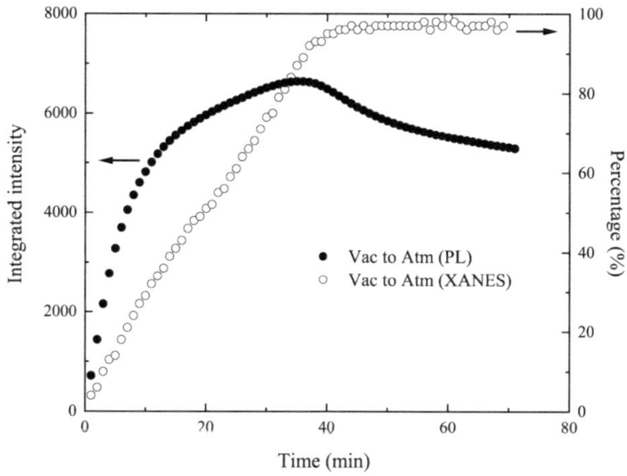

Figure 7.
Evolution of the PL intensity (solid circle) and the proportion of the XAFS spectrum (open circle) in air introduction processes as a function of the introduction time.

shows a 20% change after about 7 min, at which time the PL intensity had reached about 60% of the maximum intensity.

The **Table 1** presents the structural parameters obtained from the curve-fitting to Ag *K*-edge XAFS. Especially, we specifically examine the coordination number ($N_{Ag\text{-}Ag}$) to Ag-Ag atom in the zeolite.

The values of $N_{Ag\text{-}Ag}$ of Ag-X zeolite in air and vacuum are, respectively, 1.55 and 2.07. The Ag species which might be present in Ag-X include the following: isolated Ag ($N_{Ag\text{-}Ag} = 0$), pairs ($N_{Ag\text{-}Ag} = 1$), linear (or bent) triples ($N_{Ag\text{-}Ag} = 1.33$), triangles ($N_{Ag\text{-}Ag} = 2$), tetrahedrons ($N_{Ag\text{-}Ag} = 3$), and octahedrons ($N_{Ag\text{-}Ag} = 4$).

Now we consider the models of Ag clusters produced in the zeolite cavity in air and vacuum deduced from EXAFS results. Several possible combinations can satisfy both the coordination number of Ag-X in the atmosphere ($N_{Ag\text{-}Ag} = 1.55$) and the coordination number in vacuum ($N_{Ag\text{-}Ag} = 2.07$), such as "pair and triangle", "pair and tetrahedron", and "triple and triangle". Among them, it is considered that the change from the triple ($N_{Ag\text{-}Ag} = 1.33$) to the triangle ($N_{Ag\text{-}Ag} = 2$) can simply represent the change from 1.55 to 2.07. However, no direct conversion occurs from triple to triangle. For example, if one atom of a triple forms a triangle with another triple, a pair is also formed simultaneously, which is inconsistent with the existence of an isosbestic point. Based on the findings presented above, the simple combination most likely to satisfy the coordination number of Ag-X zeolite is considered as the combination of a pair and tetrahedron. When two pairs are twisted and the distance among the four atoms is coordinated, a tetrahedron is formed. A schematic drawing is presented in **Figure 8**.

Unheated sample	r_{O1}(Å)	N_{O1}	σ_{O1}(Å)	r_{O2}(Å)	N_{O2}	σ_{O2}(Å)	r_{Ag}(Å)	N_{Ag}	σ_{Ag}(Å)
in atmosphere	2.37	3.6	0.14	3.05	0.5	0.08	2.97	1.55	0.15
in vacuum	2.29	3.4	0.14	2.94	0.4	0.1	2.84	2.07	0.15

Table 1.
Structural parameters r, N and σ for Ag_{86}-X for unheated samples in atmosphere and vacuum.

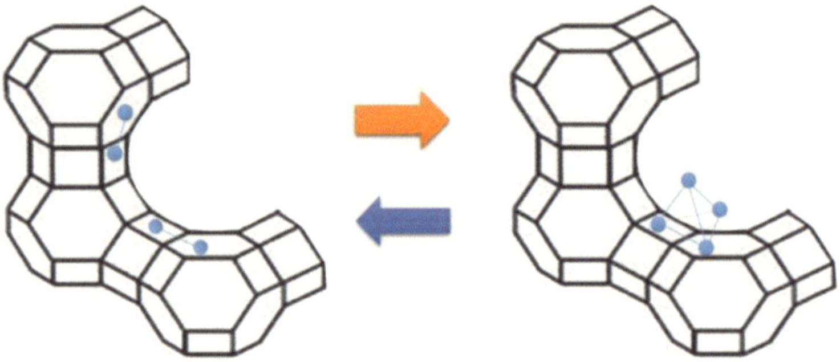

Figure 8.
Schematic drawing of mutual conversion model between pairs and tetrahedron.

Gonzales et al. assigned the double positively charged $Ag_4 (H_2O)_4$ and $Ag_4 (H_2O)_2$ clusters to the PL emission species by XEOL measurement [28]. However, the light observed by XEOL is an emission band excited by an energy of about 25.5 keV. No evidence exists that it is the same species as PL species excited by UV–Vis light. We measured PL and XAFS simultaneously and clarified that the formed cluster was not a PL luminescent species.

Assuming that the ratio of the number of pairs to tetrahedra is 31:6, then the average coordination number is 1.558, which is approximately equal to the coordination number in the atmosphere of 1.55. However, if the ratio of the number of pairs to tetrahedrons is 21:11, then the average coordination number is 2.023, which is close to 2.07 in a vacuum. From these results, a model is conceivable by which evacuation increases the tetrahedral clusters from 6 to 11. The pairs decrease from 31 to 21: about 10 pairs (20 Ag ions) of Ag change into tetrahedrons when Ag clusters are formed. Results presented in **Figures 5** and **6** show that only 2 Ag pairs (4 Ag ions) form clusters, causing a rapid decrease in PL. In addition, during the collapse process, only one tetrahedron becomes two pairs. The PL increases rapidly. This model, which can clearly represent the change between the two states of Ag cluster formation and collapse, is called the "pair and tetrahedron model" (PTM).

Importantly, PL is lost when clusters are formed by evacuation (**Figure 6**). This result indicates that the cluster formed by the evacuation is not a PL luminescent species. In Ag-A, when Ag clusters are formed by heat treatment and are then collapsed, the PL intensity is increased considerably, but no change in local structure is observed by XAFS [21, 22]. These facts suggest that a "metastable site contributing to light emission" exists very close to the "stable site". Unlike Ag-X, Ag-A and Ag-Y show almost no PL when not heated [29]. This phenomenon is explainable by considering that Ag-X has 86 Ag per unit cell, which is higher than other types (Ag-A = 12, Ag-Y = 51.2), and that Ag is located not only at stable sites but also at metastable sites. Then, in the cluster formation process, Ag that is coordinated to an unstable metastable site is consumed preferentially for cluster formation. Consequently, PL is lost. However, in the cluster collapse process, PL recovers rapidly because the cluster preferentially coordinates to metastable sites as the cluster collapses (**Figure 7**, 10 min). It is conceivable that the two pairs coordinated to the metastable site move to the more "stable site" over time. Subsequently, the collapsed Ag coordinates to this vacant metastable site. The Ag exchange is considered to occur at the metastable site. In other words, the increase in Ag at the metastable site becomes slower by the amount of Ag moving to the stable site. Accordingly, the PL intensity increases gradually (**Figure 7**, 10–35 min). The change in XAFS has stopped in 40 min. The collapse of the Ag cluster is completed in 40 min. The decrease in PL intensity after 40 min is explainable by considering that Ag is moving gradually from the metastable site to the stable site. Considering the existence of metastable sites contributing to PL, one can explain the phenomenon by which PL disappears because of the formation of a small amount of Ag cluster. The PL intensity increases considerably because of the slight collapse.

5. Conclusion

This study was conducted to elucidate the PL mechanism of Ag-type zeolite by the construction of a model that can explain the PL expression mechanism. For that purpose, XAFS measurements that can analyze local structure are combined with PL measurements. Simultaneous operando time-resolved measurements of PL and transmission Quick-XAFS during evacuation and introduction to the atmosphere were taken to elucidate the formation and collapse processes of Ag clusters. Based on those findings, we attempted to ascertain the PL mechanism. From results of this experiment, the formation and collapse processes of Ag clusters in Ag-X zeolite are explainable by Ag species of two types: a pair and a tetrahedron. Then we confirm that the Ag cluster is not related directly to light emission. We predict the existence of metastable sites in zeolite as PL emission species. Furthermore, we propose a PTM that can explain only two pairs of Ag (4 Ag) that contribute to PL.

Acknowledgements

Synchrotron radiation experiments were conducted at the Photon Factory at KEK under Proposals 2016G056, 2018G070, and 2020G070. This work was supported by JSPS KAKENHI Grant Number JP17K05026, JP21K04858 and JP20K05295.

Data availability

The data that support the findings of this study are available from the corresponding author upon reasonable request.

Author details

Yushi Suzuki*, Takafumi Miyanaga, Kazuma Yamauchi, Saya Okita, Yoshiki Oka and Reki Nakamura
Department of Mathematics and Physics, Graduate School of Science and Technology, Hirosaki University, Hirosaki, Aomori, Japan

*Address all correspondence to: uc@hirosaki-u.ac.jp

References

[1] Kim Y, Seff K. The octahedral hexasilver molecule. Seven crystal structures of variously vacuum-dehydrated fully silver(1+)-exchanged zeolite A. Journal of the American Chemical Society. 1978;**100**:6989

[2] Miyanaga T, Hoshino H, Endo H. Local structure of silver clusters in the channels of zeolite 4A. Journal of Synchrotron Radiation. 2001;**8**:557

[3] Miyanaga T, Hoshino H, Endo H, Sakane H. XAFS study of silver clusters in zeolites. Journal of Synchrotron Radiation. 1999;**6**:442

[4] Sun T, Seff K. Silver clusters and chemistry in zeolites. Chemical Reviews. 1994;**94**:857

[5] Corma A, Corell C, Perez-Pairente J. Synthesis and characterization of the MCM-22 zeolite. Zeolites. 1995;**15**:2

[6] Gomez RS, Li X, Yson RL, Patterson HH. Zeolite-supported silver and silver–iron nanoclusters and their activities as photodecomposition catalysts. Research on Chemical Intermediates. 2011;**37**:729

[7] Mohamed RM, Mkhalid LA, Abdel Salam M, Barakat MA. Zeolite Y from rice husk ash encapsulated with Ag-TiO_2: Characterization and applications for photocatalytic degradation catalysts. Desalination and Water Treatment. 2013;**51**:7562

[8] Sherry HS. Ion Exchange, in Handbook of Zeolite Science and Technology. In: Auerbach SM, Carrado KA, Dutta PK, editors. New York: Marcel Dekker; 2003. p. 1007

[9] Zhang SG, Ariyuki M, Mishima H, Higashimoto S, Yamashita H, Anpo M. Photoluminescence property and photocatalytic reactivity of V-HMS mesoporous zeolites Effect of pore size of zeolites on photocatalytic reactivity. Microporous and Mesoporous Materials. 1998;**21**:621

[10] Lin H, Imakita K, Fujii M. Reversible emission evolution from Ag activated zeolite Na-A upon dehydration/hydration. Applied Physics Letters. 2014;**105**:211903

[11] Johan E, Yamauchi Y, Matsue N, Itagaki Y, Aono H. Preparation of rare-earth-free luminescent material from partially Ag^+-exchanged zeolite X. Journal of the Ceramic Society of Japan. 2016;**124**(1):70-73

[12] Gonzalez EC, Baelelant W, Grandjean D, Roeffaer MBJ, Fron E, Aghakhani MS, et al. Thermally activated LTA(Li)–Ag zeolites with water-responsive photoluminescence properties. Journal of Materials Chemistry C. 2015;**3**:11857

[13] Gui SCR, Lin H, Bao W, Wang W. Effect of Annealing Temperature on Broad Luminescence of Silver-Exchanged Zeolites Y and A. Journal of Applied Spectroscopy. 2018;**85**:232

[14] Lee SH, Kim Y, Seff K. Weak Ag^+–Ag^+ bonding in zeolite X. Crystal structures of $Ag_{92}Si_{100}Al_{92}O_{384}$ hydrated and fully dehydrated in flowing oxygen. Microporous and Mesoporous Materials. 2000;**41**:49

[15] Lin H, Imakita K, Fujii M, Prokof'ev VY, Gordina NE, Said B, et al. Visible emission from Ag^+ exchanged SOD zeolites. Nanoscale. 2015;7:15665

[16] Lin H, Imakita K, Fujii M. Reversible emission evolution from Ag activated

zeolite Na-A upon dehydration/hydration. Applied Physics Letters. 2014;**105**:211903

[17] Seifert R, Rytz R, Calzaferri G. Colors of Ag^+-Exchanged Zeolite A. The Journal of Physical Chemistry. 2000;**104**:7473

[18] Seifert R, Kunzmann A, Calzaferri G. The yellow color of silver-containing zeolite A. Angewandte Chemie, International Edition. 1998;**37**:1521

[19] Borsella E, Cattaruzza E, De Marchi G, Gonella F, Mattei G, Mazzoldi P, et al. Synthesis of silver clusters in silica-based glasses for optoelectronics applications. Journal of Non-Crystalline Solids. 1999;**245**:122

[20] Mishra YK, Mohapatra S, Kabiraj D, Mohanta B, Lalla NP, Pivinc JC, et al. Synthesis and Characterization of Ag Nanoparticles in Silica Matrix by Atom Beam Sputtering. Scripta Materialia. 2007;**56**:629

[21] Nakamura A, Narita M, Narita S, Suzuki Y, Miyanaga T. In-situ XAFS study of Ag clusters in Ag-type zeolite-A. Journal of Physics: Conference Series. 2014;**502**:012033

[22] Narita S, Miyanaga T, Suzuki Y. IR and XAFS Studies of Photoluminescent Ag-type Zeolite-A. Advances in Applied Physics. 2016;**4**:13

[23] Sakane H, Miyanaga T, Watanabe I, Matsubayashi N, Ikeda S, Yokoyama Y. Reproducibility Tests of Extended X-Ray Absorption Fine Structure for Aqua and Ammine Complexes of First Transition Metals in Solid and Aqueous Solution. Japanese Journal of Applied Physics. 1993;**32**:4641

[24] Ankudinov AL, Rehr JJ. Real-space multiple-scattering calculation and interpretation of x-ray-absorption near-edge structure. Physical Review B. 1998;**58**:7565

[25] Hoshino H, Sannohe Y, Suzuki Y, Azuhata T, Miyanaga T, Yaginuma K, et al. Photoluminescence of the dehydrated Ag-type zeolite a packed under air. Journal of the Physical Society of Japan. 2008;**77**:064712

[26] Miyanaga T, Suzuki Y, Matsumoto N, Narita S, Ainai T, Hoshino H. Formation of Ag clusters in zeolite X studied by in situ EXAFS and infrared spectroscopy. Microporous and Mesoporous Materials. 2013;**168**:213

[27] Sayah E, La Fontaine C, Briois V, Brouri D, Massiania P. Silver species reduction upon exposure of Ag/Al_2O_3 catalyst to gaseous ethanol: An in situ Quick-XANES study. Catalysis Today. 2012;**189**:55

[28] Grandjean D, Gonzalez EC, Cuong NT, Fron E, Baekelant W, Aghakhani S, et al. Origin of the bright photoluminescence of few-atom silver clusters confined in LTA zeolites. Science. 2018;**361**:686

[29] Suzuki Y, Miyanaga T, Yamauchi K, Mori N, Nakamura R. Photoluminescence of Ag-Loaded A, X, and Y Type Zeolites Heat-Treated in Atmosphere. Advances in Applied Physics. 2019;**7**:19

Chapter 2

Some New Progress in the Experimental Measurements on Electrical Property of Main Minerals in the Upper Mantle at High Temperatures and High Pressures

Lidong Dai, Haiying Hu, Yu He and Wenqing Sun

Abstract

In this chapter, we present the recent progress in the experimental studies of the electrical conductivity of dominant nominally anhydrous minerals in the upper mantle of the deep Earth interior, namely, olivine, pyroxene, and garnet. The influences from pressure, oxygen partial pressure, and anisotropic orientation on hydrous and anhydrous electrical conductivities of minerals and rocks have been already explored detailedly. There are two main electric conduction mechanisms in Fe-bearing mantle minerals, for example, small proton and proton hopping conditions, which are well distinguished by the magnitude of activation enthalpy at high temperature and high pressure. Likewise, the conduction mechanisms are efficiently characterized by these obtained positive and negative effects from the oxygen fugacity on electrical conductivities of corresponding dry and wet Fe-bearing silicate minerals at the regions of the upper mantle under conditions of different oxygen partial pressures. On the base of high-pressure laboratory-based conductivity measurements for these nominally anhydrous minerals (e.g., olivine, pyroxene, and garnet), the water content will be estimated within the depth range of the upper mantle. In comprehensive considerations of filed geophysical magnetotelluric results, the electrical conductivity measurements of dominant upper-mantle minerals can thoroughly disclose the distribution, storage state, and migration conduction in the deep Earth interior.

Keywords: electrical property, upper-mantle minerals, water content, oxygen fugacity, high temperatures and high pressures

1. Introduction

Global and regional field magnetotelluric (MT) and geomagnetic deep sounding (GDS) results revealed that there existed many high electrical conductivity layers

(HCL) in various geotectonic units in the deep Earth's interior (the magnitude of electrical conductivity range is 10^{-2}–10^{0} S/m) [1, 2]. To investigate the cause of all of these available high conductivity layers, it is crucial to measure the electric transport properties of minerals and rocks at certain high-temperature and high-pressure conditions. As one of the crucial physical parameters of minerals, electrical conductivity (EC) is highly sensitive to temperature, pressure, and depth, which is strongly dependent on the physical and chemical environments in the deep Earth and other planetary interiors [3–5]. In particular, EC is dependent on several factors such as diffusion coefficients of alkali ion [6, 7], trace elemental contents [8], the spin transition of the electron [9, 10], anisotropic crystal orientation [11–15], contents of water and other volatile elements [16–18], partial melting [19–21], dehydration (or dehydrogenation) effects of minerals [22–24], impurity of high-conductivity phase [25, 26], salinity-bearing (or water-bearing) fluids [27, 28], and structural phase transformation (amorphization or metallization) [29–34].

In the recent 20 years, with the development of measuring techniques and experimental methods of electrical conductivity in the *AC* electrical impedance spectroscopy (EIS) technique and multi-anvil high-pressure apparatus, there is a large number of research results on the electrical properties of minerals and rocks to be reported in the upper-mantle and mantle transition zone. Some international famous research administrations have successfully set up the experimental platform and measurement system of minerals and rocks at high temperatures and high pressures, such as the Key Laboratory of High-temperature and High-pressure Study of the Earth's Interior (HTHPSEI), Institute of Geochemistry, Chinese Academy of Sciences, the People's Republic of China; the Karato High-pressure Laboratory, Department of Earth and Planetary Sciences, Yale University, United States; the Laboratoire Magmas et Volcans, Université Clermont Auvergne, French National Centre for Scientific Research, France; the Scripps Institution of Oceanography, University of California San Diego, United States; University of Bayreuth, Germany; and Okayama University, Japan.

As we know, previously available classic "Pyrolite" mineralogical models have already confirmed that the nominally anhydrous minerals (NAMs, e.g., olivine, pyroxene, and garnet) are dominant mineralogical composition in the upper mantle of the deep Earth interior. In light of the FTIR result, these NAMs can contain a certain amount of structural water rather than absolutely "dry." Whereas, it is general that the trace structural water in NAMs stably exists as a form of hydroxyl point defect of the crystalline site in these minerals. Due to the presence of trace structural water in NAMs, many physical and chemical properties of NAMs have been thoroughly changed accordingly, such as electrical conductivity [3, 8, 13–18, 35], diffusivity [36, 37], plastic deformation [38, 39], seismic wave attenuation [40, 41], grain growth [42, 43], and kinetic recrystallization [44, 45]. In the world, by virtue of the theoretical calculations of Nernst-Einstein equation between the electrical conductivity and coefficient in mineral, Professor Shun-ichiro Karato from the Karato High-pressure Laboratory, Department of Earth and Planetary Sciences, Yale University, United States firstly brought forward the viewpoint that the trace structural water in hydrous olivine can enhance several orders of magnitude in the EC of upper-mantle mineral, which can be used to reasonably explain the observed high conductivity anomaly in the region of asthenosphere [46]. In the following 20 years, as a research hotpoint in the field of solid Earth science, a large amount of research work of electrical conductivity of minerals and rocks from the laboratory high-pressure experiments and theoretical calculations investigated have been conducted to focus on this hypothesis of water for the NAMS in the upper-mantle zone (olivine: [8, 13–15, 47–50], pyroxene: [51], and garnet [52–54]). In the year 1998, it is first that Xu Yousheng from the Bayerisches Geoinstitut, University of Bayreuth, Germany fetched in AC electrical impedance spectroscopy (EIS) technique and applied it to report a series of electrical conductivity of minerals, such as olivine, orthopyroxene, and garnet of the upper mantle; wadsleyite of mantle transition zone; as well as the silicate perovskite of the lower mantle under conditions of high temperatures and high pressures in the multi-anvil high-pressure apparatus [55–60]. Generally, to explore the effect of water on the electrical conductivity, we need to obtain a series of starting materials of hydrous either hot-pressure sintering synthetic or natural hydrous samples. Then, at a fixed temperature and pressure condition, we can measure the electrical conductivity of hydrous minerals. Further, the functional relationship between the EC and water content can be established at HP and HT conditions, thereby providing constraints of the water content in the deep Earth's and planetary interiors.

In this chapter, we reviewed some recent progress in the electrical conductivity of the main NAMs in the region of the upper mantle, that is, olivine, pyroxene, and garnet at conditions of high temperatures and high pressures. Then, some experimental methods, measurement techniques, and electrical transport conductions on the electrical conductivity of minerals are summarized in the multi-anvil high-pressure apparatus. The newest progress in the recently reported conductivity measurements is outlined in detail. Finally, some comprehensive remarks on the mineral electrical conductivity are discussed.

2. Electrical conductivity of upper-mantle minerals

The electrochemical *AC* impedance spectroscopy is the most efficient method to measure the electrical conductivity of minerals and rocks at HT and/or HP conditions [61–64]. The AC signal voltage and scanning frequency need to be designated before the sample resistance is measured. As usual, for a special mineral single specimen, the electrochemical *AC* impedance spectroscopy of samples consists of grain boundary impedance arc, and as well as the interface impedance between sample and electrode. However, for a special polycrystalline aggregate or rock, the electrochemical *AC* impedance spectroscopy of samples consists of grain boundary impedance arc, grain boundary impedance, and the interface impedance between sample and electrode. For each individual complex impedance spectroscopy, there are four parameters to be obtained at the same time, for example, real part, imaginary part, magnitude, and phase angle at the same time. The relation can be expressed as,

$$Z_r = |Z| \cos \varphi \tag{1}$$

$$Z_i = |Z| \sin \varphi \tag{2}$$

Here, Z_r stands for the real part of complex impedance spectroscopy, Z_i stands for the imaginary part of complex impedance spectroscopy, $|Z|$ stands for the magnitude of complex impedance spectroscopy, and φ stands for the phase angle of complex impedance spectroscopy. Representative complex impedance spectra for natural eclogite from the Dabie-Sulu ultrahigh-pressure metamorphic belt of eastern China are shown in **Figure 1**.

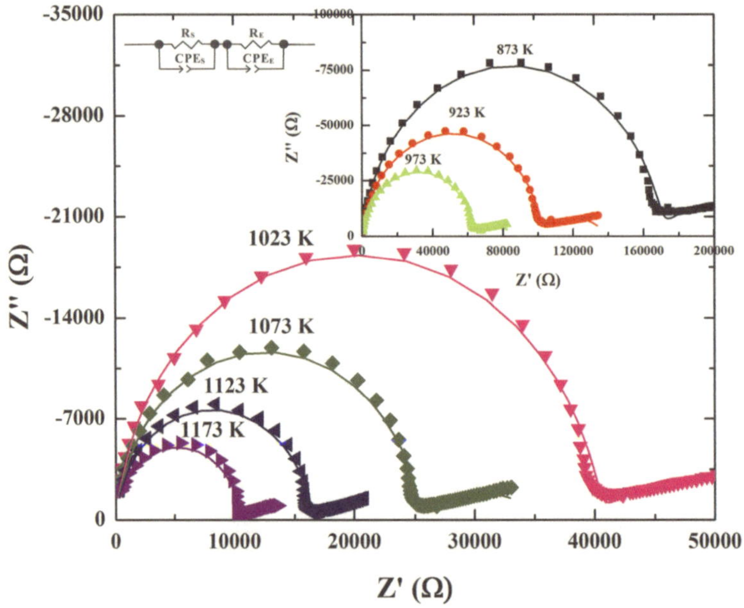

Figure 1.
Representative complex impedance spectra for natural eclogite from Dabie-Sulu ultrahigh-pressure metamorphic belt of eastern China at conditions of 3.0 GPa, 873 K–1173 K and frequency range of 10^{-1}–10^{6} Hz (reproduced with permission from Dai et al., Geochem. Geophys. Geosyst.; published by American Geophysical Union, 2016 [4]).

Detailed description of measurement theory and experimental method of impedance spectroscopy are given in our previous review chapter [9]. The equivalent electric circuit was selected to fit the impedance spectroscopy of the sample, which is composed of some fundamental electronic elements (e.g., resistor, capacitor, inductor, constant phase element (CPE), Gerischer element, Warburg element, etc.) [65–68]. After that, the electrical conductivity of the sample was obtained by the sample resistance, the calculating formula is expressed as,

$$\sigma = \frac{1}{\rho} = \frac{L}{(\mathrm{R} \times S)} \tag{3}$$

In here, σ stands for the electrical conductivity (S/m), ρ stands for the electrical resistivity (m/S), L stands for the sample height (m), and S stands for the cross-sectional area (m^2). At a given pressure condition, it is usual that the electrical conductivity of sample and temperature satisfies with an Arrhenius relation, namely,

$$\sigma = \sigma_0 \exp\left(-\frac{\Delta H}{\mathrm{k}T}\right) \tag{4}$$

In here, σ_0 stands for the pre-exponential factor (S/m), ΔH stands for the activation enthalpy (eV), k stands for the Boltzmann constant and T stands for temperature (K).

3. High-pressure apparatuses for conductivity measurements

In the recent several years, many researchers developed the high-pressure electrical property experiments of minerals and rocks by virtue of various high-pressure experimental apparatuses. From lower to higher pressure conditions, some typical high-pressure apparatuses on the laboratory-based electrical conductivity measurements are mainly included autoclave, piston-cylinder, multi-anvil press, and diamond anvil cell. In this counterpart, we focus on two types of multi-anvil apparatuses—(i) YJ-3000 t multi-anvil press is equipped in the Key Laboratory of High-temperature and High-pressure Study of the Earth's Interior (HTHPSEI), Institute of Geochemistry, Chinese Academy of Sciences, the People's Republic of China and (ii) Kawai-1000 t multi-anvil Press is equipped in the Karato High-pressure Laboratory, Department of Earth and Planetary Sciences, Yale University, United States.

3.1 YJ-3000 t multi-anvil press

Early on half a century ago, Xie Hongsen and his coworkers successfully set up one multi-anvil press of the YJ-3000 t in the Key Laboratory of HTHPSEI, Chinese Academy of Sciences, People's Republic of China. All of these available high-pressure measurement methods including the direct current, single frequency *AC*, multi-frequency electrical bridge, and electrochemical AC impedance spectroscopy are widely adopted in the past several years. *In situ* high-pressure EC results on minerals and rocks have been published by many previous researchers using this high-pressure apparatus [69–87]. Dai Lidong and his collaborator [88–91] have developed the HP-HT electrical conductivity platform of minerals and rocks in HTHPSEI, as shown in **Figure 2**. It is composed of three main counterpart pieces of equipment, namely, (a) the pressure-generated apparatus of the YJ-3000 t multi-anvil press; (b) the

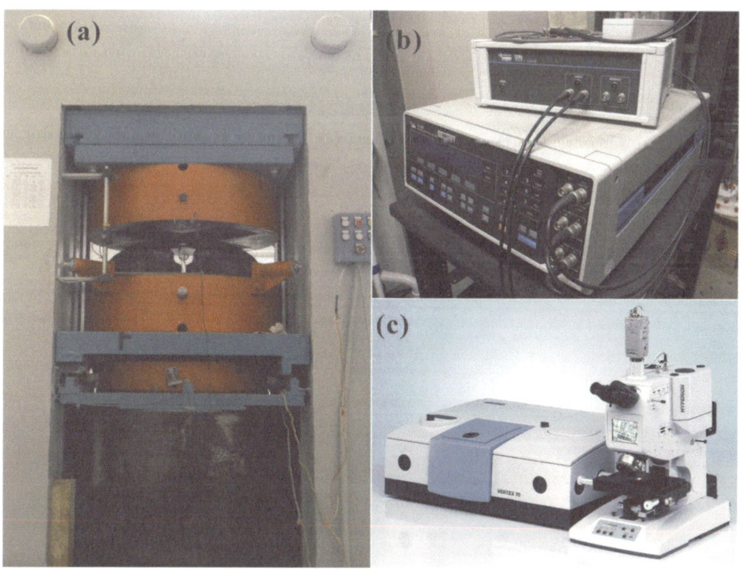

Figure 2.
High-pressure conductivity measurement platform and experimental setup in the YJ-3000 t multi-anvil press is equipped in the Key Laboratory of High-temperature and High-pressure Study of the Earth's interior (HTHPSEI), Institute of Geochemistry, Chinese Academy of Sciences, the People's Republic of China. (a) The YJ-3000 t multi-anvil apparatus; (b) the Solartron-1260 and Solarton-1296 interface impedance spectroscopy analyzer operating in the two-electrodes configuration for complex EIS measurements in the frequency range 10^{-4} Hz–10^{7} Hz; (c) the vertex-70v vacuum Fourier-transform infrared spectroscopy (FT-IR) analyzer to check the water content of sample.

Solartron-1260 and Solarton-1296 interface impedance spectroscopy analyzer operating in the two-electrodes configuration for complex EIS measurements in the frequency range 10^{-4} Hz–10^{7} Hz; and (c) the Vertex-70v vacuum Fourier-transform infrared spectroscopy (FT-IR) analyzer to check the water content of the sample. The influential ingredients include temperature, pressure, frequency, oxygen fugacity, water content, iron content, crystallographic anisotropy, grain boundary state, the content of alkali metallic elements, etc. on the electrical characterizations of minerals and rocks have already been explored using this high-pressure conductivity measurement platform in details.

In addition to the *in situ* EC measurements, it has recently become possible to measure some other high pressure-dependent physical properties of minerals and rocks by using the YJ-3000 t multi-anvil press, such as the ultrasonic elastic wave velocity, thermal conductivity, thermal diffusivity, and kinetics of grain growth. [92–100]. Except for wide application in the field of high-pressure mineral physics, the YJ-3000 t multi-anvil press is one of the indispensable tools in some other aspects of high-pressure material science and high-pressure condensed physics.

3.2 Kawai-1000 t multi-anvil press

A representative high-pressure EC experimental assemblage in the Kawai-1000 t multi-anvil press installed in the Karato High-pressure Laboratory, Department of Earth and Planetary Sciences, Yale University, United States [3, 8, 13, 14, 47–49, 51, 52] is displayed in **Figure 3**. Eight cubic WC anvils (it is corresponding to each edge length $26 \times 26 \times 26$ mm^3) with the 3–18 mm truncation was adopted to provide a high-temperature and high-pressure quasi-hydrostatic environment. Pressure calibrations were performed by pressure-induced structural phase transformations of some representative semiconducting materials (e.g., pure metallic bismuth, ZnTe, GaP, GaAs,

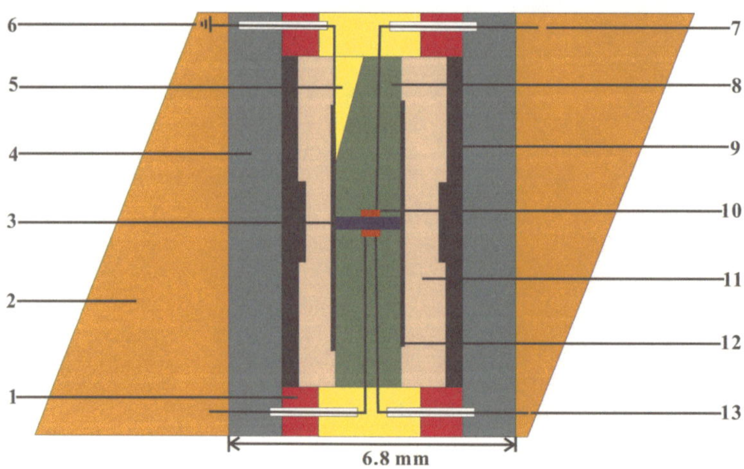

Figure 3.
Sample assembly for the electrical conductivity measurement of minerals and rocks in the Kawai-1000 t multi-anvil press from Shun-ichiro Karato's group from the Karato high-pressure laboratory, Department of Earth and Planetary Sciences, Yale University, United States: (1) metallic Mo ring; (2) MgO octahedral pressure medium with its edge length of 14 mm; (3) sample; (4) zirconia; (5) Al_2O_3 cement; (6) electric grounding; (7) lead wire of metallic electrode and Al_2O_3 insulation tube; (8) insulation tube made of four hole alumina; (9) heater of lanthanum chromite; (10) two symmetric buffer electrodes; (11) MgO insulation tube; (12) metallic shielding case made of Ni, Fe, Re or Mo foil; and (13) thermocouple and Al_2O_3 insulation tube.

etc.) at atmospheric pressure, the structural phase transformation between α-quartz, β-quartz, coesite, and stishovite, and as well as phase transitions between olivine, wadsleyite, ringwoodite, and bridgmanite. The pressure-transmitting medium is made of pure or Cr_2O_3-doped or Co-doped octahedral magnesium oxide. Two symmetric metallic discs are employed as electrodes of electrical conductivity measurement. A layer of metallic foil was installed to remove the signal disturbance of the measurement electric circuit and environmental noise. As usual, the ordinary heater, such as stainless steel slice, graphite, tantalum slice, and rhenium slice, is reasonably selected on the base of target temperature during a given electrical conductivity measurement. The temperature in the sample chamber is precisely measured by the B-type $Pt_{70\%}Rh_{30\%}$-$Pt_{94\%}Rh_{6\%}$ thermocouple or K-type thermocouple. The experimental errors of pressure and temperature are not more than 0.5 GPa and 10 K, respectively.

4. Electrical conductivity of major minerals in the upper mantle

As one crucial cycling of deep Earth interior at the depth range from 80 km to 410 km, the upper mantle mainly contains three main dominant rock-forming minerals, that is olivine, pyroxene, and garnet. Previous available high-pressure experimental results have already confirmed that the electrical conductivity of upper-mantle minerals is highly sensitive to all of these influence ingredients, such as high temperature, high pressure, oxygen partial pressure, trace structural water, iron-bearing content, grain boundary state, graphite layer of grain boundary, magnetite-bearing impurity, titanium-bearing content, chromite-bearing content, sulfur-bearing content, the orientation of crystallographic axis, and partial melting, which is most concerned in the recent several years. Because the crucial effect of water content on EC of minerals and rocks in these representative regions of upper-mantle and mantle transition was neglected by other research groups, in the following paragraph, we mainly pay attention to some related results from Dai Lidong's group from the high-pressure conductivity results from the Key Laboratory of High-temperature and High-pressure Study of the Earth's Interior (HTHPSEI), Institute of Geochemistry, Chinese Academy of Sciences, the People's Republic of China, and as well as Shun-ichiro Karato's group from the Karato High-pressure Laboratory, Department of Earth and Planetary Sciences, Yale University, United States.

4.1 Electrical conductivity of olivine

As a major rock-forming silicate mineral and nominally anhydrous mineral, olivine occupies ~60% of the volume proportion of upper-mantle minerals. Therefore, most of the previously available profile between the electrical conductivity and depth was successfully constructed by the laboratory-based electrical conductivity of olivine data at high-temperature and high-pressure conditions. Electrical conductivities on those of natural olivine single crystal, polycrystalline olivine aggregates, and hot-pressed sintered synthetic olivine have already been performed, especially considering the effects of pressure, oxygen partial pressure, iron-bearing content, and orientation of crystallographic axis on the EC of hydrous olivine.

In an early 1990 year, Roberts and Tyburczy [62] reported the room-pressure electrical conductivity of polycrystalline olivine aggregates using impedance electrochemical impedance spectroscopy (EIS) method under conditions of frequency of 10^{-4}–10^{4} Hz and 1073 K–1673 K. They discussed the influence of anisotropic thermal expansion-induced porosity and microfracture on the high-temperature electrical conductivity of polycrystalline olivine aggregates. At their measured olivine aggregates with the volume percentage of 2–8% inter- and intra-granular porosity and microfracturing, its effect becomes very feeble for the total electrical conductivity of olivine. In light of their findings, subsequent investigations on the electrical properties of upper-mantle polycrystalline samples do not need to consider the effects of porosity and microcracking. The previous investigation of pressure influence on the EC of olivine single crystal was reported by Xu et al. [55] using the EIS method in the Kawai-1000 t multi-anvil press. Detailed experimental conditions are controlled by pressures of 4–10 GPa, temperatures of 1273–1673 K, frequencies of 10^{-1}–10^{6} Hz, and as well as the solid oxygen buffer of molybdenum and molybdenum dioxide. The acquired pre-exponential factor (σ_0) of ~200 S/m, the activation energy (ΔE) of 144.7 kJ/mol and the activation volume (ΔV) of 0.6 cm^3/mole are well characterized by fitting Arrhenius relation at upper-mantle temperatures and pressures [60]. A negative pressure-dependent EC of olivine single crystal was also observed, and the electrical conduction mechanism of small polaron hopping between the ferrous and ferric irons of crystalline position was proposed. However, unfortunately, in all of those published results on the EC of olivine and its correspondent high-pressure polymorph in the regions of the upper mantle, mantle transition zone, and lower mantle, Xu et al., did not provide any available information on the water content of the selected experimental samples [55–60].

The first research work for the effect of water on the EC of upper-mantle olivine was found by Karato [46] on the basis of the theoretical calculating of Nernst-Einstein equation, who brought forward that the trace structural water of nominally hydrous minerals plays a vital role in the EC of olivine in the upper mantle. To check this theoretical calculating hypothesis, the EC of hydrous synthetic polycrystalline olivine compacts was firstly conducted by Wang et al. [17] in Shun-ichiro Karato's group from the Karato High-pressure Laboratory, Department of Earth and Planetary Sciences, Yale University, the United States at conditions of a pressure of 4.0 GPa, temperature ranges from 873 K to 1273 K and water content ranges from 100 ppm wt% to 800 ppm wt%, using the Kawai-1000 t multi-anvil press and the Solarton-1260 EIS analyzer (Schlumberger, Houston, TX, USA). It is the first time that the ionized reaction model in hydrous synthetic polycrystalline olivine compacts is originated from the free proton-dominated conduction mechanism, as follows,

$$(2H)_M^{\times} = H'_M + H^{\bullet} \quad (5)$$

where, in the Kröger-Vink notation, $(2H)_M^{\times}$ represents two hydrogens of crystalline lattice in the position of metallic iron or magnesium, H'_M represents the hydrogen vacancy in the position of metallic iron or magnesium, and $H^{\bullet}$ represents the free proton.

Furthermore, a series of hydrous electrical conductivity measurements on single-crystal olivine with different crystallographic orientations, synthetic polycrystalline olivine compacts, and synthetic hot-pressed polycrystalline olivine aggregates have been extensively studied by virtue of Kawai-1000 t multi-anvil press and the EIS methods in recent several years [8, 13, 14, 47–49]. **Figure 4** shows the influences of temperature, pressure, oxygen fugacity, iron-bearing content and water-bearing content on hydrous synthetic polycrystalline olivine compacts, and hydrous hot-pressed synthetic polycrystalline olivine aggregates at the temperature ranges from 873 K to 1273 K and pressure ranges from 4.0 GPa to 10.0 GPa.

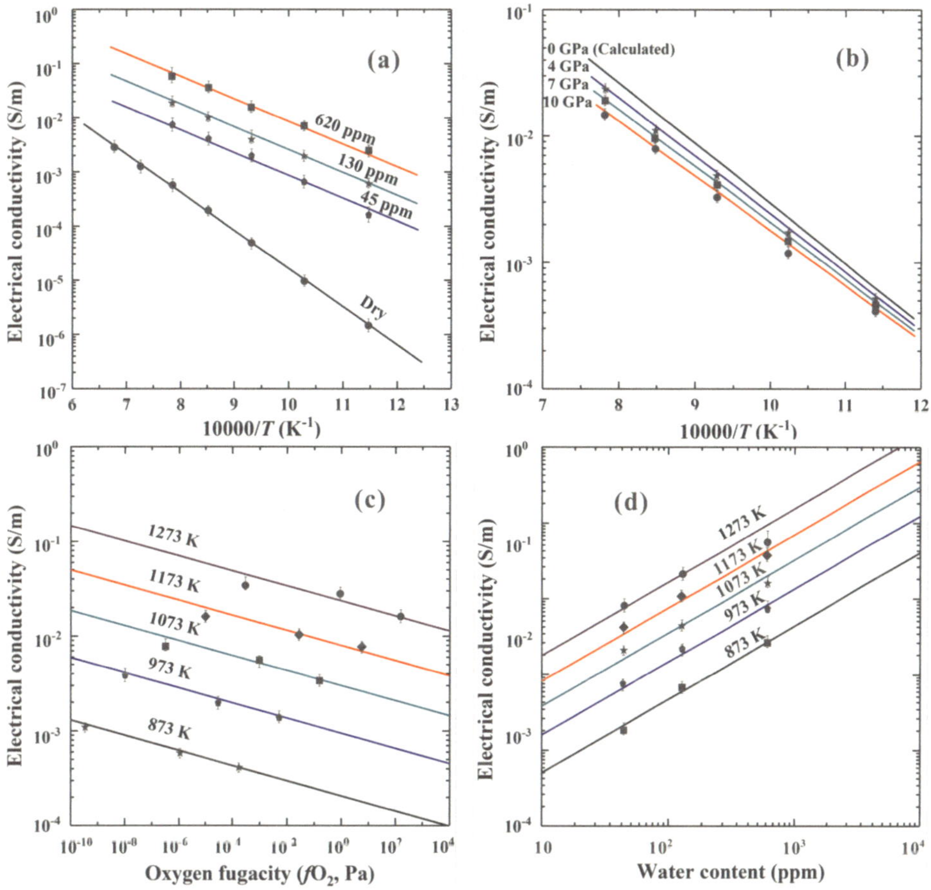

Figure 4.
The effects of (a) water content on the EC of X_{Fe} (Fe/(Fe + Mg)) = 41.2% hydrous hot-pressed synthetic polycrystalline olivine aggregates, (b) pressure on the EC of hydrous synthetic polycrystalline olivine compacts, (c) oxygen fugacity on the EC of hydrous synthetic polycrystalline olivine compacts, and (d) water-bearing content and temperature on the EC of X_{Fe} (Fe/(Fe + Mg)) = 41.2% hydrous hot-pressed synthetic polycrystalline olivine aggregates at the temperature ranges from 873 K to 1273 K and pressure ranges from 4.0 GPa to 10.0 GPa. Three oxygen buffers including Ni-NiO, Mo-MoO_2, and Re-ReO_2 were selected to oxygen fugacity during the process of high-pressure electrical conductivity measurements (reproduced with permission from Dai and Karato, Phys. Earth Planet. Inter.; published by Elsevier, 2009 [47–49]).

For hydrous hot-pressed synthetic polycrystalline olivine aggregates with fixed iron-bearing content, X_{Fe} (= (Fe/(Fe + Mg))) = 41.2% (molar ratio percentage), the electrical conductivity of the sample increases with the rise of water contents from 45 ppm wt% to 620 ppm wt%, and one relatively a fixed activation enthalpy (ΔH = 80 kJ/mole) is observed under conditions of different water contents and temperatures (refer to **Figure 4a** and **d**). The activation enthalpy for hydrous hot-pressed synthetic polycrystalline olivine aggregates is lower than that of the ΔH of dry sample (136 kJ/mole) with the fixed iron content [47]. The influence of pressure on the EC of hydrous synthetic polycrystalline olivine compacts at conditions of 873–1273 K and the oxygen buffer of Ni-NiO controlled oxygen fugacity is displayed in **Figure 4b**. It makes clear that the EC of hydrous synthetic polycrystalline olivine compacts decreases with the pressure increasing from 4.0 GPa to 10.0 GPa, and at the same time, the activation enthalpy and pre-exponential factor of the sample also reduce, accordingly [49]. **Figure 4c** displays the influence of oxygen fugacity on the EC of hydrous synthetic polycrystalline olivine compacts under conditions of 4.0 GPa, temperatures range from 873 K to 1273 K and fixed water content of 280 ppm wt%. Three oxygen buffers including Ni-NiO, Mo-MoO_2, and Re-ReO_2 were selected to oxygen fugacity during the process of high-pressure electrical conductivity measurements. An available negative electrical conductivity of sample dependence on the oxygen fugacity is observed at each given temperature point at 4.0 GPa, which is attributed to the hydrogen-related defects in hydrous synthetic polycrystalline olivine compacts at conditions of high temperature and high pressure [48]. In all these listed previous cases, the relationship between the logarithm of EC of hydrous (or dry) polycrystalline olivine and the inverse temperature follows a good Arrhenius relation, which is clearly suggesting only one single conduction mechanism controlling the electrical conductivity of hydrous (or dry) sample at high pressure. Furthermore, as pointed out by Dai and Karato [47–49], the dominant conduction mechanisms in the hydrous and anhydrous polycrystalline olivines are corresponding to the hydrogen-related defects (e.g., free proton) and iron-related defects (e.g., small polaron), respectively.

On the other hand, Dai and Karato [13, 14] also measured the EC of hydrous San Carlos single-crystal olivine along with [001, 010, 100] three different crystallographic orientations at conditions of 573–1373 K and 4.0 GPa, as illustrated in **Figure 5** in details. It is clearly observed that at lower temperatures range from 573 K to 900 K, one relatively feeble anisotropic EC in hydrous San Carlos single-crystal olivine with a lower activation enthalpy value (ΔH = 74 kJ/mole) is obtained, which is consistent

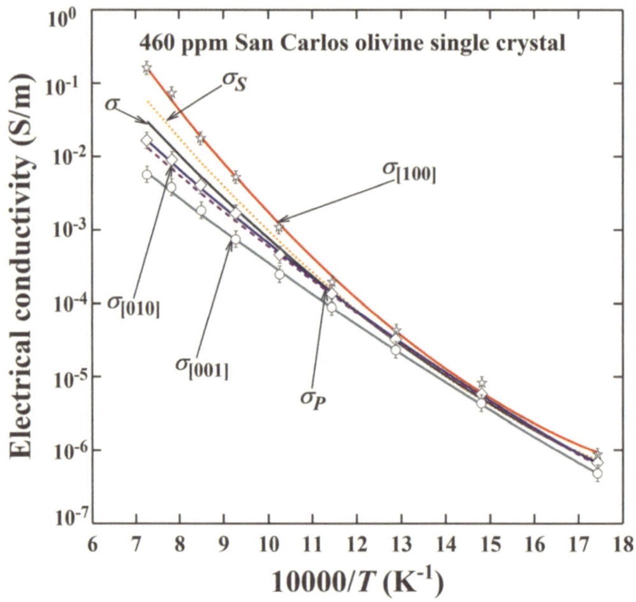

Figure 5.
The influence of anisotropy on the EC of hydrous San Carlos single crystal olivine along [001, 010, 100] three different crystallographic orientations at conditions of 573–1373 K and 4.0 GPa. Three different average schemes of series ($\sigma_S = (\sigma_{[100]} + \sigma_{[010]} + \sigma_{[001]})/3$), *parallel* ($\sigma_P = 3\left(\frac{1}{\sigma_{[100]}} + \frac{1}{\sigma_{[010]}} + \frac{1}{\sigma_{[001]}}\right)^{-1}$) *and effective medium models* ($\langle\sigma\rangle = \left[\sigma_S + \sigma_P + \sqrt{(\sigma_S + \sigma_P)^2 + 32\sigma_S\sigma_P}\right]/8$) *were applied. (Reproduced with permission from Dai and Karato, Earth Planet. Sci. Lett; published by Elsevier, 2014 [13]).*

with previously published EC results [11, 101–103]. In contrast, at a higher temperature ranges from 900 K to 1373 K, the EC of hydrous San Carlos single-crystal olivine with a higher activation enthalpy value (ΔH = ~130–150 kJ/mole) shows an obvious anisotropy along with [001, 010, 100] three different crystallographic orientations, which is good agreement with the theoretical calculating electrical conductivity results by virtue of high-temperature H–D inter-diffusion method in hydrous single-crystal olivine along with [001, 010, 100] three dominant crystallographic orientations at conditions [104].

In comprehensive considerations of geophysical field observations and geochemical models, the acquired EC results revealed that the high and highly anisotropic EC at the corresponding asthenospheric temperature and pressure conditions is reasonably explained by the high-water content in the region of the asthenosphere (100 ppm). On the other hand, the influence of the interconnected high conductive impurity phases (graphite, magnetite, chromite, sulfide impurity, etc.) and saline fluids (e.g., Ol-NaCl-H_2O, Ol-KCl-H_2O, Ol-$CaCl_2$-H_2O, etc.) on the EC of olivine has been also explored in details [25–28, 105–107].

Except for high-pressure experimental measurements on the electrical conductivity of hydrous olivine, some important progress from the first-principles calculations based on density functional theory (DFT) has already been performed in order to deeply explore the microscopic electrical transport conduction within an atomic scale in the deep Earth interior. Recently, He et al. [15] firstly designed several point defect models (e.g., $(2H)_{Mg}^{\times}$, $(4H)_{Si}^{\times}$, $\left[(Fe)_{Mg}^{\bullet}(H)_{Mg}'\right]^{\times}$, etc.) by using the ab-initio molecular dynamics (AIMD) method, and applied them to efficiently provide a robust constraint on different contributions of hydrogen and iron within a wide temperature range of 700–2000 K at 5.0 GPa. Some crucial findings were obtained—(i) the mobility of hydrogen associated with Mg vacancy and Fe^{3+} defects change significantly with temperature; (ii) at high temperature, hydrogen is able to escape from the associated defect to form "free proton," which is able to diffuse freely in olivine lattice shown high proton conductivity; (iii) the diffusion of ionized protons is highly anisotropic with extraordinary high diffusion coefficient along with [100] direction, and (V) hydrogen associated with Si vacancy is stable even at 2000 K. It suggests that different kind of hydrogen defect present different mobility at high temperature with great influence on ionic conductivity of hydrous olivine. AIMD simulations on hydrous wadsleyite and ringwoodite also suggest a higher diffusion rate of hydrogen at Mg vacancy in comparison with hydrogen at Si vacancy [108]. The migration barrier enthalpies of hydrogen along different lattice directions were calculated by using the climbing image nudged elastic band (CINEB) method [109]. The calculation results support the anisotropic diffusion behavior of hydrogen in hydrous olivine. The activation enthalpies for proton migration along with [100] direction in different models is about 180–240 kJ/mole, which is similar to the experimental results obtained by conductivity (140 ± 6 kJ/mole) and diffusion coefficient (229 ± 18 kJ/mole) measurements [13, 37]. Based on all of these obtained calculation results, an ionization-hopping proton conduction mechanism is proposed and the conductivity is calculated, as shown in **Figure 6**. Obviously, an approximately 80–160 ppm wt water content (correlated with Mg/Fe defects) in olivine is sufficient to produce the high and highly anisotropic conductivity anomaly (~10^{-1} S/m). The results are consistent with previous experimental studies and provide a good constraint on the conduction mechanism within an atomic scale for hydrous olivine. It also suggests previous conductivity measurements at relatively low temperatures ignored the contribution of ionized

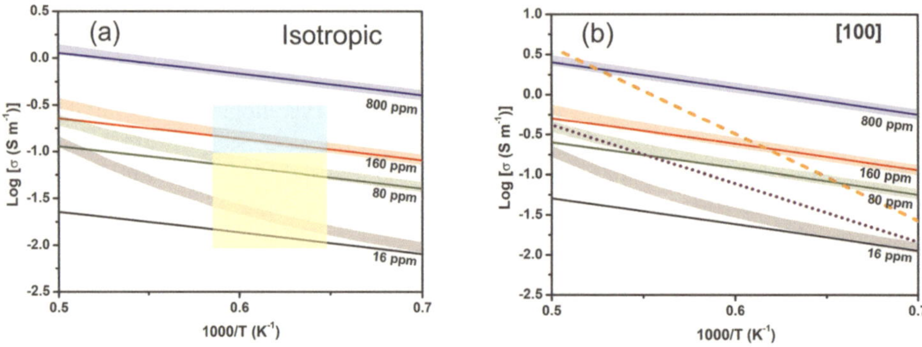

Figure 6.
The extrapolated proton and total conductivities of Fe-bearing hydrous olivine as the function of inverse temperature and water content compared with experimental results [13, 15, 110]. In here, the vertical axis stands for the logarithic electrical conductivity (the corresponding unit is S/m) and the horizontal axis stands for the inverse temperature (the corresponding unit is Kelvin), respectively. The conductivities with 16, 80, 160, and 800 ppm wt water content associated with V''_{Mg} *or* Fe^{3+}*. Thick-light curves are the total conductivities counting on the measured conductivities of dry olivine [111]. Light yellow and cyan areas show the conductivity range representing conductivity structures at the asthenosphere (Reproduced with permission from He et al., J. Geophys. Res. Solid Earth; published by American Geophysical Union, 2121 [15]).*

hydrogen, thus the extrapolated data underestimates the proton conduction for the electrical conductivity of hydrous olivine at high temperatures of the asthenosphere.

4.2 Electrical conductivity of pyroxene

As a secondary rock-forming mineral in the upper mantle, the volume percentage of pyroxene is close to the average content of 20–40% in a typical upper-mantle region. Thus, high-pressure electrical conductivity measurement on pyroxene is crucial to deeply explore the high conductivity anomalies in the regions of the upper mantle and mantle asthenosphere. For anhydrous clinopyroxene, Dai et al. [111] measured EC of dry single-crystal diopside along with [001, 010, 100] three main crystallographic orientations at 1.0–4.0 GPa, 1073–1373 K, and Ni-NiO oxygen buffer-controlled oxygen fugacity in the YJ-3000 t multi-anvil apparatus and the Solartron-1260 impedance spectroscopy analyzer. A feeble influence of anisotropy on the electrical conductivity of dry single-crystal diopside was observed. Similarly, the influence of oxygen fugacity on the electrical conductivity of dry orthopyroxene single crystal by Dai et al. [112] was performed at 1.0–4.0 GPa, 1073–1423 K, and different oxygen fugacities. The oxygen fugacity in the sample chamber of EC measurements is controlled by four solid buffers of nickel and nickel oxide, iron and magnetite, and as well as molybdenum and molybdenum dioxide. A positive dependence of oxygen fugacity on the EC of orthopyroxene single crystal is observed at a given temperature range from 1073 K to 1423 K and 2.0 GPa, which is highly related to the conduction mechanism of small polaron hopping in the anhydrous iron-bearing silicate mineral at high pressure.

On the other hand, EC measurements of anhydrous and hydrous orthopyroxene single crystals along with three main crystallographic orientations were performed by Dai and Karato [51] at conditions of 8 GPa, temperatures of 873–1273 K, and as well as the oxygen fugacity controlled by the molybdenum and molybdenum dioxide using the Kawai-1000 t multi-anvil apparatus and the Solartron-1260 impedance spectroscopy analyzer. Detailed experimental results were illustrated in **Figure 7**. According to

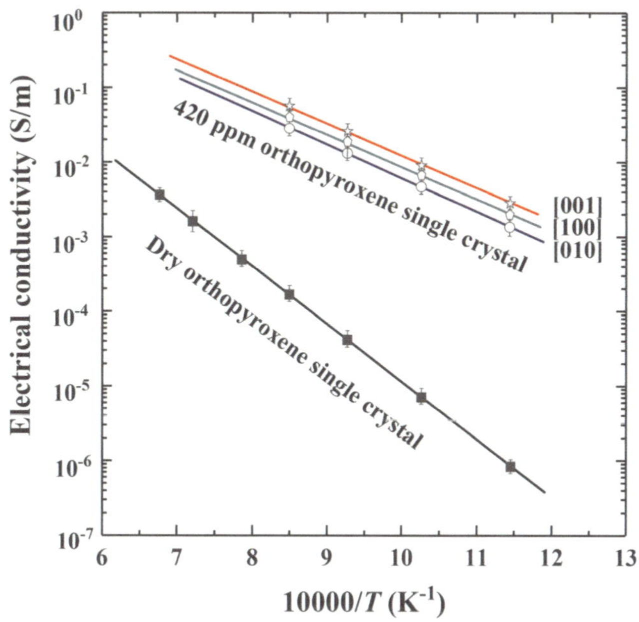

Figure 7.
The influence of anisotropy on the electrical conductivity of hydrous single-crystal orthopyroxene along [001, 010, 100] crystallographic orientations at conditions of 8 GPa, temperatures of 873–1273 K, and as well as the oxygen fugacity controlled by the molybdenum and molybdenum dioxide. The EC of dry single-crystal orthopyroxene is also included (modified from Dai and Karato [51]).

the FT-IR spectroscopy results, the hydrous orthopyroxene single crystals contain 420 ppm wt% water. They found that trace structural water can enhance several orders of magnitude in the EC of orthopyroxene single crystals. The effect from the anisotropy of hydrous orthopyroxene on the EC of the sample can be neglected, which is consistent with the above-mentioned dry single crystal diopside. As two main conduction mechanisms, small polaron and free proton play a crucial role in the electrical conductivity of anhydrous and hydrous orthopyroxene single crystals at high-temperature and high-pressure conditions.

In the recent days, Sun et al. [27] performed electrical conductivities of the polycrystalline clinopyroxene aggregates + sodium chloride + water system under conditions of 1 GPa, 673–973 K, the various salinity degree of fluid (5, 10, 15, 20, and 25%), and the volume fraction of fluid (5, 10, 15, 20, and 25%) using the YJ-3000 t multi-anvil apparatus and the Solartron-1260 impedance spectroscopy analyzer. They found that the electrical conductivity of polycrystalline clinopyroxene aggregates containing the salinity-bearing fluid with a certain salinity degree and volume fraction of fluid can be applied to explain the unusually high conductivities in some regional geotectonic units, such as southern Tibetan plateau, Dabie orogen, Grenville province, and central New Zealand.

4.3 Electrical conductivity of garnet

As a type of important constituent silicate mineral, garnet can stably exist over a wide depth from lower crust Earth to the topmost lower mantle. With increasing temperature and pressure in the deep Earth interiors, the form of pyrope-rich garnet in the lower Earth crust gradually transforms into the high-pressure phase of majorite garnet in the mantle transition zone, which is of the complex chemical composition and stable crystalline structure. Furthermore, the mineralogical content of garnet in the deep Earth interiors will gradually increase with the rise of depth [113]. Therefore, when we try to construct the profile between electrical conductivity and depth in the deep mantle of Earth's interior, it is indispensable to comprehensively assess the influence of electrical conductivity of garnet at high temperatures and high pressures.

Dai and Karato [52] conducted the EC measurements of anhydrous and water-rich single-crystal pyrope-rich garnet [its corresponding chemical composition is close to 73 mole% of pyrope (Py), 14 mole% of almandine (Alm), and 13 mole% of grossular (Gr)] at conditions of temperature range from 873 K to 1473 K, pressure ranges from 4 GPa to 16 GPa, frequency ranges from 10^{-2} Hz to 10^6 Hz and water-bearing content range from less than 10 H/10^6Si to 7000 H/10^6Si using the Kawai-1000 t multi-anvil press installed in the Karato High-pressure Laboratory, Department of Earth and Planetary Sciences, Yale University, United States. **Figure 8** shows the effect of pressure on the dry and water-bearing (160 ppm wt) single crystal pyrope-rich garnet at temperatures of 873–1473 K. In comparisons with water-free water of pyrope-rich garnet, the electrical conductivity of the hydrous sample is higher, and the activation enthalpy of pyrope-rich garnet (ΔH = ~70 kJ/mole) is lower than that of the anhydrous sample (ΔH = ~128 kJ/mole).

In subsequent investigations, a series of EC experiments on garnet with different chemical compositions ($Py_{20}Alm_{76}Grs_4$, $Py_{30}Alm_{67}Grs_3$, $Py_{56}Alm_{43}Grs_1$, $Py_{73}Alm_{14}Grs_{13}$, and $Alm_{82}Py_{15}Grs_3$) were conducted by Dai et al. [53, 54] using the YJ-3000 t multi-anvil apparatus and the Solartron-1260 impedance spectroscopy analyzer to explore the influences of oxygen fugacity and mineralogical composition on the EC of garnet at 1.0–4.0 GPa and 873–1273 K. Five different solid buffers including Fe_3O_4-Fe_2O_3, Ni-NiO, Fe-Fe_3O_4, Fe-FeO, and Mo-MoO_2 were employed to control the oxygen fugacity of the high-pressure sample chamber. Based on these obtained

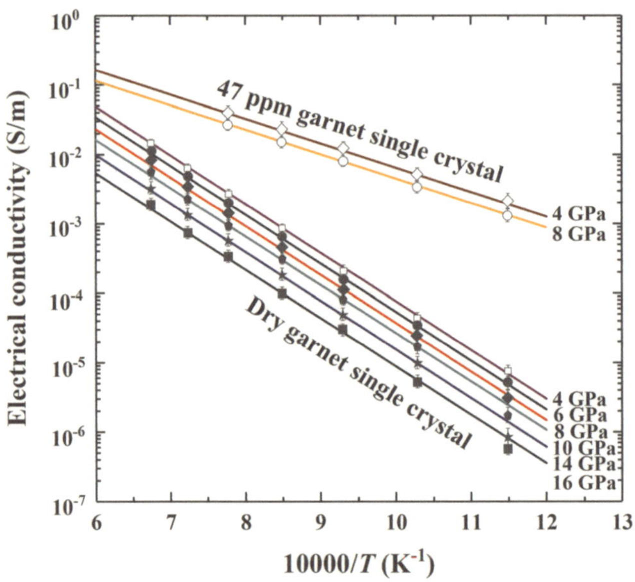

Figure 8.
The influence of pressure on the dry and water-bearing (160 ppm wt) single crystal pyrope-rich garnet at temperatures of 873–1473 K (reproduced with permission from Dai and Karato, Phys. Earth planet. Inter.; published by Elsevier, 2009 [52]).

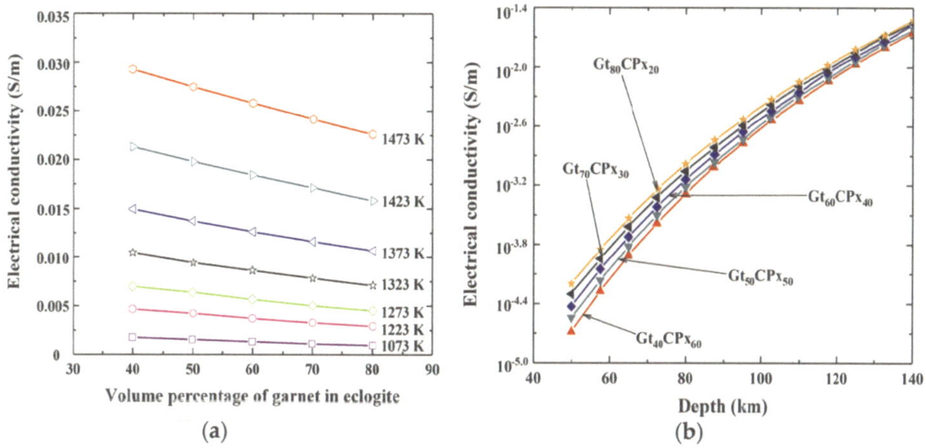

Figure 9.
(a) The influence of volume percentage on the electrical conductivity of eclogite on the base on our obtained electrical conductivity results for single crystal pyrope-rich garnet at high temperatures and high pressures (modified from Dai et al. [53]), (b) the profile between the electrical conductivity of eclogite and depth in conjunction with previously available electrical conductivity of clinopyroxene results, the average models of rock and as well as the regionally geothermal gradient is established from the EC of garnet series at high temperatures and high pressures (reproduced with permission from Dai et al., Contrib. Mineral. Petrol.; published by Springer Nature, 2012 [53]).

experimental results, some absolutely new models of electrical conductivity of garnet series as functions of the variation of oxygen fugacity and chemical compositions have already been constructed, as illustrated in **Figure 9(a)** in detail. Furthermore, in conjunction with previously available electrical conductivity of clinopyroxene results, the average models of rock and as well as the regionally geothermal gradient, the profile between the electrical conductivity of eclogite and depth is established from the EC of garnet series at high temperatures and high pressures, as shown in **Figure 9(b)** in detail. All of these acquired laboratory-based electrical conductivity profiles can be widely applied to systematically disclose and deeply explore the cause of the observable high conductivity anomalies from global and regional field magnetotelluric and geomagnetic deep sounding results in those representatively geotectonic units, for example, stable mid-lower continental Earth crust, Dabie-Sulu ultrahigh-pressure metamorphic belt of eastern China, Tibet plateau, and North China craton.

5. Conclusions

High-pressure laboratory-based measurement results on electrical properties of several dominant minerals (olivine, orthopyroxene, clinopyroxene, garnet, etc.) in the upper mantle regions are highly sensitive to some influential factors including temperature, pressure, oxygen fugacity, water content, crystallographic orientation, trace element of titanium content and iron content. The proper determination of EC of upper-mantle minerals in a broad temperature and pressure range requires the utilization of complex electrical electrochemical impedance spectroscopy measurements. The iron-related hopping of small polaron and the hydrogen-related defects are possibly the two dominant conduction mechanisms in anhydrous and hydrous Fe-bearing silicate minerals within the depth range of the upper mantle. The trace structural water of mantle minerals plays a crucial role in explaining the high conductivity anomaly and the water distribution in the deep mantle. In comprehensive considerations of some newest results from high-pressure laboratory-based conductivity measurements, geophysical field observations, and first-principles theoretical calculations, high and highly anisotropic EC at the corresponding asthenospheric temperature and pressure conditions can be used to reasonably explain by the 100 ppm wt% of water content for the high conductivity anomaly in the asthenospheric region. In addition, the influence of the interconnected high conductive impurity phases (graphite, magnetite, chromite, sulfide impurity, etc.) and saline fluids (e.g., Ol-NaCl-H_2O, Ol-KCl-H_2O, Ol-$CaCl_2$-H_2O, etc.) on the EC of olivine also needed to be considered in a special geotectonic environment. With the development and advancement of measurement techniques and experimental methods, more and more laboratory-based high-pressure conductivity experimental results will be indispensable to systematically disclose the cycle discontinuity and high conductivity anomaly in the upper-mantle region.

Acknowledgements

Almost all listed original high-pressure electrical conductivity data have already been obtained by Lidong Dai and his collaborators in the Chinese Academy of Sciences and Yale University. In here, please accept my best honest thanks and greetings to Professor Heping Li in the Key Laboratory of High-temperature and High-pressure Study of the Earth's Interior (HTHPSEI), Institute of Geochemistry, Chinese Academy of Sciences, the People's Republic of China, and Professor Shun-ichiro Karato in the Karato High-pressure Laboratory, Department of Earth and Planetary Sciences, Yale University, United States. This work was financially supported by the National Natural Science Foundation of China (Grant numbers 42072055, 41774099, and 41772042), Youth Innovation Promotion Association of Chinese Academy of Sciences (Grant number 2019390), Special Fund of the West Light Foundation of Chinese Academy of Sciences and as well as Special Fund from Shandong Provincial Key Laboratory of Water and Soil Conservation and Environmental Protection.

Author details

Lidong Dai*, Haiying Hu*, Yu He and Wenqing Sun
Key Laboratory of High-Temperature and High-Pressure Study of the Earth's Interior, Institute of Geochemistry, Chinese Academy of Sciences, Guiyang, Guizhou, China

*Address all correspondence to: dailidong@vip.gyig.ac.cn; huhaiying@vip.gyig.ac.cn

References

[1] Yang B, Egbert G, Zhang H, Meqbel N, Hu X. Electrical resistivity imaging of continental United States from three-dimensional inversion of EarthScope USArray magnetotelluric data. Earth and Planetary Science Letters. 2021;**576**: 117244. DOI: 10.1016/j.epsl.2021.117244

[2] Förster M, Selway K. Melting of subducted sediments reconciles geophysical images of subduction zones. Nature Communications. 2021;**12**:1-7

[3] Dai L, Karato S. Electrical conductivity of wadsleyite at high temperatures and high pressures. Earth and Planetary Science Letters. 2009;**287**:277-283

[4] Dai L, Hu H, Li H, Wu L, Hui K, Jiang J, et al. Influence of temperature, pressure, and oxygen fugacity on the electrical conductivity of dry eclogite, and geophysical implications. Geochemistry, Geophysics, Geosystems. 2016;**17**:2394-2407

[5] Dai L, Hu H, Jiang J, Sun W, Li H, Wang M, et al. An overview of the experimental studies on the electrical conductivity of major minerals in the upper mantle and transition zone. Materials. 2020;**13**:408. DOI: 10.3390/ma13020408

[6] Hu H, Li H, Dai L, Shan S, Zhu C. Electrical conductivity of albite at high temperatures and high pressures. American Mineralogist. 2011;**96**: 1821-1827

[7] Hu H, Li H, Dai L, Shan S, Zhu C. Electrical conductivity of alkali feldspar solid solutions at high temperatures and high pressures. Physics and Chemistry of Minerals. 2013;**40**:51-62

[8] Dai L, Karato S. Electrical conductivity of Ti-bearing hydrous olivine aggregates at high temperature and high pressure. Journal of Geophysical Research—Solid Earth. 2020;**125**:e2020JB020309. DOI: 10.1029/2020JB020309

[9] Lin J, Weir S, Jackson D, Evans W, Vohra Y, Qiu W, et al. Electrical conductivity of the lower-mantle ferropericlase across the electronic spin transition. Geophysical Research Letters. 2007;**34**:L16305. DOI: 10.1029/2007GL030523

[10] Ohta K, Onoda S, Hirose K, Sinmyo R, Shimizu K, Sata N, et al. The electrical conductivity of post-perovskite in Earth's D″ layer. Science. 2008;**320**: 89-91

[11] Wang D, Li H, Yi L, Matsuzaki T, Yoshino T. Anisotropy of synthetic quartz electrical conductivity at high pressure and temperature. Journal of Geophysical Research—Solid Earth. 2010;**115**:B09211. DOI: 10.1029/2009JB006695

[12] Guo X, Yoshino T, Katayama I. Electrical conductivity anisotropy of deformed talc rocks and serpentinites at 3 GPa. Physics of the Earth and Planetary Interiors. 2011;**188**:69-81

[13] Dai L, Karato S. High and highly anisotropic electrical conductivity of the asthenosphere due to hydrogen diffusion in olivine. Earth and Planetary Science Letters. 2014;**408**:79-86

[14] Dai L, Karato S. Reply to comment on "High and highly anisotropic electrical conductivity of the asthenosphere due to hydrogen diffusion in olivine" by Dai and Karato [Earth Planet. Sci. Lett. 408 (2014) 79–86]. Earth and Planetary Science Letters. 2015;**427**:300-302

[15] He Y, Dai L, Kim D, Li H, Karato S. Thermal ionization of hydrogen in hydrous olivine with enhanced and anisotropic conductivity. Journal of Geophysical Research—Solid Earth. 2021;**126**:e2021JB022939. DOI: 10.1029/2021JB022939

[16] Huang X, Xu Y, Karato S. Water content in the transition zone from electrical conductivity of wadsleyite and ringwoodite. Nature. 2005;**434**:746-749

[17] Wang D, Mookherjee M, Xu Y, Karato S. The effect of water on the electrical conductivity of olivine. Nature. 2006;**443**:977-980

[18] He Y, Hou M, Jang B, Sun S, Zhuang Y, Deng L, et al. Superionic iron oxide–hydroxide in Earth's deep mantle. Nature Geoscience. 2021;**14**: 174-178

[19] Manthilake M, Matsuzaki T, Yoshino T, Yamashita S, Ito E, Katsura T. Electrical conductivity of wadsleyite as a function of temperature and water content. Physics of the Earth and Planetary Interiors. 2009;**174**:10-18

[20] Pommier A, Leinenweber K, Tasaka M. Experimental investigation of the electrical behavior of olivine during partial melting under pressure and application to the lunar mantle. Earth and Planetary Science Letters. 2015;**425**: 242-255

[21] Yoshino T, Matsuzaki T, Yamashita S, Katsura T. Hydrous olivine unable to account for conductivity anomaly at the top of the asthenosphere. Nature. 2006;**443**:973-976

[22] Hu H, Dai L, Li H, Hui K, Sun W. Influence of dehydration on the electrical conductivity of epidote and implications for high-conductivity anomalies in subduction zones. Journal of Geophysical Research—Solid Earth. 2017;**122**:2751-2762

[23] Sun W, Dai L, Li H, Hu H, Wu L, Jiang J. Electrical conductivity of mudstone (before and after dehydration at high PT) and a test of high conductivity layers in the crust. American Mineralogist. 2017;**102**: 2450-2456

[24] Hu H, Dai L, Li H, Sun W, Li B. Effect of dehydrogenation on the electrical conductivity of Fe-bearing amphibole: Implications for high conductivity anomalies in subduction zones and continental crust. Earth and Planetary Science Letters. 2018;**498**: 27-37

[25] Dai L, Hu H, Sun W, Li H, Liu C, Wang M. Influence of high conductive magnetite impurity on the electrical conductivity of dry olivine aggregates at high temperature and high pressure. Minerals. 2019;**9**:44. DOI: 10.3390/min9010044

[26] Sun W, Jiang J, Dai L, Hu H, Wang M, Qi Y, et al. Electrical properties of dry polycrystalline olivine mixed with various chromite contents: Implications for the high-conductivity anomalies in subduction zones. Geoscience Frontiers. 2021;**12**:101178. DOI: 10.1016/j.gsf.2021.101178

[27] Sun W, Dai L, Li H, Hu H, Jiang J, Wang M. Electrical conductivity of clinopyroxene-NaCl-H_2O system at high temperatures and pressures: Implications for high-conductivity anomalies in the deep crust and subduction zone. Journal of Geophysical Research—Solid Earth. 2020;**125**:e2019JB019093. DOI: 10.1029/2019JB019093

[28] Sun W, Dai L, Hu H, Jiang J, Wang M, Hu Z, et al. Influence of saline fluids on the electrical conductivity of

olivine aggregates at high temperature and high pressure and its geological implications. Frontiers in Earth Science. 2021;**9**:749896. DOI: 10.3389/feart.2021.749896

[29] Dai L, Pu C, Li H, Hu H, Liu K, Yang L, et al. Characterization of metallization and amorphization for GaP under different hydrostatic environments in diamond anvil cell up to 40.0 GPa. The Review of Scientific Instruments. 2019;**90**:066103. DOI: 10.1063/1.5093949

[30] Dai L, Liu K, Li H, Wu L, Hu H, Zhuang Y, et al. Pressure-induced irreversible metallization with phase transitions of Sb_2S_3. Physical Review B. 2018;**97**:024103. DOI: 10.1103/PhysRevB.97.024103

[31] Dai L, Zhuang Y, Li H, Wu L, Hu H, Liu K, et al. Pressure-induced irreversible amorphization and metallization with a structural phase transition in arsenic telluride. Journal of Materials Chemistry C. 2017;**5**:12157-12162

[32] Hong M, Dai L, Hu H, Yang L and Zhang X. Pressure-induced structural phase transitions in natural kaolinite investigated by Raman spectroscopy and electrical conductivity. American Mineralogist. 2022. In press. DOI: 10.2138/am-2021-7863

[33] Yang L, Dai L, Li H, Hu H, Hong M, Zhang X, et al. High-pressure investigations on the isostructural phase transition and metallization in realgar with diamond anvil cells. Geoscience Frontiers. 2021;**12**:1031-1037

[34] Yang L, Jiang J, Dai L, Hu H, Hong M, Zhang X, et al. High-pressure structural phase transition and metallization in Ga_2S_3 under non-hydrostatic and hydrostatic conditions up to 36.4 GPa. Journal of Materials Chemistry C. 2021;**9**:2912-2918

[35] Freitas D, Manthilake G. Electrical conductivity of hydrous silicate melts: Implications for the bottom-up hydration of Earth's upper mantle. Earth and Planetary Science Letters. 2019;**523**: 115712. DOI: 10.1016/j.epsl.2019.115712

[36] Fei H, Wiedenbeck M, Yamazaki D, Katsura T. Small effect of water on upper-mantle rheology based on silicon self-diffusion coefficients. Nature. 2013; **498**:213-216

[37] Novella D, Jacobsen B, Weber P, Tyburczy J, Ryerson F, Du Frane W. Hydrogen self-diffusion in single crystal olivine and electrical conductivity of the Earth's mantle. Scientific Reports. 2017;7:5344. DOI: 10.1038/s41598-017-05113-6

[38] Jung H, Karato S. Water-induced fabric transitions in olivine. Science. 2001;**293**:1460-1463

[39] Kang H, Jung H. Lattice-preferred orientation of amphibole, chlorite, and olivine found in hydrated mantle peridotites from Bjørkedalen, southwestern Norway, and implications for seismic anisotropy. Tectonophysics. 2019;**750**:137-152

[40] Wei S, Wiens D, Zha Y, Plank T, Webb S, Blackman D, et al. Seismic evidence of effects of water on melt transport in the Lau back-arc mantle. Nature. 2015;**518**:395-398

[41] Cline IIC, Faul U, David E, Berry A, Jackson I. Redox-influenced seismic properties of upper-mantle olivine. Nature. 2018;**555**:355-358

[42] Nishihara Y, Maruyama G, Nishi M. Growth kinetics of forsterite reaction rims at high-pressure. Physics of the

Earth and Planetary Interiors. 2016;**257**: 220-229

[43] Cerpa N, Wada I, Wilson C. Fluid migration in the mantle wedge: Influence of mineral grain size and mantle compaction. Journal of Geophysical Research—Solid Earth. 2017;**122**:6247-6268

[44] Masotta M, Mollo S, Nazzari M, Tecchiato V, Scarlato P, Papale P, et al. Crystallization and partial melting of rhyolite and felsite rocks at Krafla volcano: A comparative approach based on mineral and glass chemistry of natural and experimental products. Chemical Geology. 2018;**483**:603-618

[45] Peslier A, Hervig R, Yang S, Humayun M, Barnes J, Irving A, et al. Determination of the water content and D/H ratio of the Martian mantle by unraveling degassing and crystallization effects in nakhlites. Geochimica et Cosmochimica Acta. 2019;**266**:382-415

[46] Karato S. The role of hydrogen in the electrical conductivity of the upper mantle. Nature. 1990;**347**:272-273

[47] Dai L, Karato S. Influence of FeO and H on the electrical conductivity of olivine. Physics of the Earth and Planetary Interiors. 2014;**237**:73-79

[48] Dai L, Karato S. Influence of oxygen fugacity on the electrical conductivity of hydrous olivine: Implications for the mechanism of conduction. Physics of the Earth and Planetary Interiors. 2014;**232**: 57-60

[49] Dai L, Karato S. The effect of pressure on the electrical conductivity of olivine under the hydrogen-rich conditions. Physics of the Earth and Planetary Interiors. 2014;**232**:51-56

[50] Katsura T, Baba K, Yoshino T, Kogiso T. Electrical conductivity of the oceanic asthenosphere and its interpretation based on laboratory measurements. Tectonophysics. 2017; **717**:162-181

[51] Dai L, Karato S. Electrical conductivity of orthopyroxene: Implications for the water content of the asthenosphere. Proceedings of the Japan Academy. 2009;**85**:466-475

[52] Dai L, Karato S. Electrical conductivity of pyrope-rich garnet at high temperature and high pressure. Physics of the Earth and Planetary Interiors. 2009;**176**:83-88

[53] Dai L, Li H, Hu H, Shan S, Jiang J, Hui K. The effect of chemical composition and oxygen fugacity on the electrical conductivity of dry and hydrous garnet at high temperatures and pressures. Contributions to Mineralogy and Petrology. 2012;**163**:689-700

[54] Dai L, Li H, Hu H, Jiang J, Hui K, Shan S. Electrical conductivity of $Alm_{82}Py_{15}Grs_3$ almandine-rich garnet determined by impedance spectroscopy at high temperatures and high pressures. Tectonophysics. 2013;**608**:1086-1093

[55] Xu Y, Poe B, Shankland T, Rubie D. Electrical conductivity of olivine, wadsleyite, and ringwoodite under upper-mantle conditions. Science. 1998; **280**:1415-1418

[56] Xu Y, McCammon C, Poe B. The effect of alumina on the electrical conductivity of silicate perovskite. Science. 1998;**282**:922-924

[57] Poe B, Xu Y. In situ complex impedance spectroscopy of mantle minerals measured at 20 GPa and 1400 °C. Phase Transitions. 1999;**68**:453-466

[58] Xu Y, Shankland T. Electrical conductivity of orthopyroxene and its

high pressure phases. Geophysical Research Letters. 1999;**26**:2645-2648

[59] Xu Y, Shankland T, Poe B. Laboratory-based electrical conductivity in the Earth's mantle. Journal of Geophysical Research—Solid Earth. 2000;**105**:27865-27875

[60] Xu Y, Shankland T, Duba A. Pressure effect on electrical conductivity of mantle olivine. Physics of the Earth and Planetary Interiors. 2000;**118**:149-161

[61] Laštovičková M. Laboratory measurements of electrical properties of rocks and minerals. Geophysical Surveys. 1983;**6**:201-213

[62] Roberts J, Tyburczy J. Frequency dependent electrical properties of polycrystalline olivine compacts. Journal of Geophysical Research. 1991;**96**: 16205-16222

[63] Nover G. Electrical properties of crustal and mantle rocks–A review of laboratory measurements and their explanation. Surveys in Geophysics. 2005;**26**:593-651

[64] Yoshino T. Laboratory electrical conductivity measurement of mantle minerals. Surveys in Geophysics. 2010; **31**:163-206

[65] Saltas V, Vallianatos F, Gidarakos E. Charge transport in diatomaceous earth studied by broadband dielectric spectroscopy. Applied Clay Science. 2013;**80–81**:226-235

[66] Vallianatos F, Saltas V. Application of the cBΩ model to the calculation of diffusion parameters of He in olivine. Physics and Chemistry of Minerals. 2014;**41**:181-188

[67] Karato S, Wang D. Electrical conductivity of minerals and rocks. In: Physics and Chemistry of the Deep Earth, 1st ed. John Wiley & Sons, Ltd: Hoboken, NJ, USA; 2013

[68] Pommier A, Leinenweber K. Electrical cell assembly for reproducible conductivity experiments in the multi-anvil. American Mineralogist. 2018;**103**: 1298-1305

[69] Xie H. Introduction to Science of the Earth Interior Material. Beijing, China: Peking Science Press; 1997

[70] Zheng H, Xie H, Xu Y, Song M, Guo J, Zhang Y. The electrical conductivity of H_2O at 0.21–4.18 GPa and 20–350 °C. Scientific Bulletin. 1997; **42**:969-976

[71] Zheng H, Xie H, Xu Y, Song M, Guo J, Zhang Y. Measurement of electrical conductivity of 0.001 mol NaCl solution under high pressures. Scientific Bulletin. 1997;**42**:1563-1566

[72] Wang D, Li H, Yi L, Shi B. The electrical conductivity of upper-mantle rocks: Water content in the upper mantle. Physics and Chemistry of Minerals. 2008;**35**:157-162

[73] Wang D, Guo Y, Yu Y, Karato S. Electrical conductivity of amphibole-bearing rocks: Influence of dehydration. Contributions to Mineralogy and Petrology. 2012;**164**:17-25

[74] Dai L, Li H, Liu C, Su G, Shan S. Experimental measurement on the electrical conductivity of pyroxenite at high temperature and high pressure under different oxygen fugacities. High Pressure Research. 2006;**26**: 193-202

[75] Dai L, Li H, Deng H, Liu C, Su G, Shan S, et al. In-situ control of different oxygen fugacity experimental study on the electrical conductivity of lherzolite at

high temperature and high pressure. Journal of Physics and Chemistry of Solids. 2008;**69**:101-110

[76] Dai L, Li H, Li C, Hu H, Shan S. The electrical conductivity of dry polycrystalline olivine compacts at high temperatures and pressures. Mineralogical Magazine. 2010;**74**: 849-857

[77] Dai L, Hu H, Li H, Jiang J, Hui K. Influence of temperature, pressure, and chemical composition on the electrical conductivity of granite. American Mineralogist. 2014;**99**:1420-1428

[78] Dai L, Hu H, Li H, Hui K, Jiang J, Li J, et al. Electrical conductivity of gabbro: The effects of temperature, pressure and oxygen fugacity. European Journal of Mineralogy. 2015;**27**:215-224

[79] Dai L, Jiang J, Li H, Hu H, Hui K. Electrical conductivity of hydrous natural basalt at high temperatures and high pressures. Journal of Applied Geophysics. 2015;**112**:290-297

[80] Dai L, Sun W, Li H, Hu H, Wu L, Jiang J. Effect of chemical composition on the electrical conductivity of gneiss at high temperatures and pressures. Solid Earth. 2018;**9**:233-245

[81] Hu H, Dai L, Li H, Jiang J, Hui K. Electrical conductivity of K-feldspar at high temperature and high pressure. Mineralogy and Petrology. 2014;**108**: 609-618

[82] Hu H, Dai L, Li H, Hui K, Li J. Temperature and pressure dependence of electrical conductivity in synthetic anorthite. Solid State Ionics. 2015;**276**: 136-141

[83] Hui K, Zhang H, Li H, Dai L, Hu H, Jiang J, et al. Experimental study on the electrical conductivity of quartz andesite at high temperature and high pressure: Evidence of grain boundary transport. Solid Earth. 2015;**6**:1037-1043

[84] Hui K, Dai L, Li H, Hu H, Jiang J, Sun W, et al. Experimental study on the electrical conductivity of pyroxene andesite at high temperature and high pressure. Pure and Applied Geophysics. 2017;**174**:1033-1041

[85] Sun W, Dai L, Li H, Hu H, Jiang J, Hui K. Effect of dehydration on the electrical conductivity of phyllite at high temperatures and pressures. Mineralogy and Petrology. 2014;**111**:853-863

[86] Sun W, Dai L, Li H, Hu H, Liu C. Effect of temperature, pressure and chemical composition on the electrical conductivity of granulite and geophysical implications. Journal of Mineralogical and Petrological Sciences. 2019;**114**:87-98

[87] Sun W, Dai L, Li H, Hu H, Liu C, Wang M. Effect of temperature, pressure and chemical compositions on the electrical conductivity of schist: Implications for electrical structures under the Tibetan plateau. Materials. 2019;**12**:961. DOI: 10.3390/ma12060961

[88] Dai L, Li H, Hu H, Shan S. Experimental study of grain boundary electrical conductivities of dry synthetic peridotite under high-temperature, high-pressure, and different oxygen fugacity conditions. Journal of Geophysical Research—Solid Earth. 2008;**113**:B12211. DOI: 10.1029/2008JB005820

[89] Dai L, Li H, Hu H, Shan S. Novel technique to control oxygen fugacity during high-pressure measurements of grain boundary conductivities of rocks. The Review of Scientific Instruments. 2009;**80**:033903. DOI: 10.1063/1.3097882

[90] Dai L, Li H, Hu H, Shan S. In-situ control of oxygen fugacity for laboratory measurements of electrical conductivity of minerals and rocks in multi-anvil press. Chinese Physics B. 2011;**20**: 049101. DOI: 10.1088/1674-1056/20/4/049101

[91] Dai L, Hu H, Li H, Sun W, Jiang J. Influence of anisotropy on the electrical conductivity and diffusion coefficient of dry K-feldspar: Implications for the mechanism of conduction. Chinese Physics B. 2018;**27**:028703. DOI: 10.1088/1674-1056/27/2/028703

[92] Xu J, Zhang Y, Hou W, Xu H, Guo J, Wang Z, et al. Measurements of ultrasonic wave velocities at high temperature and high pressure for window glass, pyrophyllite, and kimberlite up to 1400 °C and 5.5 GPa. High Temperatures—High Pressures. 1994;**26**:375-384

[93] Liu Y, Xie H, Zhou W, Guo J. A method for experimental determination of compressional velocities in rocks and minerals at high pressure and high temperature. Journal of Physics. Condensed Matter. 2002;**14**:11381-11384

[94] Xie H, Zhou W, Zhu M, Liu Y, Zhao Z, Guo J. Elastic and electrical properties of serpentinite dehydration at high temperature and high pressure. Journal of Physics. Condensed Matter. 2002;**14**:11359-11363

[95] Song W, Liu Y, Wang Z, Gong C, Guo J, Zhou W, et al. Measurement method for sound velocity of melts in large volume press and its application to liquid sodium up to 2.0 GPa. The Review of Scientific Instruments. 2011;**82**: 086108. DOI: 10.1063/1.3625267

[96] Zhou W, Fan D, Liu Y, Xie H. Measurements of wave velocity and electrical conductivity of an amphibolite from southwestern margin of the Tarim Basin at pressures to 1.0 GPa and temperatures to 700 °C: Comparison with field observations. Geophysical Journal International. 2011;**187**:1393-1404

[97] Miao S, Li H, Chen G. The temperature dependence of thermal conductivity for lherzolites from the North China Craton and the associated constraints on the thermodynamic thickness of the lithosphere. Geophysical Journal International. 2014;**197**:900-909

[98] Miao S, Li H, Chen G. Temperature dependence of thermal diffusivity, specific heat capacity, and thermal conductivity for several types of rocks. Journal of Thermal Analysis and Calorimetry. 2014;**115**:1057-1063

[99] Miao S, Zhou Y, Li H. Thermal diffusivity of lherzolite at high pressures and high temperatures using pulse method. Journal of Earth Science. 2019; **30**:218-222

[100] Zhang B, Hu X, Asimow P, Zhang X, Xu J, Fan D, et al. Crystal size distribution of amphibole grown from hydrous basaltic melt at 0.6–2.6 GPa and 860–970 °C. American Mineralogist. 2019;**104**:525-535

[101] Yang X. Orientation-related electrical conductivity of hydrous olivine, clinopyroxene and plagioclase and implications for the structure of the lower continental crust and uppermost mantle. Earth and Planetary Science Letters. 2012;**317**:241-250

[102] Poe B, Romano C, Nestola F, Smyth J. Electrical conductivity anisotropy of dry and hydrous olivine at 8 GPa. Physics of the Earth and Planetary Interiors. 2010;**181**:103-111

[103] Karato S. Some remarks on hydrogen-assisted electrical conductivity

in olivine and other minerals. Progress in Earth and Planetary Science. 2019;**6**:55. DOI: 10.1186/s40645-019-0301-2

[104] Du Frane W, Tyburczy J. Deuterium–hydrogen interdiffusion in olivine: Implications for point defects and electrical conductivity. Geochemistry, Geophysics, Geosystems. 2012;**13**: Q03004. DOI: 10.1029/2011GC003895

[105] Watson H, Roberts J, Tyburczy J. Effect of conductive impurities on electrical conductivity in polycrystalline olivine. Geophysical Research Letters. 2010;**37**:L02302. DOI: 10.1029/2009GL041566

[106] Wang D, Karato S, Jiang Z. An experimental study of the influence of graphite on the electrical conductivity of olivine aggregates. Geophysical Research Letters. 2013;**40**:2028-2032

[107] Huang Y, Guo H, Nakatani T, Uesugi K, Nakamura M, Keppler H. Electrical conductivity in texturally equilibrated fluid-bearing forsterite aggregates at 800 °C and 1 GPa: Implications for the high electrical conductivity anomalies in mantle wedges. Journal of Geophysical Research—Solid Earth. 2021;**126**:e2020JB021343. DOI: 10.1029/2020JB021343

[108] Caracas R, Panero W. Hydrogen mobility in transition zone silicates. Progress in Earth and Planetary Science. 2017;**4**:9. DOI: 10.1186/s40645-017-0119-8

[109] Henkelman G, Uberuaga B, Jonsson H. A climbing image nudged elastic band method for finding saddle points and minimum energy paths. The Journal of Chemical Physics. 2000;**113**: 9901-9904

[110] Gardés E, Gaillard F, Tarits P. Toward a unified hydrous olivine electrical conductivity law. Geochemistry, Geophysics, Geosystems. 2014;**15**:4984-5000

[111] Dai L, Li H, Liu C, Shan S, Cui T, Su G. Experimental study on the electrical conductivity of orthopyroxene at high temperature and high pressure under different oxygen fugacities. Acta Geologica Sinica—English Edition. 2005; **79**:803-809

[112] Dai L, Li H, Liu C, Su G, Cui T. In situ control of oxygen fugacity experimental study on the crystallographic anisotropy of the electrical conductivities of diopside at high temperature and high pressure. Acta Petrologica Sinica. 2005;**21**: 1737-1742

[113] Yoshino T, Nishi M, Matsuzaki T, Yamazaki D, Katsura T. Electrical conductivity of majorite garnet and its implications for electrical structure in the mantle transition zone. Physics of the Earth and Planetary Interiors. 2008;**170**: 193-200

Chapter 3

Mineral Physics

Robert Cooper Liebermann

Abstract

Mineral physics emerged as an independent discipline of Earth sciences in the middle of the 20th century, drawing together geophysics and mineralogy. Using the principles of condensed matter physics and solid-state chemistry, it focuses on exploring how physical properties of minerals depend on atomic structure. With the advent of new experimental tools (e.g., automated X-ray diffractometers, electron microscopes, various spectrometric techniques, digital computers and synchrotron X-radiation sources) in the past 70 years, geophysicists and mineralogists began to talk with one another.

Keywords: mineral physics, geophysics, mineralogy, high pressure, high temperature, applications to earth

1. Introduction

In the 1960s, Orson Anderson established a new laboratory at Columbia University's Lamont Geological Observatory and chose to name it "Mineral Physics." Experiments in his laboratory measured mineral properties over the wide range of pressures, temperatures and chemical compositions seen in the interior of Earth and other terrestrial planets. These studies included properties of minerals, but all materials related to natural minerals (e.g., structural analogs, but also glasses, melts and fluids). According to Robert Hazen [1], "mineral physics is the study of mineralogical problems through the application of condensed matter physics". In reality, mineral physicists use not only physics but also solid-state chemistry. Knowledge of these properties is essential to interpretations of seismic data accurately and performing realistic geodynamic simulations. Today, mineral physics is widely considered one of the three pillars of geophysics, along with geodynamics and seismology.

Today, scientists approach these problems through a combination of experimental and computational methods. Precise information at lower pressures and temperatures is provided by experiments, and detailed information at conditions difficult to re-created in the laboratory is provided by computational work. Bulk material properties are vital to understanding the behavior of planets, but atomistic inspection of these complex materials provides a connection to planetary-scale phenomena. Theoretical mineral physicists are in a unique to position to illuminate this connection [2] (see also paragraph by Taku Tsuchiya below).

In the past half century since the first scientific conference focused on mineral physics was held in 1977, mineral physics has matured into an independent scientific discipline firmly in the mainstream of geosciences. Now mineral physicists are faced with many challenging problems and many exciting opportunities for research; in large measure, this is now possible due to access to synchrotron X-radiation facilities throughout the world. This evolution is highlighted in this chapter (see also our earlier paper [3]).

2. History prior to the 1950s

Long before seismology became a recognized discipline early in the 20th century, many scientists and philosophers had speculated on the nature of the interior of Earth. One of the earliest of these was Athanasius Kircher (1602–1680) whose view of Earth (**Figure 1**) was published in 1665 and entitled: Mundus Subterraneus [4].

Over the subsequent three centuries as more detailed information became available from improved scientific methods, especially in seismology, our view of the interior of Earth has evolved, but major features have remained unchanged (**Figure 2**).

Harvard University faculty dominated investigations of the problems related to mineral physics in the first half of the 20th century. The first was Percy Bridgman whose investigations of the properties if matter under high pressure earned him the Nobel Prize in Physics in 1946. Among the graduate students of Bridgman was Francis Birch, who focused on high-pressure studies relevant to geophysical problems. In his classical paper in 1952 [5], he demonstrated that the mantle is predominately composed of silicate minerals and that the upper mantle and lower mantle regions, each essentially homogeneous but of somewhat differing compositions, are separated by a thin transition zone associated with silicate phase transitions (see green band at 670 km in **Figure 3**). Birch also concluded that the inner and outer core are alloys of crystalline and molten iron, respectively, which contrasted with the prevailing views at the time. Although a few refinements have become necessary in light of subsequent research, the essential details of this model are still valid. For this paper and other contributions, Birch is widely acknowledged as the "father of mineral physics" (though he never used that term).

Figure 1.
View of earth in 1665 by Athanasius Kircher (1602–1680).

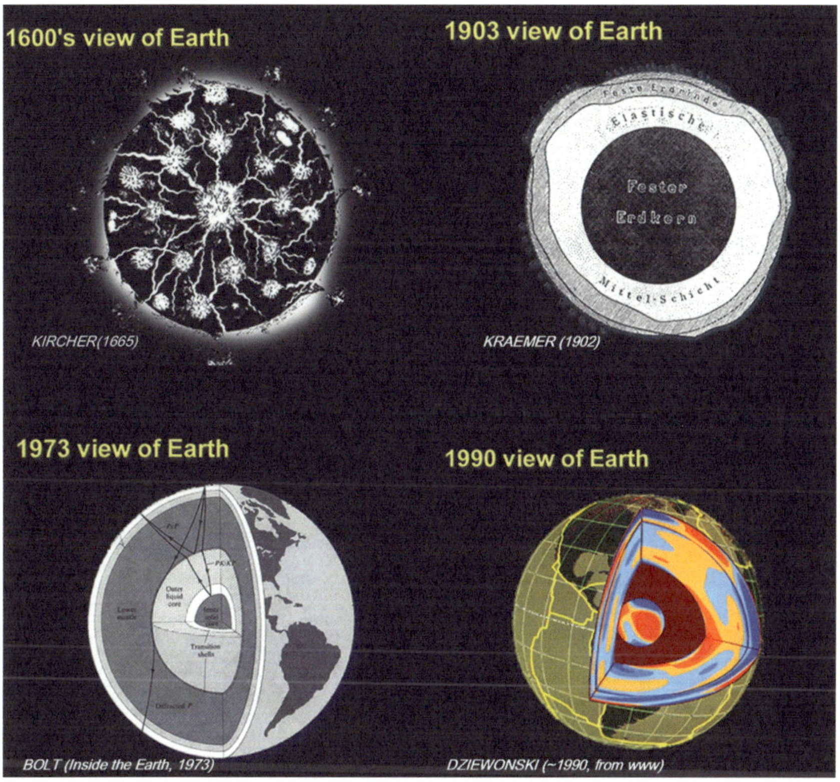

Figure 2.
Views of the interior of the earth from 1665 to 1990. Courtesy of Ed Garnero.

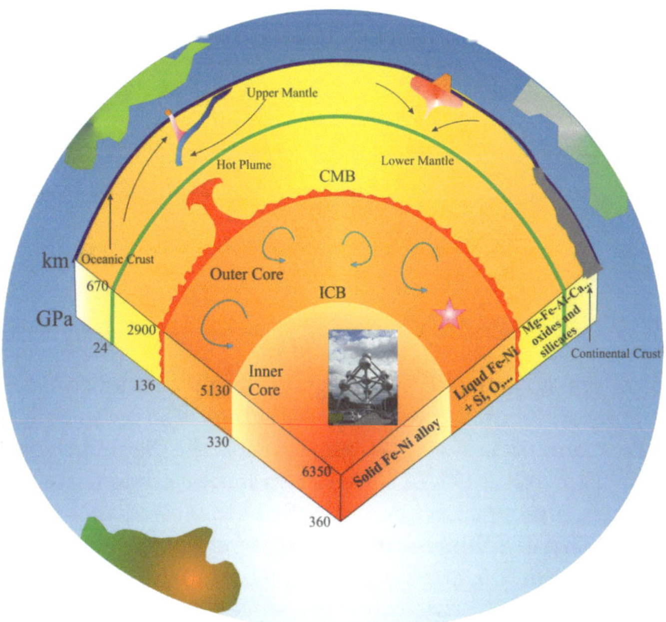

Figure 3.
Internal structure and composition of the Earth's interior.

3. Post 1950s developments

Over the last four decades of the 20th century, many laboratories in the United States, Japan, Australia and France began to conduct experiments on the physical properties of minerals at high pressures and temperatures. These laboratories focused their research on understanding the role of fundamental properties of minerals and thus provided links to other disciplines of the geosciences such as seismology, tectonophysics, volcanology, geochemistry and petrology. These links are graphically illustrated by **Figure 4**.

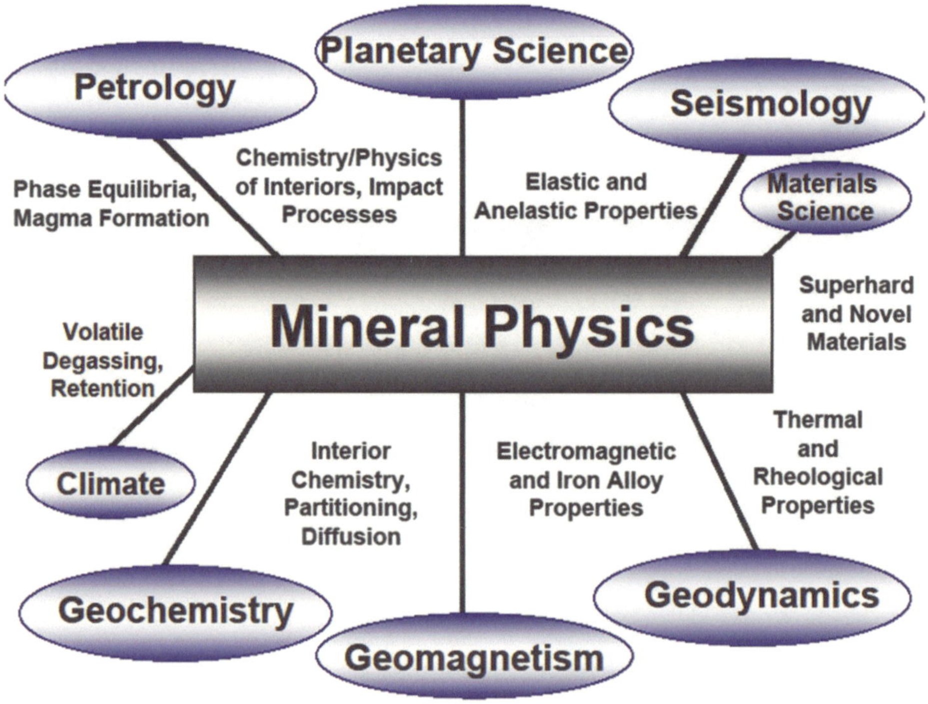

Figure 4.
Williams-Lattimore diagram reflecting the relationship of mineral physics to other disciplines of geosciences (copyright R. C. Liebermann).

4. Inventions of DACs and MAA

In the past 50+ years, two major static high-pressure techniques have been developed in mineral physics laboratories by earth scientists desiring to replicate in the laboratory the P–T conditions of Earth's deep interior: the diamond-anvil cell (DAC) and the multi-anvil apparatus (MAA). These two static technique are both useful and very complementary, although have occasionally been viewed as competitive. Higher pressures and large sample volumes can be achieved either through the use of larger tonnage hydraulic rams in MAAs or by increasing the culet size of DACs, and thus experimental conditions of the two techniques will eventually merge.

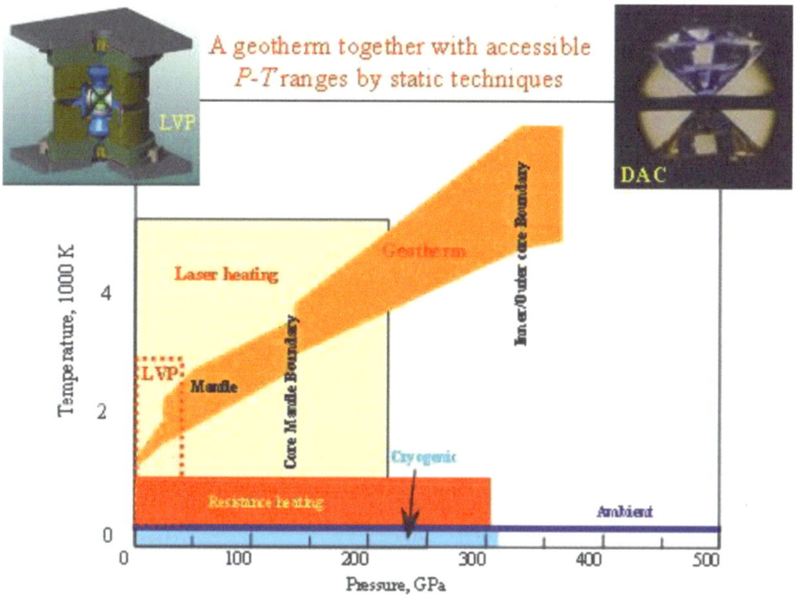

Diagram from G. Shen

(a)

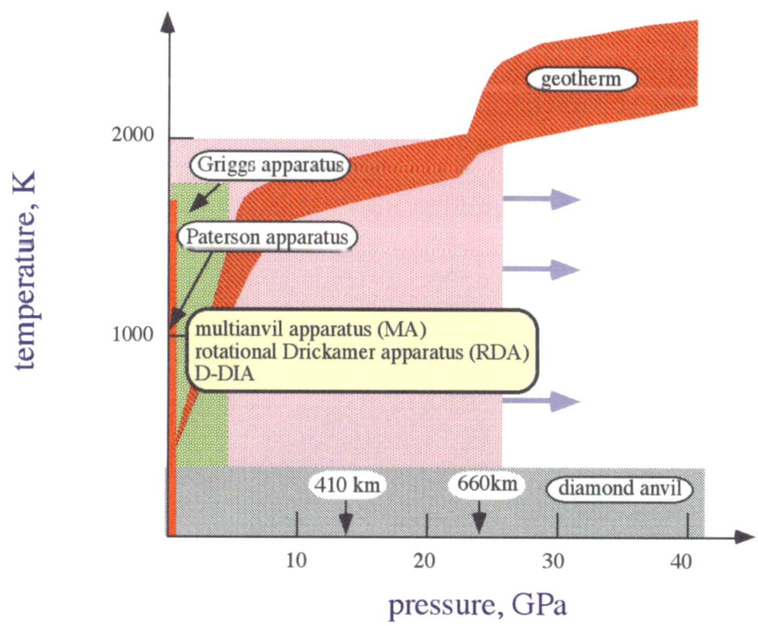

Diagram from S. Karato

(b)

Figure 5.
a. Geotherm with P–T ranges compared with the P–T regions achievable with diamond anvil cells (and various types of heating) and with LVP (large-volume presses, aka multi-anvil apparatus). b. Temperature and pressures achievable in difference types of large-volume apparatus.

The diamond-anvil cell (DAC) was invented at the National Bureau of Standards [NBS] under the direction of Alvin Van Valkenburg and his colleagues in 1958. For the subsequent developments of the DAC, see the excellent review by William Bassett [6]. In the same time period, Tracy Hall invented the first multi-anvil apparatus, a tetrahedral-anvil machine; such MAA have evolved progressively [7] and can now achieve pressures close to 100 GPa at high temperatures. With the advent of synchrotron radiation facilities in the early 1980s, many of these DAC and MAA devices have been utilized in conjunction with *in situ* X-ray diffraction.

In **Figure 5a** and **b**, we illustrate the pressure and temperature ranges achievable with the diamond anvil cells and multi-anvil apparatus and compare those with the P–T ranges of the geotherm of the Earth.

In **Figure 6**, the regions of the Earth's interior (Crust, Upper mantle, Lower mantle, Outer core and Inner core) accessible to difference types of high-pressure apparatus: Belt, Split-Sphere, DIA and DAC are illustrated.

With the aid of high-pressure devices such as diamond anvil cells and multi-anvil apparatus, scientists have been able to explore the crystallographic transformations which the principal minerals of the upper mantle undergo as they are buried deeper in the Earth's interior. Thus, olivine, garnet, clinopyroxene and orthopyroxene of the upper mantle evolve to mixtures of (Mg, Fe)O-ferropericlase and Ca- and (Mg, Fe, Al)-SiO_3-perovskites [the latter recently named bridgmanite] in the lower mantle. In the vicinity of the outer core, the bridgmanite transforms to a post-perovskite phase which has not yet been found in nature (see **Figure 7a**, courtesy of Stas Sinogeikin and **Figure 7b**, courtesy of Nick Schmerr and Ed Garnero).

In addition to experiments using static multi-anvil apparatus or diamond anvil cells, many scientists have utilized dynamic shock wave techniques to measure the physical properties of minerals at high pressures and temperatures. In **Figure 8**, the shock wave gun at Caltech is shown; it was originally built by Thomas Ahrens and is now under the supervision of Paul Asimow.

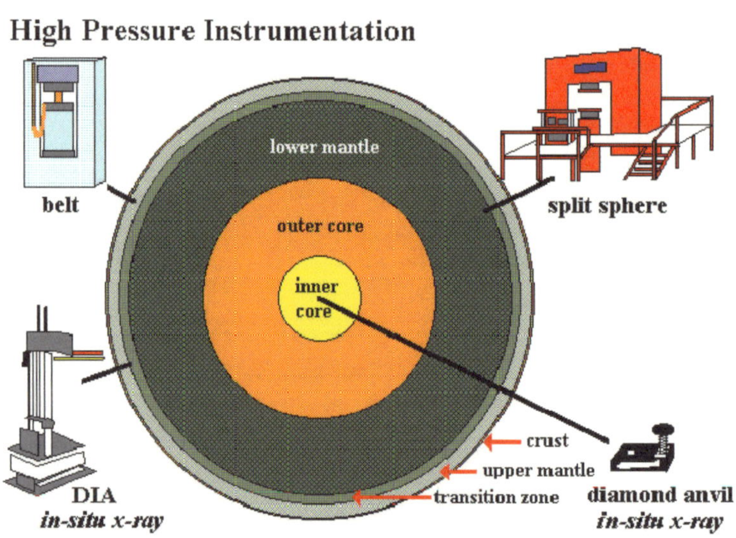

Figure 6.
Regions of the Earth's interior accessible to different types of high-pressure apparatus: Belt, Split-sphere, DIA and DAC.

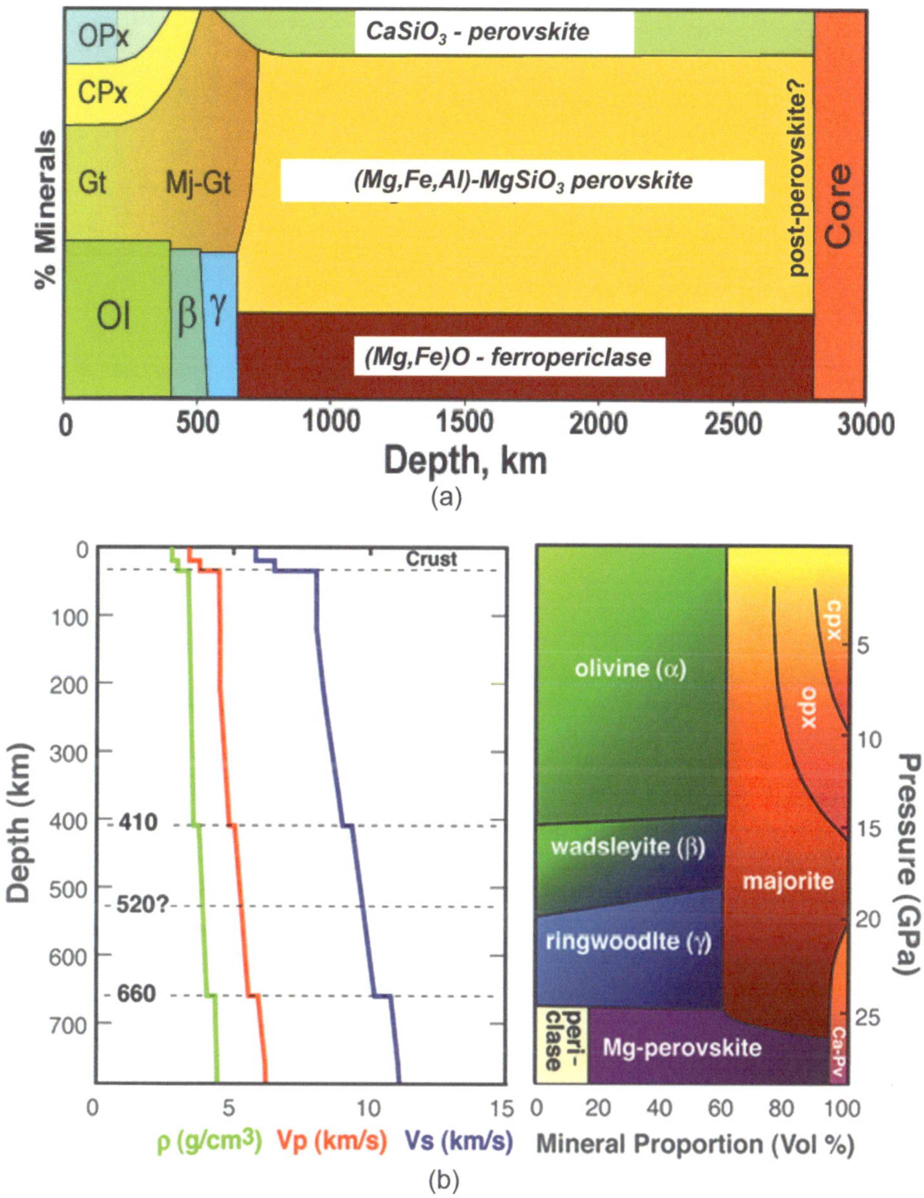

Figure 7.
a. Evolution of minerals in the earth from the upper mantle to the core. Courtesy of Stas Sinogeikin. b. Upper mantle seismic discontinuities: Mineral phase boundaries. Courtesy of Nick Schmerr and Ed Garnero.

In a paper now in press entitled "New analysis of shock-compression data for selected silicates", Thomas Duffy has illustrated one of the important uses of shock wave experiments:

"The study of minerals under shock compression provides fundamental constraints on their response to conditions of extreme pressure, temperature, and strain rate and has applications to understanding meteorite impacts and the deep Earth. The recent development of facilities for real-time *in situ* X-ray diffraction studies under gun- or laser-based dynamic compression provides new capability for understanding

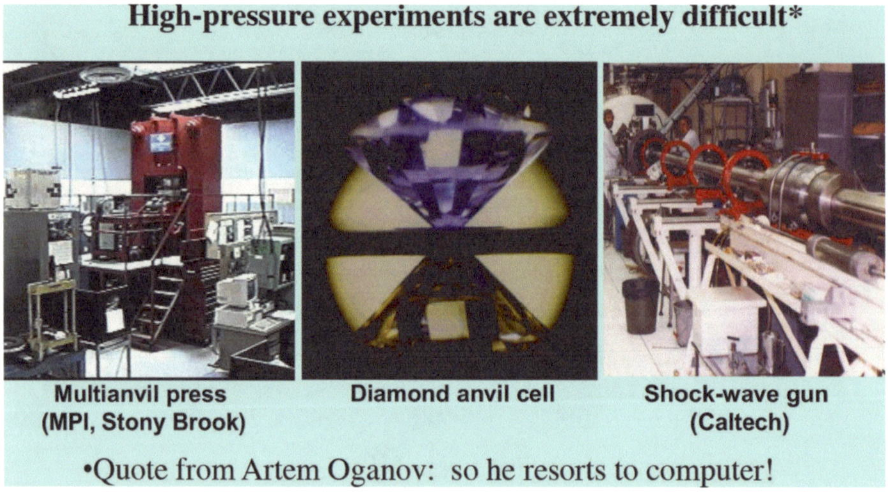

Figure 8.
Static (multi-anvil press and diamond anvil cell) high-pressure devices illustrated with the dynamic shock-wave gun at Caltech.

the atomic-level structure of shocked solids. Here traditional shock pressure-density data for selected silicate minerals (garnets, tourmaline, nepheline, topaz, and spodumene) are examined through comparison of their Hugoniots with recent static compression and theoretical studies. The results provide insights into the stability of silicate structures and the possible nature of high-pressure phases under shock loading. This type of examination highlights the potential for *in situ* atomic-level measurements to address questions about phase transitions, transition kinetics, and structures formed under shock-compression for silicate minerals."

Experiments are not the only solution to finding progress. In the figure above is a quote by Artem Oganov:

"High-pressure experiments are extremely difficult".

Thus, he resorts to the computer for theoretical calculations.

"Recent progress in theoretical mineral physics based on the *ab initio* quantum mechanical computation method has been dramatic in conjunction with the rapid advancement of computer technologies. This technique solves electronic structures and chemical bonding natures of materials highly accurately and became practical after the beginning of this century. It is now possible to predict thermodynamic stability, elasticity, and transport properties of complex minerals quantitatively with uncertainties that are comparable or even smaller than those attached in experimental data under high pressure and high temperature. These calculations under *in situ* high-pressure (P) and high-temperature (T) condition allow us to construct *a priori* mineralogical models of the deep Earth and have opened a new generation in solid geophysics and geochemistry". (Taku Tsuchiya, personal communication, 2021).

5. Experimental physical acoustics and elasticity of minerals

Experimental physical acoustics has been used to study the elastic properties of minerals. These high-precision techniques provide measurements of the velocity of sound in single crystals or polycrystalline aggregates (i.e., rocks) as functions of pressure and temperature.

There are several goals of such research in geophysics:

a. To study the relationship of the elastic properties to crystallographic structure and composition.

b. To provide the input parameters for theoretical equations of state of solids to enable these ultrasonic data to be extrapolated to higher pressures and temperatures, and

c. To deduce the composition and mineralogy of the Earth's interior from a comparison of the laboratory data with the velocity-depth profiles in the Earth derived from seismology.

Our recent paper [8] summarizes the current state-of-the-art in studies of the Earth's interior using measurements of sound velocities in minerals by ultrasonic interferometry. In that paper, we reviewed the progress of the technology of ultrasonic

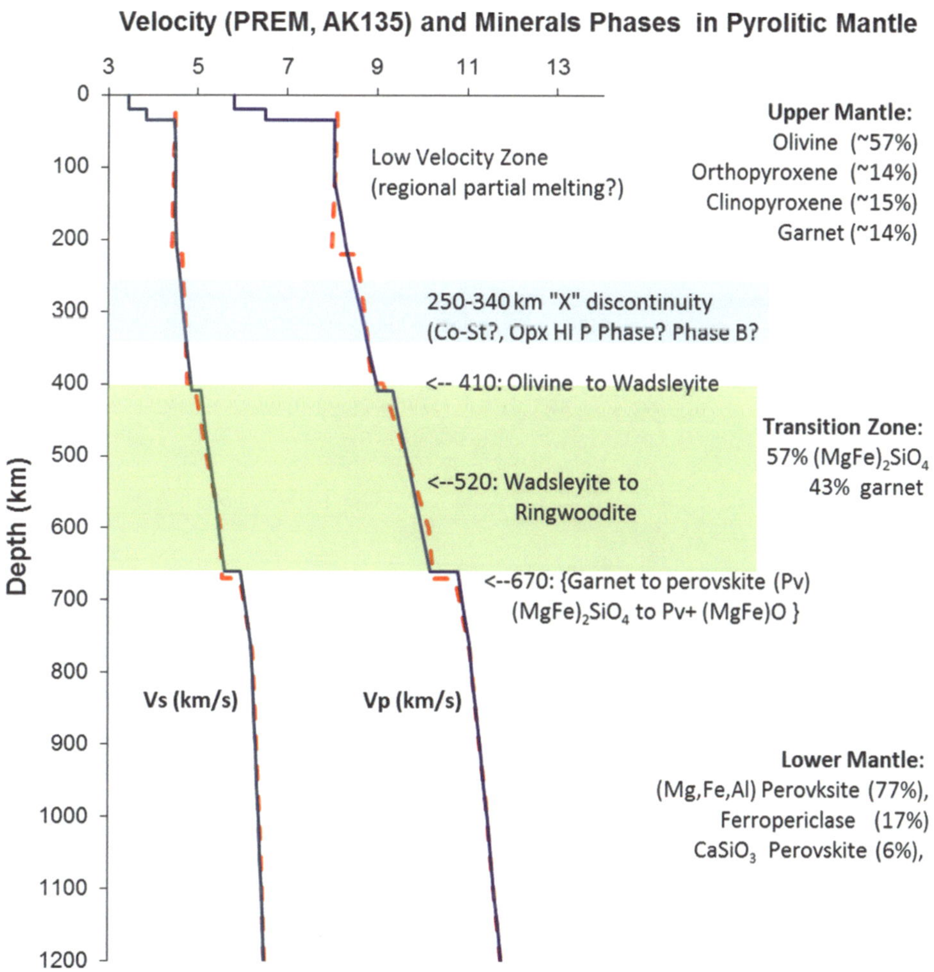

Figure 9.
Comparison of seismic models (PREM-dashed lines and AK135-solid lines) with possible mineral phase transitions in the minerals of the upper mantle (see also ***Figures 3*** *and* ***7a, b****).*

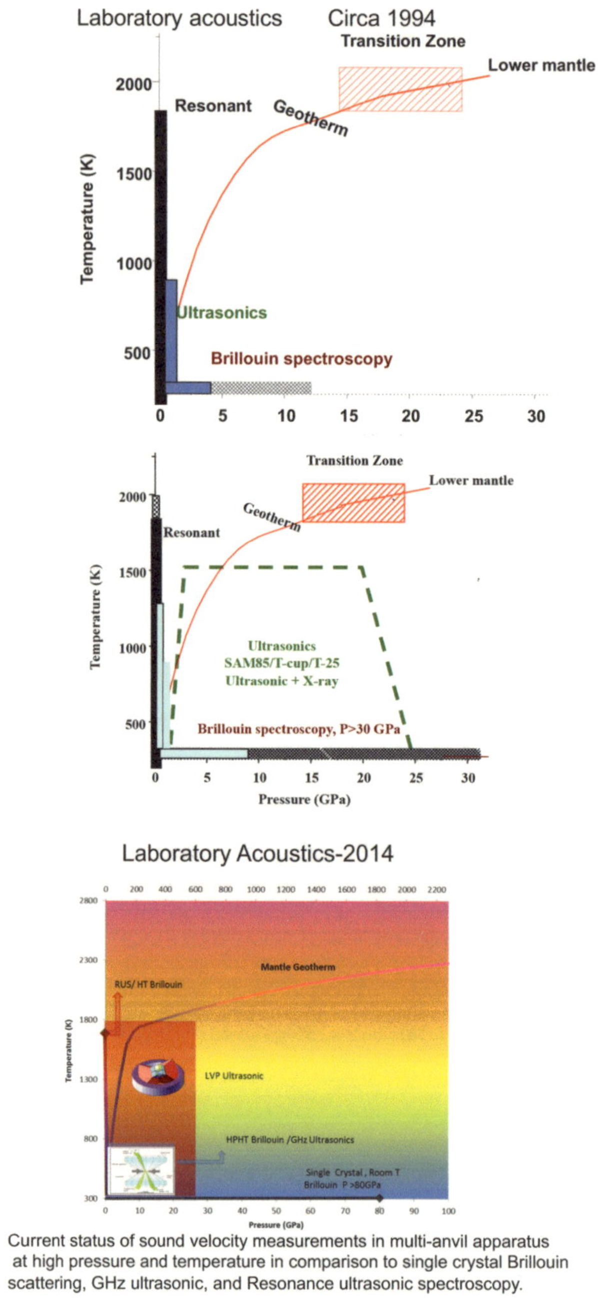

Figure 10.
The progress in the pressures and temperatures achievable in these investigations using laboratory acoustics from 1994 to 2014 is reflected in these three figures.

interferometry from the early 1950s to the present day. During this period of more than 60 years, sound wave velocity measurements have been increased from pressures less than 1 GPa and temperatures less than 800 K to conditions above 25 GPa and temperatures of 1800 K. This is complimentary to other direct methods to measure sound velocities (such as Brillouin spectroscopy and impulsive stimulated scattering) as well as indirect methods (e.g., resonance ultra- sound spectroscopy, static or shock compression, inelastic X-ray scattering). Newly-developed pressure calibration methods by Wang *et al.* [9] and data analysis procedures using a finite strain approach are described and applied to data for the major mantle minerals. These state-of-the-art ultrasonic experiments performed in conjunction with synchrotron X-radiation provide simultaneous measurements of the elastic bulk and shear moduli and their pressure and temperature derivatives with direct determination of pressure.

These sound velocity data are important in enabling scientists to interpret seismic data for the Earth's interior. Two of the most popular global seismic models are PREM and AK135 [10, 11] are plotted in **Figure 9**; these models include some distinct features including a low velocity zone around 80 km–150 km, jumps at 410- and 670-km depths, and high velocity gradients in the transition zone. In addition, regional seismic studies also revealed discontinuities at 520 km as well as at depths of 250 km–340 km (X discontinuity). Also plotted are phase transitions as possible causes of these velocity anomalies in a pyrolytic mantle compositional model (See **Figure 9** and **Figures 3** and **7a** and **b** above).

The progress in the pressures and temperatures achievable in these investigations using laboratory acoustics from 1994 to 2014 is reflected in the following figures (**Figure 10**); copyright R. C. Liebermann.

6. Current status of mineral physics research

In early 2019, I wrote a paper entitled "The Orson Anderson Era of Mineral Physics at Lamont in the 1960s", and began to explore options for its publication. When the Assistant Editor for Minerals, Ms. Jingjing Yang, agreed to consider my paper, she also inquired as to whether I would like to the be the Guest Editor for a Special Issue in honor of Orson Anderson. After asking prospective authors about the viability of such a Special Issue, I accepted her invitation, with the hope and expectation that it would be a wonderful present for his 95th birthday. This Special Issue is the result [12]. It contains original scientific papers, as well as historical reviews of the field of mineral physics (and also rock physics).

The papers in this Special Issue are grouped into four categories: Reviews, Experimental Science, Theoretical Science and Technological Developments. These papers include those from; first authors covering five generations of mineral physicists, including contemporaries of Orson, the next generation of leaders in mineral physics throughout the world, current leaders in the field, senior graduate students, and an undergraduate student (i.e., Tyler Perez). Note that Tyler, a student of Jennifer Jackson at Caltech, is an academic great-great grandson of Orson Anderson (Anderson > Liebermann > Bass > J. Jackson > Perez).

Examples of papers in all four categories of the Special Issue [12]:

7. Review

William A. Bassett.

The Takahashi–Bassett Era of Mineral Physics at Rochester in the 1960s. Reprinted from: Minerals 2020, 10, 344, doi:10.3390/min10040344

8. Experimental science

Agnés Dewaele.

Equations of State of Simple Solids (Including Pb, NaCl and LiF) Compressed in Helium or Neon in the Mbar Range, Reprinted from: Minerals 2019, 9, 684, doi:10.3390/min9110684

Tyler Perez, Gregory J. Finkelstein, Olivia Pardo, Natalia V. Solomatova and Jennifer M. Jackson.

A Synchrotron Mössbauer Spectroscopy Study of a Hydrated Iron-Sulfate at High Pressures Reprinted from: Minerals 2020, 10, 146, doi:10.3390/min10020146

Francesca Miozzi, Jan Matas, Nicolas Guignot, James Badro, Julien Siebert and Guillaume Fiquet.

A New Reference for the Thermal Equation of State of Iron, Reprinted from: Minerals 2020, 10, 100, doi:10.3390/min10020100

9. Theoretical science

J. Michael Brown and Baptiste Journaux.

Local-Basis-Function Equation of State for Ice VII–X to 450 GPa at 300 K, Reprinted from: Minerals 2020, 10, 92, doi:10.3390/min10020092

Jun Tsuchiya, Risa Nishida and Taku Tsuchiya.

First Principles Calculation of the Stability of Iron Bearing Carbonates at High Pressure Conditions, Reprinted from: Minerals 2020, 10, 54, doi:10.3390/min10010054

10. Technological developments

Tony Yu, Clemens Prescher, Young Jay Ryu, Feng Shi, Eran Greenberg, Vitali Prakapenka, Peter Eng, Joanne Stubbs, Yoshio Kono, Guoyin Shen, Heather Watson, Mark L. Rivers, Stephen R. Sutton and Yanbin Wang.

A Paris-Edinburgh Cell for High-Pressure and High-Temperature Structure Studies on Silicate Liquids Using Monochromatic Synchrotron Radiation, Reprinted from: Minerals 2019, 9, 715, doi:10.3390/min911071

Acknowledgements

I am grateful to William Bassett, Thomas Duffy and Taku Tsuchiya for their contributions to the sections on diamond anvil cells, shock waves and theoretical mineral physics, respectively. I appreciate the courtesy of many colleagues in providing figures to illustrate this paper: Ed Garnero, Quentin Williams, Guoyin Shen, Shun Karato, Nick Schmerr and Stas Sinogeikin; all colleagues have granted explicit permission to use their figures in this chapter, none of which have been published. Portions of this chapter have been adapted and modified from the author's previous publications [3, 7, 8].

Conflict of interest

The author declares no conflict of interest.

Mineral Physics
DOI: http://dx.doi.org/10.5772/intechopen.102326

Author details

Robert Cooper Liebermann
Department of Geosciences and Mineral Physics Institute, Stony Brook University, Stony Brook, NY, USA

*Address all correspondence to: robert.liebermann@stonybrook.edu

References

[1] Hazen RM. What is mineral physics? EOS. Transactions of the American Geophysical Union. 1984;**65**:746

[2] Wentzcovitch R, Stixrude L. Theoretical and computational methods in mineral physics: Geophysical applications, Reviews in Mineralogy and Geochemistry. Chantilly, Virginia: Mineralogical Society of America; 2010. Ser, 71, xvii + 484 pp

[3] Liebermann RC, Prewitt CT. From Airlie House in 1977 to Lake Morey in 2007: 30 Years of evolution of mineral physics. Phys. Earth Planet. Interiors. 2014;**228**:36-45

[4] Kircher A. Mundus subterraneus, quo universae denique naturae divitiae; 1665

[5] Birch F. Elasticity and constitution of the Earth's interior. Journal of Geophysical Research. 1952;**57**:227-286

[6] Bassett WA. Diamond anvil cell, 50th birthday. High Pressure Research. 2009;**29**:165-186

[7] Liebermann RC. Multi-anvil, high-pressure apparatus: A half century of development and progress. High Pressure Research. 2011;**31**:493-532

[8] Li B, Liebermann RC. Study of the Earth's interior using measurements of sound velocities in minerals by ultrasonic interferometry. Physics of Earth and Planetary Interiors. 2014;**233**:135-153

[9] Wang X, Chen T, Qi X, Zou Y, Kung J, Yu T, et al. Acoustic travel time gauges for in-situ determination of pressure and temperature in multi-anvil apparatus. Journal of Applied Physics. 2015;**118**. 065901, 9 pp

[10] Dziewonski AM, Anderson DL. Preliminary reference Earth model. Physics of the Earth and Planetary Interiors. 1981;**25**:297-356

[11] Kennett BLN, Engdahl ER, Buland R. Constraints on seismic velocities in the Earth from traveltimes. Geophysical Journal International. 1995;**122**:108-124

[12] Liebermann RC. [Guest Editor], Mineral Physics—In Memory of Orson Anderson, Book version of Special Issue of MINERALS. Switzerland: MDPI, Basel; 2020. pp. 628

Chapter 4

Testing and Validating Instruments for Feedstocks of Mineral Carbonation

Muhammad Imran Rashid

Abstract

Different feedstocks Dunite, Olivine and Lizardite are examined in this research using various measuring techniques such as TGA-MS, XRD and Quantitative XRD and EDS. Quantitative XRD results matched with TGA-MS results. Malvern Mastersizer, EDS and QXRD results also showed a good match regarding the individuality of results which are shown graphically. TGA-MS calibration curves example is provided. Matching the results of different measuring techniques is a key to fundamental research. Comparison of the reactivity of dunite, soaked dunite, heat-activated dunite and lizardite and raw dunite soaked has been performed. TGA-MS and QXRD results match each other. Malvern Mastersizer, EDS and QXRD results match with their individual results indicating the instrument's reliability. Semi-Quantitative XRD results authenticity is EXCELLENT. TGA-MS results match with QXRD is excellent. Mineral carbonation converts CO_2 into stable mineral carbonates. This research explores the utilisation of serpentinised dunite (which is comprised of 61% lizardite) as a potential feedstock for mineral carbonation. Heat activation, *ex-situ* regrinding and concurrent grinding techniques were employed to enhance the reaction rate and yield, and to provide information on the carbonation reaction mechanism. Silica-rich layers that appeared during reference experiments were disrupted using concurrent grinding and significantly higher magnesite yields and Mg extractions were obtained.

Keywords: CCS (carbon capture and storage), CCSU (carbon capture, storage and utilisation), mineral carbonation (MC), carbon capture (CC), materials science

1. Introduction

Greenhouse gases especially CO_2 concentration in the atmosphere has increased to a level of 419 ppm compared to a value of 280 ppm from the preindustrial revolution (1975) [1]. Reduction in greenhouse gases is a need of time. Significant research has been published regarding mineral carbonation [2–15], geological carbon dioxide storage, oceanic storage [3], carbon dioxide conversion into chemicals, carbon dioxide fixation in polymers and carbon dioxide conversion into Urea [16, 17]. Mineral carbonation is one of the forefront technologies recently proposed. Although various publications have been done in this field [2], the basic need of time is to foresee how the research efforts need to be oriented or centred on that technology [3, 4, 9–11, 13, 14, 18–20]. This article will indicate some directions for the utilisation of different feedstocks for CO_2 utilisation and fixation. Geological storage poses a threat to nearby occupants as there were thousands of killings in the Crater Lake incident. ALOHA software can be used for the estimation of such leakages if CO_2 is to be stored in geological formations. CO_2 can cause asphyxiation; hence, safety measures are at most necessity. ALOHA can estimate CO_2 vapours travelling distances and how far this gas can travel and how much concentration will be at a specified point. Oceanic storage pose threat to aquatic life and is expected to disrupt the ecosystem seriously. Increased concentration of CO_2 in oceans [21] will reduce the pH of the oceanic water, rivers or canals making them undrinkable. Seawater is used in various industries. Reduced pH will cause serious corrosion issues and may result in materials damage and or stress corrosion cracking.

Greenhouse gases are uncontrollable. Each greenhouse gas concentration increases day by day. CO_2 is recently converted to jet fuel using sunlight by Adele Peters from Fast Company [22]. Researches are not giving up. However, the Antarctic lake has disappeared in just 3 days [23]. More efforts need to be initiated. Extremism in climate shattering weather patterns is expected right now [23]. Europe has seen extreme flooding in 2021. Pakistan has seen extreme summers like what the Middle East has seen shooting of temperatures. Catalysts have been discovered to convert CO_2 into fuel [24]. A single reason why CO_2 is not controlled is that industries emit more than capture. Adam Vughan has indicated that atmosphere warming could not have been kept below 1.5°C [25]. Alas, more seriousness is required. No negative emissions drama. The cement industry is also one of the largest CO_2 emitting industries. Novel modifications are proposed to overcome this threat [26]. Coal-burning emissions and their environmental effects are also highlighted [27].

2. Analytical instruments

Dunite, different varieties of olivine and lizardite are used in this research. Proper functioning of analytical instruments is a fundamental to perform the highest level of research. Fundamental instruments operation and working is described here.

2.1 TGA-MS analysis

Thermogravimetric analysis (TGA) measures the change in mass over time as the sample is heated. These measurements provide compositions of different feeds or carbonated products. TGA is suitable to characterise different materials that display mass loss or gain due to thermal decomposition and thus enable an estimate of magnesite yields of the carbonated products to be obtained. Feed materials or carbonated products were heated in TGA (**Figure 1**) from 25 to 1000°C and mass losses due to decomposition of different phases present are identified. To identify the evolving gases generated during heating, the TGA-DSC (Setsys Evolution 1200) was coupled with a mass spectrometer (Thermostar Quadrupole). The initial loss of mass observed between 25 and 280°C corresponds to physically bound moisture present in the sample, while the second mass loss from 280 to 430°C corresponds to brucite decomposition, while the third major mass loss in the range of 430–830°C corresponds to lizardite decomposition (**Figure 2**).

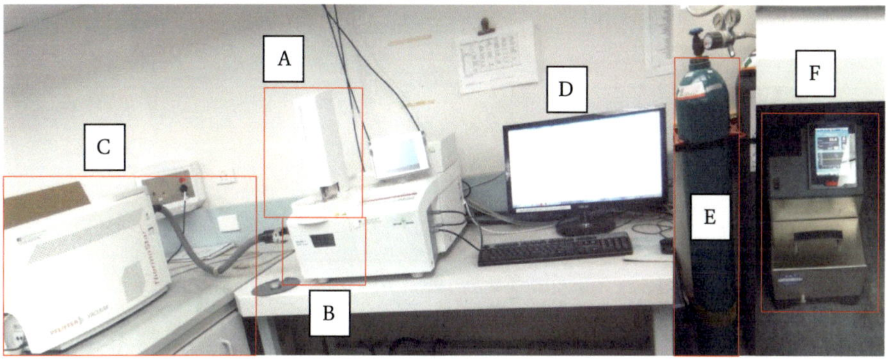

Figure 1.
Photo of TGA-MS set up. A, autosampler and small sample crucibles; B, TGA furnace where the sample is being heated; C, mass spectrometer connected with TGA furnace to receive evolved gases from TGA; D, computer for data output; E, argon cylinder for argon gas flow; F, chiller to cool down TGA furnace.

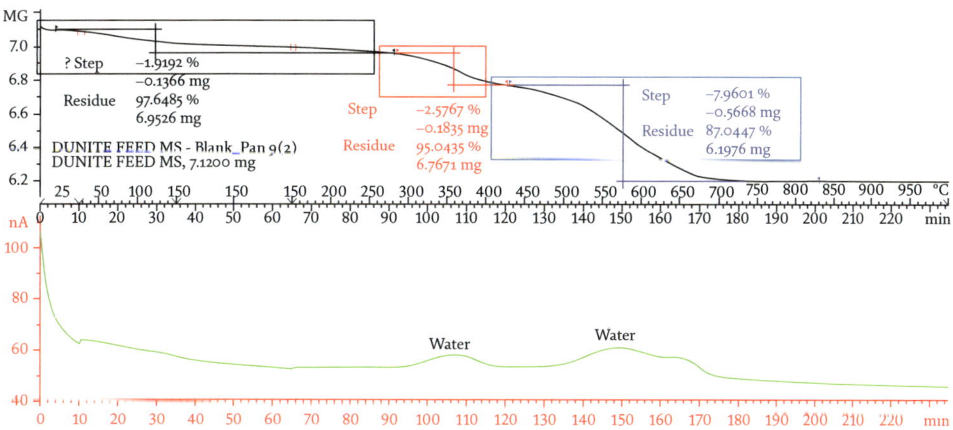

Figure 2.
Typical TGA-MS curve. The first significant loss of mass is due to moisture present in the sample. The second mass loss is due to brucite decomposition. The third mass loss is due to lizardite decomposition. All major changes in mass are due to the elimination of H_2O vapour from the sample.

Lizardite decomposes over the same temperature range (300–600°C) as the magnesite [6] and this can introduce a systematic error in magnesite yield estimation (leading to an over-estimation of the magnesite yield) unless the mass loss in this period can be quantitatively attributed to loss of H_2O or CO_2 from the sample. To distinguish between these species, the ion current from the m/z = 44 ion (CO_2^+) from mass spectrometer was calibrated using sodium bicarbonate samples and a calibration curve for CO_2 concentration was obtained, which is used to quantify CO_2 mass loss (distinguishing CO_2 production from the loss of water vapour, which occurs simultaneously) and thus render more accurate estimation of the magnesite yield. CO_2 peak areas were calculated using the mass spectrometer data and CO_2 mass response is estimated based on the reaction ($2NaHCO_3 \rightarrow Na_2CO_3 + CO_2 + H_2O$). The relationship between peak areas and CO_2 mass loss was linear.

For carbonated samples, the CO_2 peak areas were estimated using MS data and then these peak areas are used to determine CO_2 mass loss applying the CO_2

Step	Reaction [29]	Theoretical mass loss (%)	Measured mass loss (%)
1	$CaC_2O_4 \cdot H_2O \rightarrow CaC_2O_4 + H_2O$ (189°C)	12.3	12.6 ± 0.66
2	$CaC_2O_4 \rightarrow CaCO_3 + CO$ (502°C)	19.2	19.9 ± 0.98
3	$CaCO_3 \rightarrow CaO + CO_2$ (763°C)	30.1	31.2 ± 0.53

Table 1.
TGA calibration data for three runs.

calibration curve. This CO_2 mass loss was used in an equation to calculate magnesite yield. The equation is based on the Gadikota formula [28].

Three TGA runs (**Table 1**) were completed with calcium oxalate hydrate (99% pure) to calibrate the TGA response. Theoretical and measured mass loss shows good agreement (**Table 1**).

3. XRD analysis and QXRD (semi-quantitative XRD)

XRD analysis is useful in the determination of crystalline phases that exists in a powdered sample [30]. Each phase has a specific, identifiable x-ray diffraction pattern, which is used to determine different phases present in the sample. X-rays are generated from the emission of high energy electrons from hot tungsten elements, which are bombarded on a copper metal target. This bombardment causes an electron emission from target atoms, thus generating an electron vacancy which is filled by an electron from higher energy orbitals and this transition generates x-rays. Filtration of these x-rays is performed to get monochromatic radiation which is bombarded on the sample being analysed. Bragg's equation is the main law used in XRD diffraction pattern analysis [31].

$$n\lambda = 2d\sin\Theta \quad (1)$$

λ = wavelength of x-rays, n = integer, d = plane spacing, Θ = Bragg's diffraction angle.

To derive Bragg's law, consider two x-rays (A and D) impinging on the atom B and E of a crystal and the angle of incident and angle of reflectance are equal as shown in **Figure 3**. Incident waves A and D are in phase with each other although wave D has to travel an extra distance of GE + EH to remain in the same phase as wave A. This extra

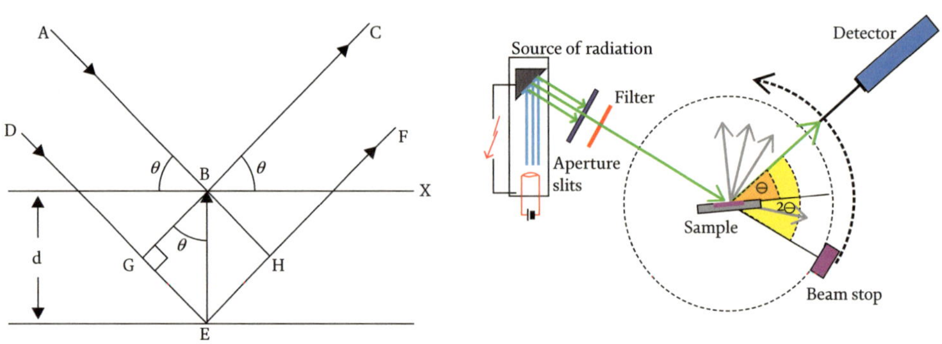

Figure 3.
Schematic of X-ray diffraction (left), Bragg–Brentano geometry (right).

distance must have been an integral (n) multiple of wavelength (λ). The length GE and EH are equal and GE equals to $d\ sinn\Theta$. Bragg–Brentano design is the most commonly used instrument geometry for high-resolution powder diffraction. The incident beam through a number of slits diverges towards the sample, the diffracted signal from the sample again converges through a number of slits towards the detector. A $\Theta/2\Theta$ rotation is employed to keep incident and diffracted wave paths in symmetry. During sample scanning, the sample rotates by Θ while the detector is rotated by 2Θ with each step [32].

In practice, finely ground feed materials and carbonated samples (up to 100 μm size particles) were put in the instrument holder for analysis. Samples were finely grounded in order to avoid intensity fluctuations and preferred orientation. XRD analyses were performed using Philips X'Pert Pro multipurpose diffractometer with Cu radiation and 2θ from 5 to 90° or 11 to 31° depending upon the sample being analysed. Collection time used was 1 s with a step size of 0.02°. The patterns from XRD were matched with the International Centre for Diffraction Data ® (ICCD) using X'Pert Highscore® in order to identify crystalline phases. A typical x-ray diffraction pattern for feed dunite is shown in **Figure 4**. Phases identified are lizardite, olivine, brucite and magnetite.

If the reference intensity ratio (RIR) of an analytical phase i (such as silicon) is known, then its concentration can be calculated by doping the original sample with the analytical phase. This can be done by the addition of a known amount of standard (silicon) of which the RIR is known. For semiquantitative method details please see below.

After obtaining the diffraction pattern of the doped sample, the concentration C_i in the original sample is calculated as follows:

$$C_i = A_x \times \left(\frac{I_i}{I_x}\right) \times \left(\frac{RIR_X}{RIR_i}\right) \qquad (2)$$

C_i = concentration of given phase i in the original sample
Axe = known amount of standard (silicon) added to the original sample

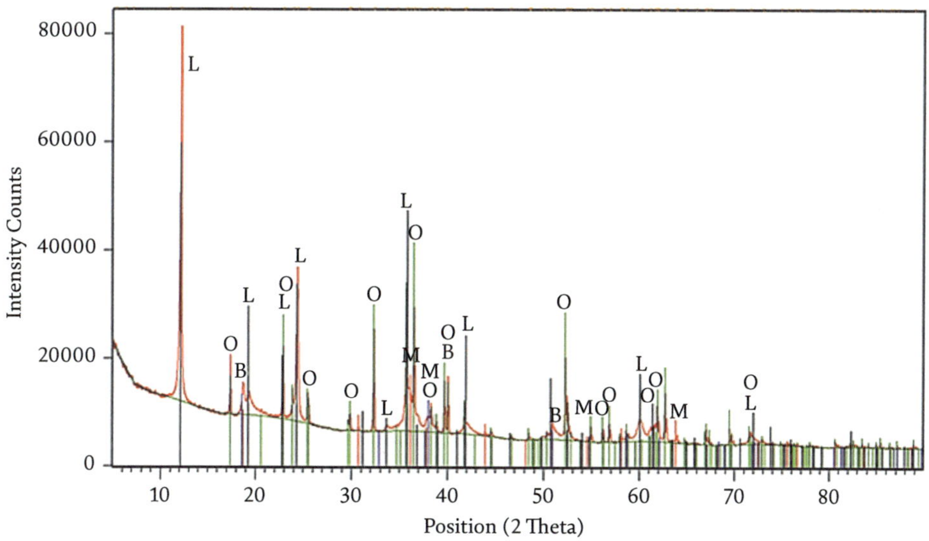

Figure 4.
Typical x-ray diffraction pattern for dunite. L, Lizardite; O, olivine; B, Brucite; M, magnetite.

Position (2Θ)	Intensity			Formulas used
x	y	Y′	Area	
11.83	15,480	15,480	2.96	(Slope) $\mathbf{M} = \frac{(y_2 - y_1)}{(x_2 - x_1)}$ (Intercept) $\mathbf{C} = y - M \times X_1$ (Derivative) $\mathbf{Y}' = M \times X_1 + C$ $\mathbf{Area} = \frac{((y_1 - Y'_1) + (y_2 - Y'_2)) \times (X_2 - X_1)}{2}$ $\mathbf{C_i} = \mathbf{A_x} \times \left(\frac{I_i}{I_x}\right) \times \left(\frac{RIR_X}{RIR_i}\right)$ (3)
11.84	15,924	15,468	11.0	
11.85	16,699	15,456	17.7	
11.87	16,940	15,445	23.0	
11.88	17,482	15,433	30.6	
11.89	18,093	15,422	38.1	

Table 2.
Semiquantitative XRD analysis.

I_i, I_x = intensities (peak areas) of phases i and x in the doped sample
RIR_i, RIR_x = reference intensity ratio values of i and x respectively
The procedure is described below in detail (for calculation detail see **Table 2**)

1. Add known amount of silicon (usually 20 wt/wt%) in the sample and scan it with XRD

2. Identify phases present in a given XRD pattern

3. Determine the area under the main peak of each phase. The peak should not overlap other peaks. To calculate the area under the main peak, first, calculate slope (M) and then intercept (C). Then calculate derivative (Y′) and use the given formula to calculate the area under the peak (please refer to **Table 2**). Use these peak areas in Eq. (2).

4. The RIR value is found from the Highscore® database

5. The concentration C_i of each phase is estimated using Eq. (2)

3.1 ICP-OES (inductively coupled plasma: Optical emission spectrometry)

The elemental composition of solid and liquid samples can be determined using ICP-OES. ICP-OES consists of two major components; the torch and optical spectrometer. The torch comprises quartz tubes [33]. To produce plasma, argon gas is normally used, which passes through the tubes around the induction coil. The argon gas is "ignited" by the Telsa unit and the ionisation process (plasma formation) is initiated. The ionisation of argon gas occurs at this stage. A plasma having approximately 7000 K temperature is generated because of collisions between neutral argon atoms and charged particles [34]. Using a peristaltic pump, an aqueous sample is continuously supplied to the nebuliser where it changes to mist and moves to the plasma envelope. The introduced sample interacts with electrons and ions in the plasma and is converted into charged ions. This causes the decomposition of different molecules into respective atoms that lose electrons to induce the emission of radiation of distinctive wavelengths of elements present inside the sample. The optical spectrometer separates these wavelengths into component wavelengths. Intensities are

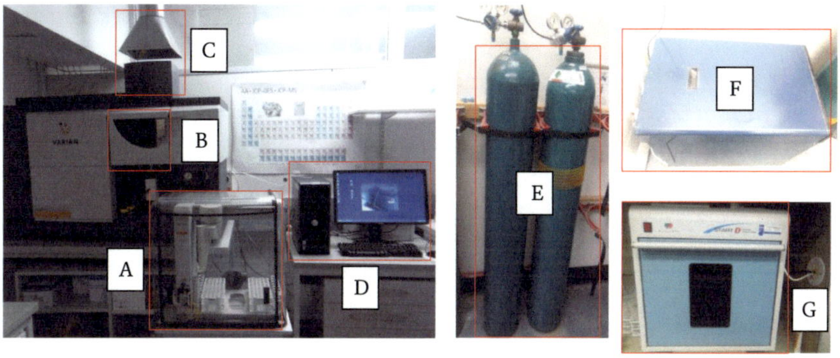

Figure 5.
Photo of the ICP-OES set-up and microwave digestion system. A, autosampler with standards and sample tube holders; B, plasma chamber; C, gases exhaust; D, computer for analysis output; E, argon gas cylinders; F, chiller; G, microwave digestion system.

compared with the intensities of standard solutions of known element compositions and elements concentrations are computed based on the calibration curves. ICP-OES set-up and microwave digestion system is shown in **Figure 5**.

Solid samples used in the present investigation were first digested in acidic solution. Dunite sample (0.1 g) was digested in a microwave oven (**Figure 5**) using a mixture of 4.5 mL HNO_3 (65%), 4.5 mL HCl (37%) and 3 mL HBF4 (tetrafluoroboric acid, 50%). Thulium (50 µL) was added as a tracking element. The volume of this mixture was increased to 20 mL by the addition of 2% nitric acid prior to its digestion in the microwave. Digestion was not required for supernatant solution samples and they are diluted using 2% nitric acid to the required level (50%/100% dilution) prior to their analysis by ICP-OES (Varian, Australia). The typical curve for ICP-OES is shown in **Figure 6**. Mg concentration drop with the passage of time due to magnesite precipitation. Si concentration increases during the first hour but then it stays constant, which is due to simultaneous silicon leaching from dunite and its precipitation in the form of silica.

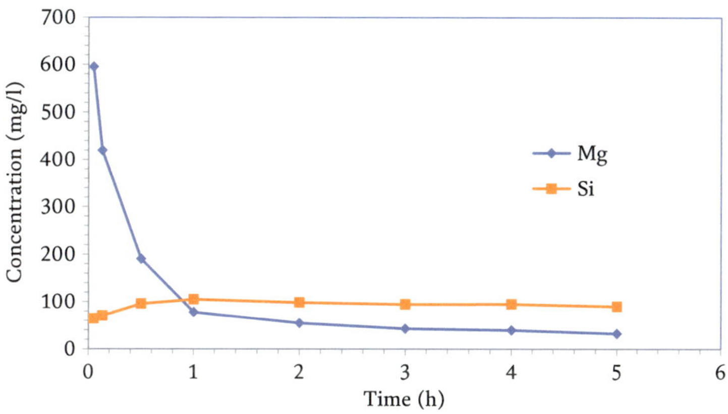

Figure 6.
Typical curve for ICP-OES. The graph represents Mg and Si concentrations variation with time for supernatant solution of sub 75 µm heat-activated dunite carbonated sample. Carbonation reaction was performed with 15% solids slurry at 185°C, 130 bar pressure and using 0.64 M $NaHCO_3$.

3.1.1 Scanning Electron microscope (SEM)/energy dispersive scattering (EDS)

Morphology, surface topography and elemental compositions of feed materials and carbonated products were determined using SEM (Zeiss Sigma VP FESEM) and EDS (Bruker). SEM scans a fine electron beam over the material being analysed and uses different detectors to reconstruct the image from signals produced from the sample [35]. SEM consists of different parts, e.g., microscope column which also includes electron gun and electron beam travels in this column; the computer that drives the microscope; ancillary equipment which analyses the composition. SEM can magnify objects from 10 times to 300,000 times. Scanning from an electron microscope can be compared with a person having a torch and looking for objects on the wall. As a person builds an image in his/her memory, SEM works in the same way and uses a fine electron beam instead of the torch to build an image.

EDS is a technique that provides information about the chemical composition of the sample. For EDS, an electron beam is focussed on the sample during SEM analysis and these electrons interact with the atoms. X-rays are produced from these interactions and an energy dispersive detector detects these x-rays and displays a signal in the form of spectrum, histogram or intensity versus x-ray energy. This makes it possible to identify elements present in the sample.

Sample preparation is important for SEM. Samples are gold (imaging) or carbon (EDS) coated prior to their analysis. Gold coating provides a thin layer to the samples and samples were coated four times at a 90° angle and fifth time from the top. A typical SEM micrograph and EDS spectrum of dunite feed sample are shown in **Figure** 7. SEM shows an image of the dunite feed and the EDS spectrum indicates intensities of the elements present in the sample. Polished resin blocks were used to study the silica-rich layers. Polished resin blocks were prepared using feed material, carbonated products and resin. Photo of polished resin blocks and sample holders is shown in **Figure 8**. The polished resin block samples preparation procedure is given in appendix 3D.

3.1.2 Transmission Electron microscope (TEM)

TEM is useful to study the structure, properties and compositions of different mineral powders, especially in the submicron range. Mineral particles should have

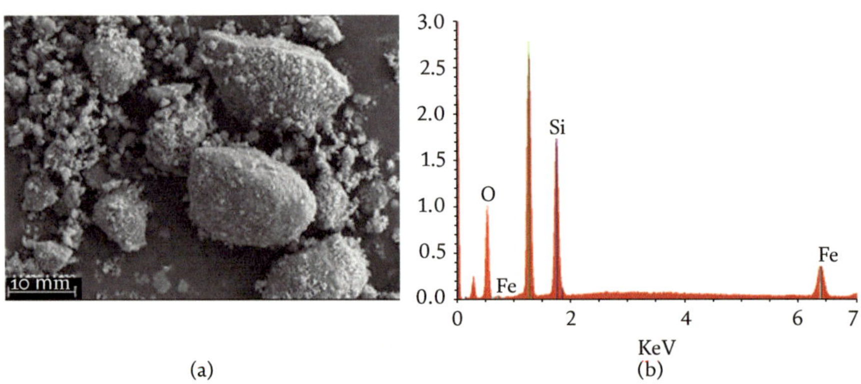

Figure 7.
SEM micrograph and EDS spectrum of dunite feed. (a) Dunite feed SEM micrograph, 10 µm is a resolution of the SEM (b) EDS spectrum of dunite feed, intensities of different elements are shown, Mg, magnesium; Si, silicon; O, oxygen; Fe, iron.

Figure 8.
Photo of the sample holders (left) and polished resin blocks (right).

been in 50–100 nanometres size to be properly analysed by TEM. Electrons transmission through the mineral particles enables detailed analysis of the particle features especially its crystal structure, orientation and chemical composition. In the present study, TEM was used to identify shell (silica-rich layers) and core part of the reacted mineral particles and study the corresponding elemental compositions and structure.

To prepare samples for TEM (JEOL 2100 TEM) analysis, 10 mg of sample powder was added to the pestle and mortar. Ethanol (4 ml) was mixed with the sample powder and contents were ground for 3 min. Ground sample was moved into a 5 ml plastic vile and sonicated for 20 min. Using pipette 1–2 drops were dropped on the TEM grid (200 mesh Cu, ProSciTech) and air-dried overnight prior to TEM analysis.

3.1.3 Fourier-transform infrared (FTIR) spectroscopy

FTIR is useful to acquire an infrared spectrum of a solid sample either feed or product. In FTIR spectroscopy, most molecules absorb light in the infra-red region of the electromagnetic spectrum and this absorption corresponds to the specific bonds present in the molecule. Measurement usually is in wave numbers typically over the range 4000–400 cm^{-1}. When the sample being analysed is exposed to radiation, some portion of the radiation is absorbed while other is transmitted. The block diagram of the FTIR spectrometer is shown in **Figure 9**. Dunite and carbonated products were analysed by FTIR (Bruker, Tensor 37 Spectrometer). Samples for FTIR spectroscopy were prepared using 99% KBr and 1% sample. Samples were thoroughly mixed and ground before making pellets. These pellets were placed in an oven under a vacuum at 150°C for 24 hours before analysis. Typical FTIR spectra of raw dunite are provided in **Figure 10**.

3.1.4 Malvern Mastersizer

Particle size distribution (PSD) of feed materials and carbonated products were determined using a particle size analyser (Mastersizer 2000, Malvern Instruments) (**Figure 11**). For particle size measurement, it is important how particles scatter and absorb light. Initially, the Fraunhofer model [36] was used to predict scattering patterns when a solid disc of particles is passed through the laser beam. This model failed to describe exact scattering as very few particles are disc-shaped. Mie theory is currently used which is able to predict the light scattering behaviour of all materials. Each size of particle has a specific scattering pattern. The particle size analyser uses the above-mentioned theories and works backwards to calculate particle

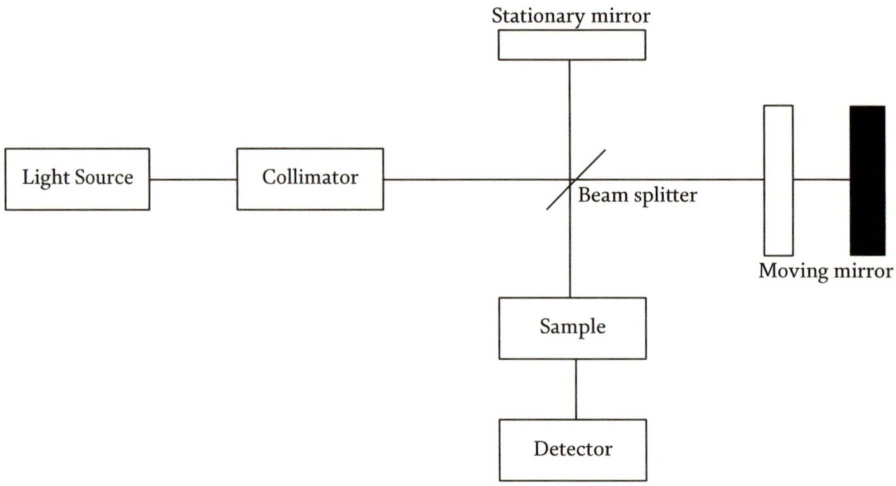

Figure 9.
Block diagram of FTIR spectrometer.

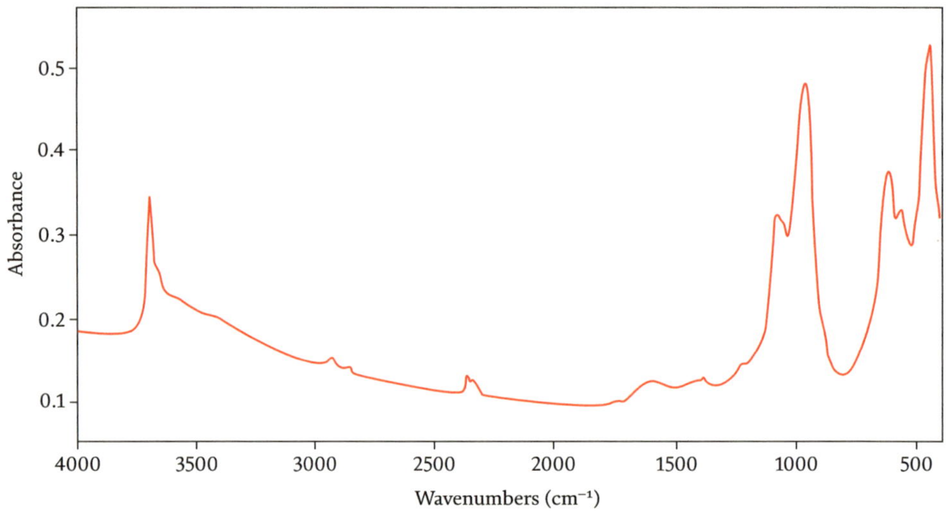

Figure 10.
Typical FTIR spectra of raw dunite, stretch present around 3690 cm^{-1} are due to presence of surface-bound OH moieties, the vibration at 1073 cm^{-1} due to out of plane vibration of Si-O, adsorption band at 970 cm^{-1} are in-plane Si-O stretching vibration, the feature at 629 cm^{-1} are due to deformation of hydroxyl groups, stretches at 564 cm^{-1} are Mg-O out of plane vibration, stretches at 450 cm^{-1} are Si-O-Si bond bending vibration (reference [9] and references therein).

size from the captured scattering pattern. Basic laser diffraction system is shown in **Figure 12**.

Three procedures are used for any particle size measurement. First, a sample is prepared and dispersed in a dispersion unit in proper concentration followed by its delivery to the optical bench. Second, a scattering pattern is captured from this sample which is also called "measurement" and is done by the optical bench. Third, raw data from measurement is analysed by instrument software to provide the PSD. Finely ground samples ($\leq$20 μm) were preferably analysed in wet solution form. These samples tend to agglomerate during drying even if dried under vacuum and at low

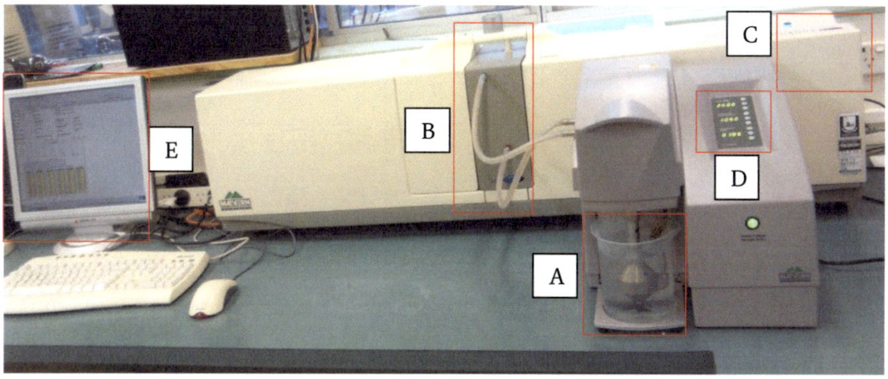

Figure 11.
Photo of Malvern mastersizer 2000. A: Sample dispersion unit and pump impeller, B: Sample cell where dispersed sample is moving and laser light pass through it, C: Laser source and laser ON indicator, D: Pump speed adjustment and ultrasound operation system, E: Computer to show output.

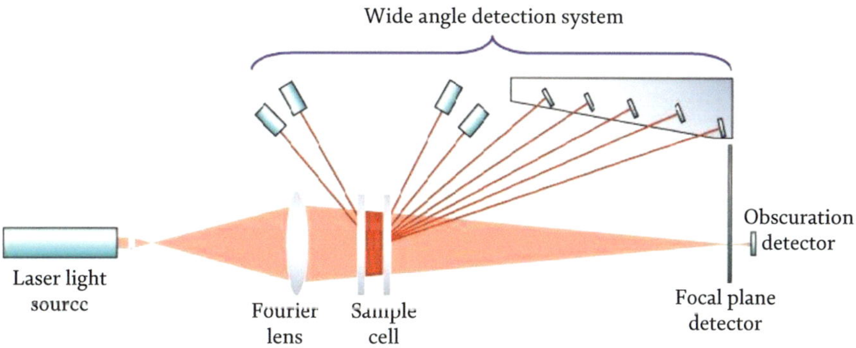

Figure 12.
Basic laser diffraction system of Mastersizer 2000.

temperatures (70°C). For these samples and heat-activated samples, a built-in ultrasonic system in the Malvern mastersizer was used to break any agglomerates present. A typical PSD from the Malvern mastersizer is shown in **Figure 13**. Size classes are represented on the x-axis in μm and the volume density of particles is represented in percentage on the y-axis. The percentiles, d_{10}, d_{50} and d_{90} are shown in **Table 3**. D_{10} means that 10% volume of particles is smaller than this size (27 μm), d_{50} means that 50% volume of particles is smaller than this size (42 μm) and d_{90} means that 90% volume of particles is smaller than this size (64 μm).

4. Results and discussion

Olivine (Netherland) and Olivine (Norway) characterisation using Quantitative XRD (X-Ray Diffraction) analysis is discussed. The supplier of these samples has promised to supply olivine; however, these appear as a mixture of olivine and a few other minerals. These samples may be rocks but still, Olivine (Norway) is probably like olivine as it has higher olivine content. This olivine content (62%) almost match lizardite content (61–62%) of the dunite [18] used primarily in my research.

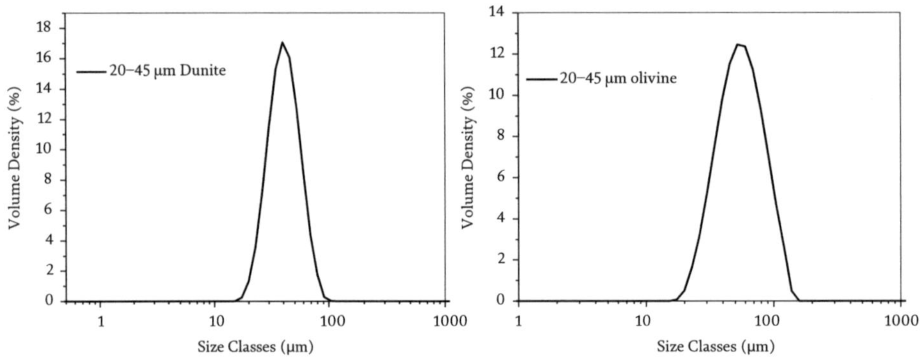

Figure 13.
PSD for 20–45 µm dunite (left), PSD for 20–45 µm olivine (right).

	d_{10} (µm)	d_{50} (µm)	d_{90} (µm)
20–45 µm dunite	27	42	64
20–45 µm olivine	26	45	76

Table 3.
PSD for 20–45 µm dunite and 20–45 µm olivine.

Component	Olivine (Norway) %	Olivine (Netherland) %
Olivine	62.01	8.40
Clinochlore	11.58	7.10
Lizardite	5.43	53.01
Orthopyroxene	9.993	Tremolite 21.73
Phlogophite+Annite	1.307	Dolomite + Pyroxene (8.15)
Halite	0.884	Magnesioferrite (1.15)
Chabazite-Ca	4.543	—
Cordierite	4.243	—
Total	100	100

Table 4.
Olivine (Norway) and olivine (Netherland) QXRD analysis.

Powders samples were prepared for QXRD analysis of olivine (Netherland) and olivine (Norway). **Table 4** shows the analysis. Please refer to earlier publications [2, 10, 13, 14, 18, 19] for detailed description of methods. TGA-MS curves are presented in **Figure 14**. TGA commonly used in mineral carbonation to obtain yields was coupled with MS (Mass Spectrometer). This enabled us to calculate yields for dunite rock as evolved CO_2 gas and water vapours were measurable. TGA-MS was routinely calibrated. One example of calibration curves is shown in **Figure 15**.

4.1 TGA-MS analysis of olivine (Norway) and olivine (Netherland)

Olivine (Norway) and Olivine (Netherland) were analysed using TGA coupled with MS (Mass Spectrometer). Results are presented in **Figure 14**.

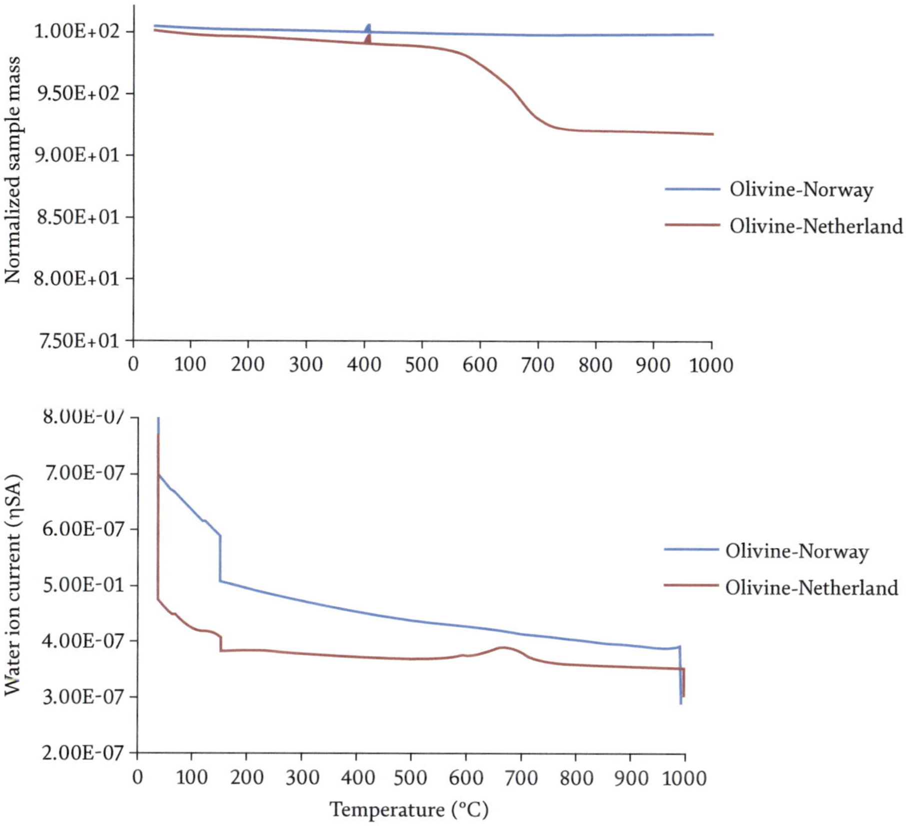

Figure 14.
TGA-MS analysis of olivine (Norway) and olivine (Netherland).

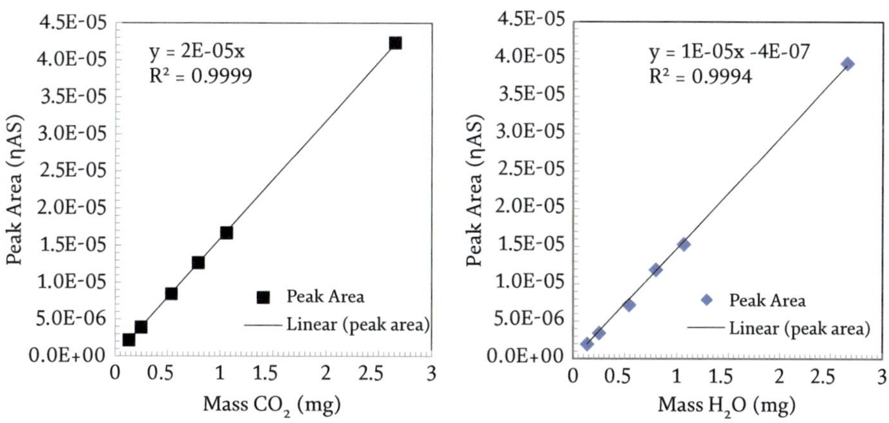

Figure 15.
Calibration curve to calculate mass loss of CO_2 (left), Calibration curve to calibrate mass loss of H_2O (right).

These results at least partially confirm that Olivine (Netherland) has more lizardite and is 53.01% as shown in **Table 4**. This verifies QXRD analysis. Dunite for this research was handpicked from Doonba deposit in vicinity of a small village/town Barbara, close to Tamworth city. **Tables 5** and **6** and other following tables present

	Lizardite	Olivine	Brucite	Magnetite
20–45 μm dunite	70	28	1.2	0.23
20–45 μm dunite	69	29	1.9	0.43
20–45 μm dunite	71	27	1.5	0.27
20–45 μm dunite	71	27	0.82	0.25
20–45 μm dunite	72	26	1.3	0.17

Table 5.
Semiquantitative XRD results (3 times XRD repeat, 2 reanalyses).

	Lizardite	Olivine	Brucite	Magnetite
−75 μm dunite	61	29	8.3	1.3
20–45 μm dunite	51	42	5.2	1.3
20–45 μm dunite repeat	51	42	5.9	1.3
45–75 μm	50	43	6	1.3
−20 μm dunite	66	24	8.6	1.3

Sr. No		Lizardite	Olivine	Brucite	Magnetite
1	−75 μm dunite QXRD	61	32	6.4	0.37
	−75 μm dunite TGA-MS	61	30 (difference)	8.3	0.37 (XRD)

Variation of results with the variation of size, e.g., −75 μm, 20–45 μm, 45–75 μm, −20 μm is interesting.

Table 6.
TGA-MS analysis results.

phases present in this dunite. Sub 75 μm dunite have 61% lizardite, 29% olivine, 8.3% brucite and 1.3% magnetite [18]. Percentage phases for other dunite fractions (20–45-micron, 45–75 micron and sub 20 micron) are presented in **Table 6**. An example of calibration curves for TGA-MS calibration is provided in **Figure 15**, routinely calibration was performed. Dunite is used here for calibration.

4.2 Yield calculation from dunite carbonated product

For magnesite yield calculation, please refer to earlier publication [10].

Carbonated products were heated in TGA from 25 to 1000°C and mass losses due to decomposition of different phases present are identified. To identify the evolving gases generated during heating, the TGA-DSC (Setsys Evolution 1200) was coupled with a mass spectrometer (Thermostar Quadrupole). Lizardite decomposes over the same temperature range (300–600°C) as the magnesite and this can introduce a slight systematic error in magnesite yield estimation (leading to an over-estimation of the magnesite yield) unless the mass loss in this period can be quantitatively attributed to loss of H_2O or CO_2 from the sample. To distinguish between these species, the ion current from the m/z = 44 ion (CO_2^+) from mass spectrometer was calibrated using sodium bicarbonate samples and a calibration curve for CO_2 concentration is obtained, which is used to quantify CO_2 mass loss (distinguishing CO_2 production from the loss of water vapour which occurs simultaneously) and thus render more accurate estimation of the magnesite yield. CO_2 peak areas were calculated using the

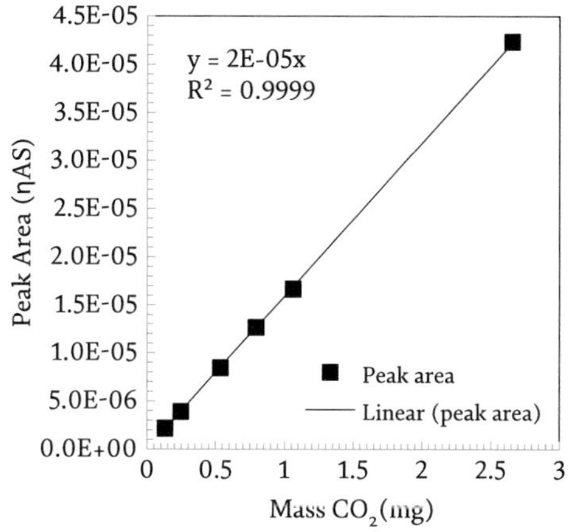

Figure 16.
Calibration curve to calculate mass loss of CO_2.

mass spectrometer data and CO_2 mass response is estimated based on the reaction ($2NaHCO_3 \rightarrow Na_2CO_3 + CO_2 + H_2O$). The relationship between peak areas and CO_2 mass loss was linear as shown in the CO_2 calibration curve (**Figure 16**).

For carbonated samples, the CO_2 peak areas were estimated using MS data and then these peak areas are used to determine CO_2 mass loss applying the CO_2 calibration curve. This CO_2 mass loss was used in Eq. (6) to calculate magnesite yield. Eq. (6) is based on the Gadikota formula (4).

Calculating fraction of magnesium (y_{Mg}) in dunite

$$\begin{aligned} \text{Yield (Rx)} &= \left[\frac{\text{Measured weight ratio of } CO_2 \text{ stored in mineral}}{\text{The residual } CO_2 \text{ storage capacity}}\right] \times 100\% \\ &= \frac{\left(\frac{W_{CO_2}}{W_{mineral}}\right)}{\frac{1}{R_{CO_2}}} \times 100\% \\ &= R_{CO_2} \times \left(\frac{\text{TGA}}{(100-\text{TGA})}\right) \times 100\% \\ &= R_{CO_2} \times \left(\frac{\text{TGA}}{(100-\text{TGA})}\right) \times 100\% \end{aligned} \tag{3}$$

$$CO_2 \text{ storage capacity of dunite} = \frac{1}{RCO_2} = \left(\frac{y_{Mg}}{MW_{Mg}} + \frac{y_{Ca}}{MW_{Ca}}\right) \times MW_{CO_2} \tag{4}$$

$$\%\text{of Mg in MgO} = \left(\frac{MW_{Mg}}{MW_{MgO}}\right) \times 100\% = \left(\frac{24.3}{40.3}\right) \times 100\% = 60.3\%$$

$$\%\text{of MgO in dunite} = 42.6\%$$

$$\%\text{of Mg in dunite} = 60.3\% \times 0.426 = 25.7\%$$

$$y_{Mg} = 0.257 \text{ (using this value in equation 5)}$$

Calculating fraction of calcium (y_{Ca}) in dunite

$$\%\text{of Ca in CaO} = \left(\frac{MW_{Ca}}{MW_{Cao}}\right) \times 100\% = \left(\frac{40}{56}\right) \times 100\% = 71.4\%$$

$$\%\text{of CaO in dunite} = 0.35\%$$

$$\%\text{of Ca in dunite} = 71.4\% \times 0.0035 = 0.25\%$$

$y_{Ca} = 0.0025$ (using this value in Eq. 5)

$$\frac{1}{RCO_2} = \left(\frac{y_{Mg}}{MW_{Mg}} + \frac{y_{Ca}}{MW_{Ca}}\right) \times MW_{CO_2} (\textit{equation } 5)$$

$$\frac{1}{RCO_2} = \left(\frac{0.257}{24.3} + \frac{0.0025}{40}\right) \times 44 = 0.468$$

$$RCO_2 = \left(\frac{1}{0.468}\right) = 2.136\ (\textit{using this value in equation } 4)$$

$$\text{Yield (Rx)} = 2.136 \times \left(\frac{TGA}{(100 - TGA)}\right) \times 100\% \quad (5)$$

W_{CO2} = Weight of CO_2 present in dunite before carbonation.
$W_{mineral}$ = Weight of dunite present before carbonation.
$1/R_{CO2}$ = CO_2 storage capacity of dunite.
y_{Mg} = Weight fraction of magnesium present in dunite which can react with CO_2.
MW_{Mg} = Molecular weight of magnesium (24.3 g/g mol).
MW_{MgO} = Molecular weight MgO (40.3 g/g mol).
y_{Ca} = Weight fraction of calcium present in dunite which can react with CO_2.
MW_{Ca} = Molecular weight of calcium (40 g/g mol).
MW_{CaO} = Molecular weight of CaO (56 g/g mol).
MW_{CO2} = Molecular weight of CO_2 (44 g/g mol).
R_{CO2} = Mass of dunite required to store unit mass of CO_2.
TGA = CO_2 mass loss from calibration curve.
R_X = Yield or extent of carbonation.

For a detailed description of materials, analytical instruments and experimental methods, please refer to Chapter 3 of the Ph.D. thesis [2]. Materials, Dunite, heat-activated dunite, heat-transformed dunite, twin sisters mountain dunite, olivine, lizardite and heat-activated lizardite are discussed. Analytical instruments, TGA-MS, XRD, Semi-Quantitative XRD (QXRD), ICP-OES, SEM, EDS, TEM, FTIR and Malvern Mastersizer are discussed. Experimental methods, acid dissolution, regrinding, single-stage carbonation, acrylic reactor testing without temperature and pressure, concurrent grinding both in situ and in operando and two-stage carbonation are discussed. Please refer to these publications for further details [2, 10, 13, 14, 18, 19].

A comparison of elemental composition of dunite by ICP-OES and XRF is provided in **Table 7**.

4.3 Magnesite yield results using different feedstocks

Magnesite yield results using various feedstocks are presented in **Table 8**.

Few of these results are presented graphically in **Figure 17**. For already published results, please refer to [2, 19] and [10, 18] and [13, 14]. Soaked dunite especially heat-activated provided the highest yields. This is not evident from literature, especially for heat-activated dunite. However, for raw dunite, some results are presented in Ph.D.

Element	ICP-OES (wt %)	XRF (wt %)
Mg	25.8 ± 1	25.7
Si	12.4 ± 0.3	16.6
Fe	6.38 ± 0.3	6.46
Ca	0.373 ± 0.07	0.251
K	0.04 ± 0.01	0.0015
Al	0.285 ± 0.04	0.171

This is a preliminary analysis comparison.

Table 7.
Comparison of elemental composition of dunite by ICP-OES and XRF.

Sr. No	Particle Size (μm)	Material	Reaction condition	Experiment condition	Reaction time (h)	Yield (%)
1	Sub 75	dunite	185°C, 130 bar, 15% solids, 0.64 M $NaHCO_3$	raw	1	24
2	Sub 75	dunite	185°C, 130 bar, 15% solids, 0.64 M $NaHCO_3$	Raw and regrinding	1.5	24
3	Sub 75	dunite	185°C, 130 bar, 15% solids, 0.64 M $NaHCO_3$	Heat-activated (630°C, 4 hrs)	1	37
4	Sub 75	dunite	185°C, 130 bar, 15% solids, 0.64 M $NaHCO_3$	Soaked 15 days	1	20
5	Sub 75	dunite	185°C, 130 bar, 15% solids, 0.64 M $NaHCO_3$	Soaked 30 days	1	22
6	Sub 75	dunite	185°C, 130 bar, 15% solids, 0.64 M $NaHCO_3$	raw	1	19
7	Sub 75	lizardite	185°C, 130 bar, 15% solids, 0.64 M $NaHCO_3$	raw	1	2.2
8	Sub 75	dunite	185°C, 130 bar, 15% solids, 0.64 M $NaHCO_3$	Heat-transformed (800°C, 3 hrs)	1	18
9	Sub 75	dunite	185°C, 130 bar, 15% solids, 0.64 M $NaHCO_3$	Heat-transformed (800°C, 3 hrs)	1	18
10	Sub 75	dunite	185°C, 130 bar, 15% solids, 0.64 M $NaHCO_3$	Heat-activated (630°C, 4 hrs)	1	37
11	Sub 75	dunite	185°C, 130 bar, 15% solids, 0.64 M $NaHCO_3$	Heat-activated (630°C, 4 hrs), 1 month soaked	1	42
12	20–45	dunite	180°C, 130 bar, 15% solids, 0.64 M $NaHCO_3$	Reference	2.5	5.3
13	20–45	dunite	180°C, 130 bar, 15% solids, 0.64 M $NaHCO_3$	Concurrent ground	2.5	31.6
14	20–45	dunite	180°C, 130 bar, 30% solids, 0.64 M $NaHCO_3$	Concurrent ground	2.5	16.4
15	20–45	dunite	180°C, 65 bar, 15% solids, 0.64 M $NaHCO_3$	Concurrent ground	2.5	12
16	20–45	dunite	155°C, 65 bar, 15% solids, 0.64 M $NaHCO_3$	Concurrent ground	2.5	22.5

Sr. No	Particle Size (μm)	Material	Reaction condition	Experiment condition	Reaction time (h)	Yield (%)
17	20–45	dunite	180°C, 65 bar, 30% solids, 0.64 M $NaHCO_3$	Concurrent ground	2.5	10.4
18	20–45	dunite	155°C, 130 bar, 30% solids, 0.64 M $NaHCO_3$	Concurrent ground	2.5	10.9
19	20–45	dunite	180°C, 130 bar, 15% solids, 0.64 M $NaHCO_3$	Concurrent ground	8	62
20	Sub 20	dunite	180°C, 130 bar, 15% solids, 0.64 M $NaHCO_3$	In operando grinding	2.5	23.6
21	20–45	dunite	180°C, 130 bar, 15% solids, 0.64 M $NaHCO_3$	reference	8	12.3
22	20–45	dunite	180°C, 130 bar, 15% solids, 0.64 M $NaHCO_3$	Effect of sampling	2.5	28.2
23	20–45	dunite	180°C, 130 bar, 15% solids, 0.64 M $NaHCO_3$	Effect of sampling	1	13
24	20–45	dunite	180°C, 130 bar, 15% solids, 0.64 M $NaHCO_3$	Effect of sampling	0.5	9.3
25	20–45	dunite	180°C, 130 bar, 15% solids, 0.64 M $NaHCO_3$	Effect of sampling	0.05	6.3
26	Sub 10	dunite	180°C, 130 bar, 15% solids, 0.64 M $NaHCO_3$	In operando grinding	8	54.6
27	20–45	dunite	180°C, 130 bar, 15% solids, 0.64 M $NaHCO_3$	Concurrent grinding	2.5	31.6
28	20–45	dunite	180°C, 130 bar, 15% solids, 0.64 M $NaHCO_3$	Concurrent grinding	2.5	27.2
29	20–45	olivine	180°C, 130 bar, 15% solids, 0.64 M $NaHCO_3$	reference	2.5	0.94
30	20–45	olivine	180°C, 130 bar, 15% solids, 0.64 M $NaHCO_3$	Concurrent grinding	2.5	34.1
31	Sub 75	lizardite	180°C, 130 bar, 15% solids, 0.64 M $NaHCO_3$	Concurrent grinding	2.5	5.3
32	45–75	dunite	180°C, 130 bar, 15% solids, 0.64 M $NaHCO_3$	Concurrent grinding	2.5	7.1
33	45–75	dunite	180°C, 130 bar, 15% solids, 0.64 M $NaHCO_3$	Concurrent grinding	2.5	38.2
34	Sub 75	olivine	180°C, 130 bar, 15% solids, 0.64 M $NaHCO_3$	reference	2.5	2.4
35	Sub 75	olivine	180°C, 130 bar, 15% solids, 0.64 M $NaHCO_3$	Concurrent ground	2.5	9.2

Table 8.
Yield results for different feedstocks under various experiment and reaction conditions.

thesis. Dunite yield calculation is very easy and straightforward, please refer to my Ph.D. thesis publication. Six times higher magnesite yields, or say an increase of 600% [10, 14], or almost two times higher yields in two-stage [13] were achieved using concurrent grinding. Olivine does not accept this much, but still shows some increased trend.

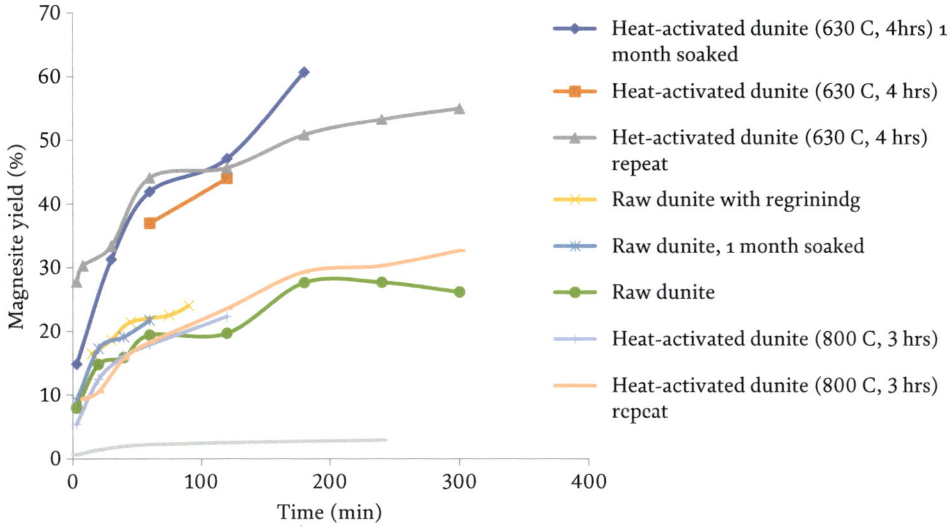

Figure 17.
Comparison of reactivity of dunite, soaked dunite, heat-activated dunite and lizardite and raw dunite soaked.

4.4 Semiquantitative XRD results authenticity

Authenticity of QXRD is shown in **Table 5**.

4.5 TGA-MS results authenticity

TGA-MS results authenticity is excellent. Please see the consistency of magnesite results, which are constant. However, these results have variations as per variation of size fraction.

Error in Brucite calculation due to slight peak overlap.

Error and second option of calculation is shown in **Figure 18**.

Sr.No. 1. QXRD Calculation details.

Csi	Cliz	Coli	Cbru	Cmag	Total
20.3	48.89	25.287	5.117	0.298	99.91
Exclude silicon	61.4	31.77	6.42	0.374	100

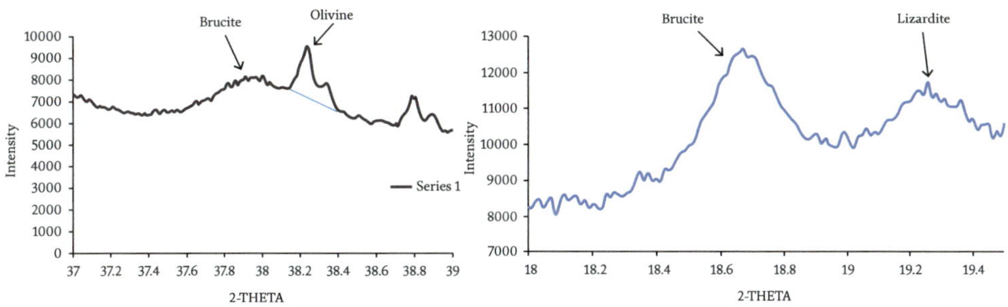

Figure 18.
The left-hand side image graph shows overlap. The right-hand side shows an alternative option.

More results related to this TGA-MS matching with QXRD.

Sr.		Lizardite	Olivine	Brucite	Magnetite
	20–45 μm dunite TGA-MS	52	42 (difference)	5.9	0.2 (XRD)
	20–45 μm dunite TGA-MS repeat	51	43.6(difference)	5.27	0.2 (XRD)
1	20–45 μm dunite QXRD	52	42	5.8	0.2
2	20–45 μm dunite QXRD repeat	52.6	42	5.27	0.19
3	20–45 μm dunite QXRD repeat	53	40	6.1	0.23

A very good match between TGA-MS and QXRD results was obtained when using an olivine peak at 17.3°. Brucite also shows a good match. The same peak points were used for all 3 XRD patterns and they are also similar to the −75 μm dunite analysis which provides more confidence in results. Points are slightly changed for 20–45 μm dunite for olivine peaks as these peaks show a slight variation. Results authenticity is excellent.

4.6 Validation of Malvern Mastersizer results

Relationship between d_{80} (mean particle size) and Malvern mastersizer RPM for olivine (**Figure 19**). The minimum RPM required for Malvern mastersizer based on feed mean size is given below.

4.7 Validation of olivine yields through QXRD measurements and matched TGA-MS

4.7.1 20: 45 μm olivine crushed carbonated sample QXRD

20–45 μm olivine crushed carbonated reference sample is mixed with 20% silicon and the sample is then scanned for semiquantitative analysis for 3 hrs. This is a reference experiment in which grinding media is not used. The reaction was done at 180°C and 130 bar. **Table 9** shows QXRD results matched with TGA-MS results.

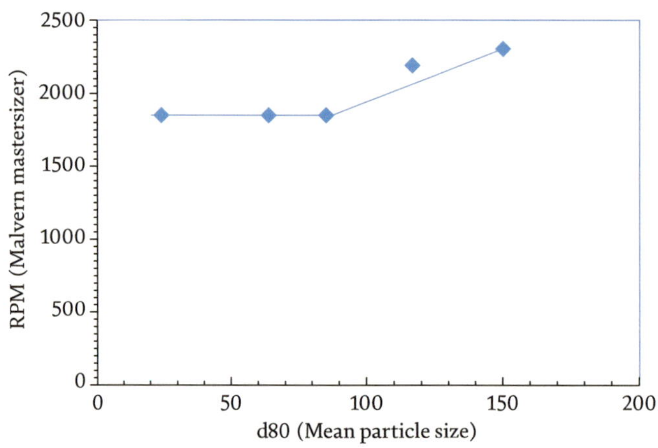

Figure 19.
Relationship between mean particle size and Malvern Mastersizer minimum RPM.

	Olivine	Lizardite	Clinochlore	Enstatite	Talc	Magnesite	Yield
20–45 μm olivine Reference Experiment	91%	0.68%	1.7%	0.36%	0.49%	5.3%	2.76%

Table 9.
QXRD analysis and yield from QXRD compared with yield from TGA-MS.

4.8 Olivine yield calculation

$$\text{Mass of } MgCO_3 = 5.3\%(\text{from QXRD}) = 0.053 \text{ g.}$$
$$= 0.053/84 = 0.000628 \text{ moles}$$

$MgCO_3$ ➔ $MgO + CO_2$.
Moles CO_2 = 0.000628 moles.
Mass CO_2 = 0.000628 * 44 = 0.02766.
Yield = (0.02766/1) * 100 = 2.76% ≥ (1%) from TGA-MS.

4.8.1 20–45 μm olivine crushed concurrent ground sample QXRD

20–45 μm olivine crushed concurrent ground sample is mixed with 20% silicon and the sample is then scanned for semiquantitative analysis for 3 hrs. This is a concurrent grinding experiment in which grinding media is used. The reaction was done at 180°C and 130 bar. **Table 10** shows QXRD results matched with TGA-MS results.

4.9 Yield calculation

$$\text{Mass of } MgCO_3 = 76\%(\text{from QXRD}) = 0.76 \text{ g}$$
$$= 0.76/84 = 0.009 \text{ moles}$$

$MgCO_3$ ➔ $MgO + CO_2$.
Moles CO_2 = 0.009 moles.
Mass CO_2 = 0.009 * 44 = 0.396.
Yield = (0.396/1) * 100 = 39.6% ≥ (34.1%) from TGA-MS, reference [14].

4.9.1 Validation of EDS for 20–45 μm dunite resin embedded samples

Various particles analysis indicate the authenticity of EDS analysis (**Figures 20–22**).

From our earlier articles, a significant difference in morphology of silica-rich layers, especially core and shell part is visible [14]. However, EDS analysis especially silicon shows no significant difference as depicted in above Figures (20–22). However,

	Olivine	Lizardite	Clinochlore	Enstatite	Talc	Magnesite	Yield
20–45 μm olivine concurrent ground experiment	19%	0.15%	0.53%	0.1%	3.6%	76%	39.6%

Table 10.
QXRD analysis and yield from QXRD compared with yield from TGA-MS.

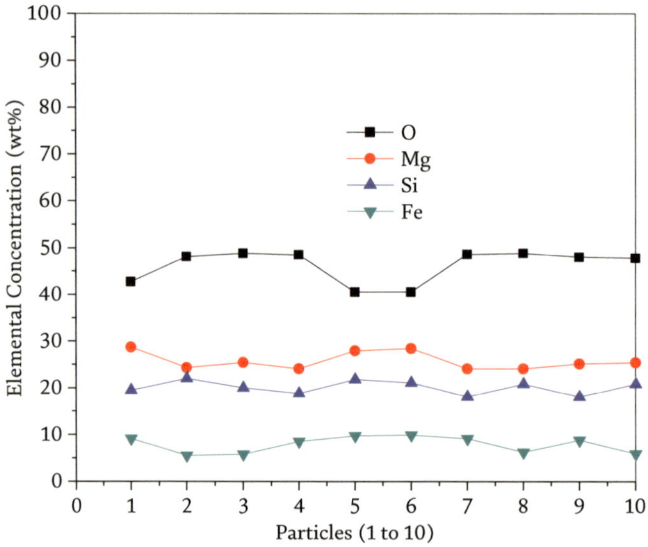

Figure 20.
20–45 μm dunite sample (embedded in resin) ten particles analysis.

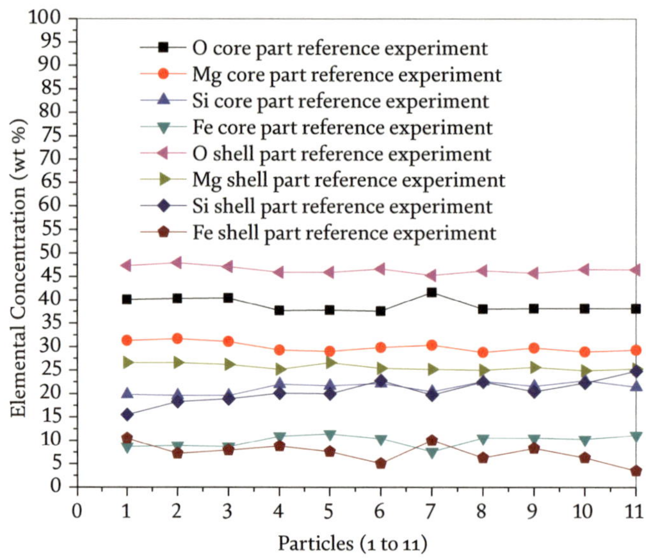

Figure 21.
20–45 μm dunite reference carbonated (8 h) sample (embedded in resin) eleven particles analysis and consistency of EDS analysis.

as reported earlier, Mg/Si ratio difference [13, 14] is there to confirm the presence of silica-rich layers. This may be taken as one of the key findings of this chapter.

5. Conclusions and recommendations

Suppliers may give wrong materials, but a variety of analyses will determine this. Semiquantitative XRD (QXRD) results authenticity is excellent. TGA-MS results

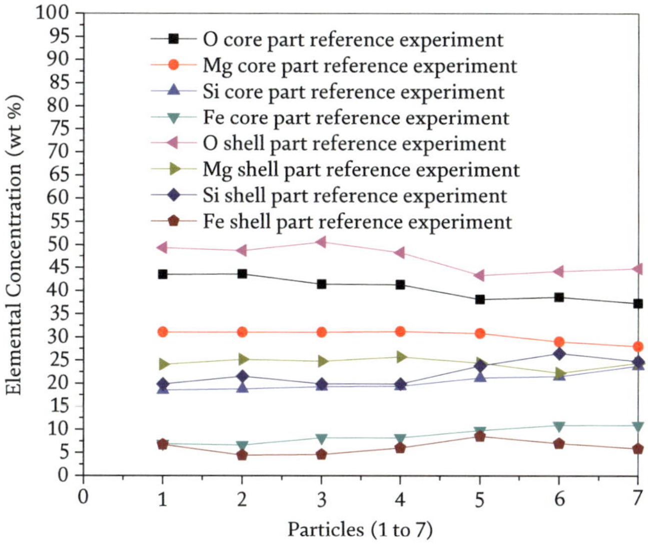

Figure 22.
20–45 μm dunite concurrent ground carbonated (8 h) sample (embedded in resin) seven particles analysis and consistency of EDS analysis.

authenticity is excellent. No doubt left on TGA-MS and QXRD results matching. ICP-OES results match with XRF results is excellent. EDS results graphically shown are excellent. Routine calibration of measuring instruments must be performed. This very instrument to instrument. Reputed researchers will know the frequency of calibration. The key to calibration is that calibration results match with standard calibration figures/charts or numbers provided by the supplier of the instrument. I recommend contacting the supplier directly or indirectly if calibration curves results are not matching as per intended results. I recommend using pure standard materials for calibration of TGA, MS, ICP-OES, XRD, SEM, EDS, TEM, Malvern Mastersizer, ATR, TPD and other measuring instruments.

Acknowledgements

Muhammad Imran Rashid acknowledges The University of Newcastle, (UoN), Newcastle, NSW, Australia for awarding a Postgraduate scholarship and enabling to use all research facilities especially from Research Division, EMX unit, University of Newcastle, Newcastle, Callaghan Campus, NSW, Australia. Mineral Carbonation International support is beyond imagination. Ms. Kitty Tang support in Malvern Master sizer analysis, Ms. Jennifer Zobec, Ms. Yun Lin and Mr. Huiming provided inevitable support in XRD, QXRD, SEM/EDS/EDX, Simulated QXRD, and TEM. Mr. Scott Molloy support in equipment installations, modifications and difficult amendments is unforgettable. Mr. Glenn Bryant support in ATR analysis and other analyses is highly appreciated. Ms. Jane Hamson support in initial trials of TGA and ICP-OES results is fundamentally acknowledged. Mr. Scott Molloy also helped with TGA-MS especially MS calibrations. His support keeps my work continuous. My advisor Dr. Geoff Brent imaged me with his industry knowledge and capabilities. Fellow students are also acknowledged. Everyone supported me from the heart. I enjoyed cultural diversity and especially food and particularly Lamb.

Author details

Muhammad Imran Rashid[1,2]

1 Discipline of Chemical Engineering, The University of Newcastle, NSW, Australia

2 Department of Chemical, Polymer and Composite Material Engineering, University of Engineering and Technology, Lahore, Pakistan

*Address all correspondence to: muhammadimran.rashid@uon.edu.au

References

[1] Tans P, Keeling R. Trends in Atmospheric Carbon Dioxide: Mauna Loa, Hawaii. Washington, DC: National Oceanic and Atmospheric Administration, U.S. Department of Commerce; 2010

[2] Rashid MI. Mineral Carbonation of CO2 using Alternative Feedstocks. Australia: The University of Newcastle; 2019

[3] Rashid MI, Benhelal E, Rafiq S. Reduction of greenhouse gas emissions from gas, oil, and coal power plants in Pakistan by carbon capture and storage (CCS): A review. Chemical Engineering & Technology. 2020;**43**(11):2140-2148

[4] Emad Benhelal JH, Zhao G, Rashid MI, Oliver T, Rayson M, Brent G, et al. Insights into chemical stability of Mg-silicates and silica in aqueous systems using 25Mg and 29Si solid-state MAS NMR spectroscopy: Applications for CO2 capture and utilisation. Chemical Engineering Journal. 2020

[5] Farhang F, Oliver TK, Rayson M, Brent G, Stockenhuber M, Kennedy E. Experimental study on the precipitation of magnesite from thermally activated serpentine for CO2 sequestration. Chemical Engineering Journal. 2016;**303** (Supplement C):439-449

[6] Julcour C, Bourgeois F, Bonfils B, Benhamed I, Guyot F, Bodénan F, et al. Development of an attrition-leaching hybrid process for direct aqueous mineral carbonation. Chemical Engineering Journal. 2015;**262**:716-726

[7] Oliver TK, Farhang F, Hodgins TW, Rayson MS, Brent GF, Molloy TS, et al. CO2 capture modeling using heat-activated serpentinite slurries. Energy & Fuels. 2019;**33**(3):1753-1766

[8] Benhelal E, Oliver TK, Farhang F, Hook JM, Rayson MS, Brent GF, et al. Structure of silica polymers and reaction mechanism for formation of silica-rich precipitated phases in direct aqueous carbon mineralization. Industrial & Engineering Chemistry Research. 2020; **59**(15):6828-6839

[9] Benhelal E, Rashid MI, Holt C, Rayson MS, Brent G, Hook JM, et al. The utilisation of feed and byproducts of mineral carbonation processes as pozzolanic cement replacements. Journal of Cleaner Production. 2018;**186**:499-513

[10] Rashid MI, Benhelal E, Farhang F, Oliver TK, Rayson MS, Brent GF, et al. Development of concurrent grinding for application in aqueous mineral carbonation. Journal of Cleaner Production. 2019;**212**:151-161

[11] Benhelal E, Rashid MI, Rayson MS, Prigge J-D, Molloy S, Brent GF, et al. Study on mineral carbonation of heat activated lizardite at pilot and laboratory scale. Journal of CO2 Utilization. 2018;**26**: 230-238

[12] Farhang F, Oliver TK, Rayson MS, Brent GF, Molloy TS, Stockenhuber M, et al. Dissolution of heat activated serpentine for CO2 sequestration: The effect of silica precipitation at different temperature and pH values. Journal of CO2 Utilization. 2019;**30**:123-129

[13] Rashid MI, Benhelal E, Farhang F, Oliver TK, Stockenhuber M, Kennedy EM. Application of a concurrent grinding technique for two-stage aqueous mineral carbonation. Journal of CO2 Utilization. 2020;**42**:101347

[14] Rashid MI, Benhelal E, Farhang F, Oliver TK, Stockenhuber M, Kennedy EM. Application of concurrent

grinding in direct aqueous carbonation of magnesium silicates. Journal of CO2 Utilization. 2021;**48**:101516

[15] Oliver TK, Dlugogorski BZ, Kennedy EM. Biologically enhanced degassing and precipitation of magnesium carbonates derived from bicarbonate solutions. Minerals Engineering. 2014;**61**(0):113-120

[16] Rashid NRMI. Fluid mechanics and heat-transfer operations combination involved in urea unit of fertilizer complex. Non-Metallic Material Science. 2019;**1**(1):5-10

[17] Muhammad NR, Rashid I. Urea Synthesis Hazard Analysis—PHA, HAZOP and Quantitative Risk Assessment, LAMBERT Academic Publishing AG & Co. KG. Germany; 2012

[18] Rashid MI, Benhelal E, Farhang F, Oliver TK, Rayson MS, Brent GF, et al. ACEME: Direct aqueous mineral carbonation of dunite rock. Environmental Progress & Sustainable Energy. 2019;**38**(3):e13075

[19] Rashid MI, Benhelal E, Farhang F, Stockenhuber M, Kennedy EM. Magnesium leachability of Mg-silicate peridotites: The effect on magnesite yield of a mineral carbonation process. Minerals. 2020;**10**:1091

[20] Benhelal E, Rashid MI, Rayson MS, Oliver TK, Brent G, Stockenhuber M, et al. "ACEME": Synthesis and characterization of reactive silica residues from two stage mineral carbonation Process. Environmental Progress & Sustainable Energy. 2019;**38**(3):e13066

[21] Shaffer G. Long-term effectiveness and consequences of carbon dioxide sequestration. Nature Geoscience. 2010; **3**(7):464-467

[22] Peters A. This startup is using sunlight and capture CO2 to make jet fuel. In: Read M, editor. Dimensional Energy. 2021

[23] Fischer EM, Sippel S, Knutti R. Increasing probability of record-shattering climate extremes. Nature Climate Change. 2021;**11**(8):689-695

[24] Ju L, Tan X, Mao X, Gu Y, Smith S, Du A, et al. Controllable CO2 electrocatalytic reduction via ferroelectric switching on single atom anchored In2Se3 monolayer. Nature Communications. 2021;**12**(1):5128

[25] Vaughan A. We are Nowhere Near Keeping Warming Below 1.5°C Despite Climate Plans Availbale from: https://www.newscientist.com/article/2269432-we-are-nowhere-near-keeping-warming-below-1-5c-despite-climate-plans/#ixzz78ZUV14CM. NewScientist; 2021

[26] Benhelal E, Shamsaei E, Rashid MI. Novel modifications in a conventional clinker making process for sustainable cement production. Journal of Cleaner Production. 2019;**221**:389-397

[27] Naseem S, Rashid MI, Ramzan N. Coal as an energy source for mitigating energy crisis in Pakistan. Journal of Engineering and Technology. 2014;**4**(2): 127-134

[28] Gadikota G, Matter J, Kelemen P, Park AHA. Chemical and morphological changes during olivine carbonation for CO2 storage in the presence of NaCl and NaHCO3. Physical Chemistry Chemical Physics. 2014;**16**(10):4679-4693

[29] Balucan RD. Thermal Studies of Magnesium Silicates from The Great Serpentine Belt in New South Wales for CO2 Sequestration by Mineral

Carbonation in. Australia The University of Newcastle; 2012

[30] Treacy M, Higgins J. Powder Pattern Simulations of Disordered Intergrowths, in Collection of Simulated XRD Powder Patterns for Zeolites. fifth ed. Amsterdam: Elsevier Science B.V.; 2007. pp. 477-485

[31] Cullity BD, Stock SR. Elements of X-ray Diffraction. Prentice Hall; 2001

[32] Lavina B, Dera P, Downs RT. Modern X-ray diffraction methods in mineralogy and geosciences. Reviews in Mineralogy and Geochemistry. 2014;**78**:1-31

[33] Rezaaiyaan GMHR, Anderson H, Kaiser H, Meddings B. Design and construction of a low-flow, low-power torch for inductively coupled plasma spectrometry. Applied Spectroscopy. 1982;**36**(6):627-631

[34] Huang M, Hieftje GM. Simultaneous measurement of spatially resolved electron temperatures, electron number densities and gas temperatures by laser light scattering from the ICP. Spectrochimica Acta Part B: Atomic Spectroscopy. 1989;**44**(8):739-749

[35] Mehta R. Interactions, imaging and spectra in SEM. In: Kazmiruk V, editor. Scanning Electron Microscopy. InTech; 2012

[36] Malvern Mastersizer 2000 Manual, 2007

Chapter 5

Normative Mineralogy Especially for Shales, Slates, and Phyllites

Hans Wolfgang Wagner

Abstract

First, an insight into normative mineralogy and the most important methods for calculating the standard or norm minerals, such as the CIPW norm, is given. This is followed by a more detailed explanation of "slatenorm" and "slatecalculation" for low and very low metamorphic rocks, such as phyllites, slates, and shales. They are particularly suitable for fine-grained rocks where the mineral content is difficult to determine. They enable the determination of a virtual mineral inventory from full chemical analysis, including the values of carbon dioxide (CO2), carbon (C), and sulfur (S). The determined norm or standard minerals include the minerals—feldspars, carbonates, micas, hydro-micas, chlorites, ore minerals, and quartz. The advantages of slatenorm and slatecalculation compared to other methods for calculating normal minerals of sedimentary rocks are discussed.

Keywords: slates, shales, phyllites, norm mineral calculation, CIPW, slatenorm, slatecalculation, micas, hydro-micas, illite, low- and very low-grade metamorphism

1. Introduction

A norm or standard mineral inventory is a fictitious inventory that, in contrast to the modal and in fact mineral inventory, is calculated from the oxides of a chemical analysis of rock. Each oxide is then divided by its molecular weight. These molecular numbers are then the basis for further calculations (**Table 1**).

The results of the calculations are given as standard or norm minerals. They are given as abbreviations and mostly in small letters such as q (or qu) = quartz or mu = muscovite. The standard minerals each have a clearly defined chemical composition. This distinguishes them from some minerals in nature that have a more complex composition, such as pyroxenes, amphiboles, or chlorites. In these cases, simplifications or end links of a mixture series are used [1–3].

Usually, standard minerals are frequently occurring rock-forming minerals. Only in exceptional cases do you have to use minerals that occur only rarely or not at all in nature. The norm mineral inventory should come as close as possible to the actual, modal mineral inventory, which is not always possible [1–3].

The CIPW (Cross, Iddings, Pirsson, Washington) norm [4] is the most common method for magmatites. The Rittman norm for igneous rocks or the Niggli norm for igneous rocks and metamorphic rocks are less common.

Oxide symbol	Approximate molecular weight
	(AMU)
SiO_2	60.1
TiO_2	79.9
Al_2O_3	102.0
Fe_2O_3	159.7
FeO	71.8
MnO	70.9
MgO	40.3
CaO	56.1
Na_2O	62.0
K_2O	94.2
P_2O_5	141.9
CO_2 (Non-carbonate C)	44.0 12.0)
SO_3 (or S)	80.1 (or 32.1)

Table 1.
The full chemical analysis for standard mineral calculations.

There are other methods for standard mineral calculations for sediments, such as SEDNORM [5], SEDMIN [6], PELNORM [7], and a linear program without a name (mainly for sandstones [8]).

The slatenorm and the slatecalculation norm are used for fine-grained sedimentary rocks and very low and low-grade metamorphic rocks, such as shales, slates, and phyllites [9]. As new methods, they are dealt with in great detail here.

2. CIPW norm

The CIPW system is one of the best-known and best-elaborated chemical classifications of igneous rocks. It was developed as early as 1902. The CIPW system (CIPW norm) is based on a normative mineral inventory. This normative mineral inventory consists of a number of standard minerals (**Table 2**).

With them, groups of substances of chemical analysis are summarized and thus the quite complex chemistry of the igneous rocks is made clearer. As before, however, the CIPW standard plays a major role for the purpose of comparing the chemical properties.

The calculation of the CIPW standard is carried out according to a specified scheme. It can be found in a number of publications [3, 4]. Computer programs have also been developed since 1965 [10]. Several websites now offer online facilities for calculating CIPW [11, 12].

There are now numerous suggestions for improvement and additions to the CIPW standard. In particular, a standard method that is consistent with a rock classification as recommended by the International Union of Geological Sciences IUGS has proven to be necessary. There will be detailed step-by-step instructions for a standard igneous

Normative mineral name	Abbreviation	Mineral formula
Quartz	Q	SiO_2
Korund	C	Al_2O_3
Orthoclase	or	$K_2O\ Al_2O_3\ 6SiO_2$
Albite	ab	$Na_2O\ Al_2O_3\ 6SiO_2$
Anorthite	an	$CaO\ Al_2O_3\ 2SiO_2$
Leucite	le	$K_2O\ Al_2O_3\ 4SiO_2$
Nepheline	ne	$Na_2O\ Al_2O_3\ 2SiO_2$
Kaliophilite	kp	$K_2O\ Al_2O_3\ 4SiO_2$
Diopside	di	$CaO\ (Mg, Fe)O\ 4SiO_2$
Wollastonite	wo	$CaO\ SiO_2$
Hypersthene	hy	$(Mg, Fe)O\ SiO_2$
Olivine	ol	$2(Mg, Fe)O\ SiO_2$
Acmite	ac	$Na_2O\ Fe_2O_3\ 4SiO_2$
Magnetite	mt	$FeO\ Fe_2O_3$
Hematite	hm	Fe_2O_3
Ilmenite	il	$FeO\ TiO_2$
Apatite	ap	$3(3CaO\ P_2O_5)$
Pyrite	pr	FeS_2
Calcite	cc	$CaO\ CO_2$

Table 2.
The standard minerals in the CIPW norm [1–3].

norm (SIN) presented [13]. The review of the main computer programs for the classification of igneous rocks in the sense of the IUGS leads to the new program "Igneous Rock Classification System" (IgRoCS) [14].

3. Niggli and Rittmann norm

The Niggli norm (or Niggli's molecular catanorm) was developed by P. Niggli in 1933 [2, 3, 15]. It was later renamed the equivalent norm. It is in contrast to the CIPW standard, flexible in the choice of minerals to be considered. On the basis of the equivalent numbers standardized to 100%, the chemical rock analysis (**Table 1**) are the so-called base molecule groups developed. These are the simplest chemical compounds of the most common elements in the earth's crust. A standard mineral inventory can now be calculated from these basic molecule groups. The standard minerals calculated in this way correspond to the minerals that are present at high temperatures and high-pressure conditions arise. This is why, this standard was given the name Katanorm. Niggli's work was taken up by Barth (1952) [3, 16] and the norm was converted into the atomic equivalents. Barth developed the other methods Mesonorm and Epinorm for lower metamorphic rocks [17]. A computer program has also been developed for this purpose [18].

Even so, the Niggli norm never really caught on. The greater flexibility of the Katanorm compared to the CIPW norm brings one greater computational effort. In addition, the Katanorm minerals do not come from the modal mineral composition of rock closer than the standard minerals to the CIPW standard [3].

Rittmann (1973) [3, 19] developed a calculation method based on the Katanorm Niggli. It is closer to nature, but also much more complicated. The calculation method is particularly important for volcanites, in which the modal mineral inventory can only be partially recognized. The minerals are often submicroscopic, only partially or not at all crystallized. The Rittmann norm uses information that remain unused in the CIPW norm and the Niggli norm, such as the degree of oxidation, the H_2O and CO_2 content, or the geological situation, and the structure of the rock. The method, therefore, includes several different standard calculation methods that take this information into account. A computer program was developed at an early stage [20].

4. SEDNORM, SEDMIN, PELNORM and other

SEDNORM [5] is a method for calculating standard minerals from unconsolidated sediments and coal ash from full chemical analysis. As with the Rittmann standard, different standard variants can be selected depending on the available data (e.g., clay chemistry). According to the authors, SEDNORM is suitable for use on sandstones, shales, and carbonate rocks. The calculation process of SEDNORM is described but is not available online as a program.

SEDMIN [6] is a standard mineral calculation for sediments that focuses on the minerals smectite, chlorite, kaolinite, illite, and ambiguous sericite. Within the full chemical analysis, TiO_2 is used to calculate kaolinite. According to the author, SEDMIN is able to predict predominant clay minerals even with atypical sample data. A calculation program for SEDMIN is available online [21].

PELNORM [7] is a calculation method for clay minerals and pelitic rocks from data as in **Table 1**. The procedure differentiates between the variants with missing smectite (A) and existing smectite (B). K_2O can be used to determine both orthoclase (K-feldspar, or) (A1, B1) or illite (ill) (A2, B2) plus orthoclase. According to the authors, the method achieves a good match between normative and modal mineral composition. The calculation steps of PELNORM are described together with program steps in FORTRAN. A computer program is not available online.

5. Slatenorm and slatecalculation norm in general

In a research project from 1989 to 1991, an attempt was made to use a full chemical analysis similar to the CIPW norm for magmatic rocks to make a rather inaccurate norm mineral evaluation for roof and wall slates ([22] and **Table 2** therein). Sericites (muscovite and paragonite), chlorites, quartz, and total carbonates were estimated as the main minerals. Ward and Gómez-Fernandez [23] used the Rietveld method-based Siroquant data processing system for X-ray powder diffraction analysis for the determination of the slate main minerals quartz, feldspar, micas, and chlorites. However, the application of the method was limited to low carbonate Spanish roofing slate. The determined feldspar (albite) values were higher than those of chlorite and likely to be too high. Jung and Wagner [24] created a calculation method similar to the CIPW norm that was ready for practical use. They managed to determine the mineral constituents, and in particular, the content of free quartz with sufficient accuracy—for the first time—to some essential practical statements.

The results of such norm calculations have already been used not only in test certificates but also in a manual [25]. Other authors cited such norm calculations together with results of other analyses and found good matches [26]. The results of more than 20 years of application of "slatenorm" [22, 25, 26–30] have shown that the inclusion of additional ore minerals, color-giving minerals, and hydro-micas (especially illite) in the new slatecalculation is reasonable.

The extended method slatecalculation presented here is based on a previous, unpublished program called "slatenorm" [24, 31] (Appendix A and B). The calculations are based on the full chemical analysis (**Table 1**). In the case of slatenorm and slatecalculation norm, a distinction is made between CO_2 (= carbonated C) and C (= non-carbonated C, approximate atomic weight = 12.0). In addition, S (approximate atomic weight = 32.1) and not SO_3 are being used as a basis. In a first step, the extended algorithm includes the distinction of sulfides. So far, pyrite was the only sulfide calculated in "slatenorm." In many higher metamorphic slates (e.g., slates from Spain), pyrrhotite is predominant and should be included in the calculation because it is more susceptible to oxidation. Therefore, in the first step, the extended algorithm differentiates various sulfides.

The basic calculations of the algorithm, fundamental norm minerals, and chemical formulas are described in a very simplified form below (quoted from [9], detailed flowchart see Appendix A):

$$P_2O_5 \rightarrow \text{ap} = \text{apatite} = 3.3\text{CaO}\ P_2O_5 \tag{1}$$

$$\begin{aligned} &\text{S(approx. Half, sofar macroscopically or microscopically determined)} \\ &\quad \rightarrow \text{pn pyrrhotite} \approx \text{FeS} \end{aligned} \tag{2}$$

$$\text{Residual S} \rightarrow \text{pt.} = \text{pyrite} = \text{FeS}_2 \tag{3}$$

Fe_2O_3 and TiO_2 → tm = titanomagnetite = FeO Fe_2O_3 TiO_2, frequently occurring mixture mineral in slates (cf. [22, 32]) At the deficit of FeO in some cases a back calculation (28–31 or 49–51) tm in ru = rutile = TiO_2 is needed.

$$TiO_2 \rightarrow \text{ilm} = \text{ilmenite} = \text{FeO}\ TiO_2 \tag{4}$$

$$Fe_2O_3 \rightarrow \text{he} = \text{hematite} = Fe_2O_3 \tag{5}$$

$$\text{C} \rightarrow \text{gr} = \text{graphite} = \text{C} \tag{6}$$

$$\begin{aligned} CO_2 \rightarrow \text{carbonates}(&\text{cc} = \text{calcite} = \text{CaO}\ CO_2, \text{dol} = \text{dolomite} \\ &= \text{CaOMgO}\ 2CO_2, \text{sid} = \text{siderite} = \text{FeO}\ CO_2) \end{aligned} \tag{7}$$

$$\text{Residual CaO} \rightarrow \text{an} = \text{anorthite} = \text{CaO}\ Al_2O_3\ 2SiO_2 \tag{8}$$

$$K_2O \rightarrow \text{mu} = \text{muskovite} = K_2O\ 3Al_2O_3\ 6SiO_2\ 2H_2O \tag{9}$$

$$Na_2O \rightarrow \text{pa} = \text{paragonite} = Na_2O\ 3Al_2O_3\ 6SiO_2\ 2H_2O \tag{10}$$

The very variable minerals of the chlorite group require more complicated considerations with regard to their composition (see subsection 7).

$$MgO \rightarrow mc = 3MgO\ 2SiO_2\ 2H_2O = mc/3\ (\text{serpentine}) \quad (11)$$
$$FeO \rightarrow fc = 3FeO\ 2SiO_2\ 2H_2O = fc/3\ (\text{greenalite}) \quad (12)$$
$$MgO \text{ and } Al_2O_3 \rightarrow mac = 2MgO\ Al_2O_3\ SiO_2\ 2H_2O = mac/2\ (\text{amesite}) \quad (13)$$
$$FeO \text{ and } Al_2O_3 \rightarrow fac = 2FeO\ Al_2O_3\ SiO_2\ 2H_2O = fac/2\ (\text{daphnite}) \quad (14)$$

With the negative rest of $Al_2O_3 \rightarrow$ (ab, or) = feldspars:

$$\text{instead of pa } Na_2O \rightarrow ab = \text{albite} = Na_2O\ Al_2O_3\ 6SiO_2 \quad (15)$$
$$\text{instead of mu} \rightarrow K_2O \rightarrow or = \text{orthoclase} = K_2O\ Al_2O_3\ 6SiO_2 \quad (16)$$

In case of high Al_2O_3 – contents the calculation of chloritoid might be necessary:

$$\begin{aligned} \text{instead of fac} \rightarrow ct &= \text{chloritoid} \\ &= FeO\ Al_2O_3\ SiO_2\ H_2O\text{Residual } SiO_2 \rightarrow qz \quad (17) \\ &= \text{quartz} = SiO_2\text{Residual } H_2O \rightarrow \text{water} \rightarrow aq \end{aligned}$$

The remaining H_2O = aq may have a positive or negative value, and this is the basis for an extended algorithm determining hydro-micas. In the input data, carbon compounds (e.g., CO_2 and organic C) and elementary sulfur (S) are not included in the glow loss (or loss of ignition (LOI), but rather subtracted. Furthermore, in the case of predominant FeO compounds, the description of the total amount of Fe as Fe_2O_3 leads to an unrealistic oxidation gain at the expense of the LOI. Only a carefully corrected LOI can be incorporated in the norm calculation as H_2O = aq but will still be less precise. Thus, the calculated data will have a higher range of variation.

6. The calculation of hydro-micas (ill = illite and br = brammallite) in slatecalculation

A positive aq value is a basis for the hydro-micas calculation (see steps of calculation 40–60 in Appendix A). The original values, as well as the results of calculation steps 9 (Al_2O_3 and SiO_2) and 17 or 19 (in Appendix A) (MgO and FeO), are used as the starting point (Quoted from [9]):

$$aq\ (\text{share } K_2O) \rightarrow ill = \text{illite} = 0.65K_2O\ 3Al_2O_3\ 6SiO_2\ 2.35H_2O \quad (18)$$
$$aq\ (\text{share } Na_2O) \rightarrow br = \text{brammallite} = 0.65Na_2O\ 3Al_2O_3\ 6SiO_2\ 2.35H_2O \quad (19)$$

The calculation of hydro-micas requires a recalculation of the remaining phyllosilicates, like micas (mu, pa, steps 41 and 42 in Appendix A) and chlorites (mac, mc, fac, fc, steps 47–59 in Appendix A).

If aq - contents >0, the calculation of limonite is necessary as well:

$$\text{instead of he} \rightarrow lm = \text{limonit} = Fe_2O_3\ H_2O \quad (20)$$
$$\text{Residual } SiO_2 \rightarrow qz = \text{quartz} = SiO_2 \quad (21)$$
$$\text{Residual } H_2O \rightarrow aq \quad (22)$$

If the ill and br values are exceptionally high, in rare cases, MgO and FeO can remain after the calculation and lead to an Al_2O_3-deficit. In this case, the following varieties may be calculated as hydro-micas:

$$aq(\text{share } K_2O) \rightarrow maill = 0.65K_2O\ 1.05MgO\ 1.95Al_2O_3\ 6SiO_2\ 2.35 \times H_2O(\text{Mol.Wt of Mineral}: 705.21) \quad (23)$$

$$aq(\text{share } K_2O) \rightarrow faill = 0.65K_2O\ 1.05FeO\ 1.95Al_2O_3\ 6SiO_2\ 2.35 \times H_2O(\text{Mol.Wt of mineral}: 738.32) \quad (24)$$

or in combination:

$$aq(\text{share } K_2O) \rightarrow mfaill = 0.65K_2O\ 0.525MgO\ 0.525FeO\ 1.95Al_2O_3\ 6SiO_2 \times 2.35H_2O(\text{Mol.Wt of mineral}: 721.77) \quad (25)$$

$$aq(\text{share } Na_2O) \rightarrow mabr = 0.65Na_2O\ 1.05MgO\ 1.95Al_2O_3\ 6SiO_2\ 2.35 \times H_2O(\text{Mol.Wt of mineral}: 684.27) \quad (26)$$

$$aq(\text{share } Na_2O) \rightarrow fabr = 0.65Na_2O\ 1.05FeO\ 1.95Al_2O_3\ 6SiO_2\ 2.35 \times H_2O(\text{Mol.Wt of mineral}: 717.39) \quad (27)$$

or in combination:

$$aq(\text{share } Na_2O) \rightarrow mfabr = 0.65Na_2O\ 0.525MgO\ 0.525FeO\ 1.95Al_2O_3 \times 6SiO_2\ 2.35H_2O(\text{Mol.Wt of mineral}: 700.83) \quad (28)$$

If the first calculation steps yield negative values or additional information about other minerals is available, further calculations may be considered (e.g., Alta-Quartzite-schist and others):

$$CaO \rightarrow epi = epidote = 4CaO\ Fe_2O_3\ 2Al_2O_3\ 6SiO_2\ H_2O \quad (29)$$

$$CaO \rightarrow wo = wollastonite = CaO\ SiO_2 \quad (30)$$

$$TiO_2 \rightarrow tit = titanite = CaO\ TiO_2\ SiO_2 \quad (31)$$

$$Fe_2O_3 \rightarrow mt = magnetite = FeO\ Fe_2O_3 \quad (32)$$

$$MnO \rightarrow sp. = spessartine = 3MnO\ Al_2O_3\ 3SiO_2 \quad (33)$$

As a final step the norm minerals are added up to minerals and/or mineral groups:

$$an + ab + or = feldspars \quad (34)$$

$$cc + dol + sid = carbonates \quad (35)$$

$$mu + pa = micas \quad (36)$$

$$ill + br = hydro - micas \quad (37)$$

$$mac + fac + mc + fc = chlorites \quad (38)$$

$$pn + pt. = sulfides \quad (39)$$

$$pn + pt. + tm + ilm + ru + he\ (+mt) = ore - minerals \quad (40)$$

$$qz + an + ab + or\ (+sp) = rigid\ minerals \quad (41)$$

$$mu + pa + ill + br + mc + mac + fc + fac = elastic\ minerals \quad (42)$$

7. The chlorites in both methods

Due to the great variability of the chlorites, 3 versions (I, II, and "Grundversion" = GV) were used as a basis for the calculation of slatenorm. Later only the "Grundversion" was used as a basis for slatenorm and slatecalculation. Jung & Wagner (1996–2000) [24] determined for chlorite porphyroblasts in roof slate relatively high iron and aluminum contents with Fe/(Fe + Mg) ratios between 0.6 and 0.7 and a replacement of silicon by aluminum [IV] of 35–40 (Weight-%). According to the classification of the chlorites, these are aphrosiderites in this case (**Figure 1**). The three versions concern the inherently variable proportion of aluminum that replaces tetravalent silicon (Si). The Mg-Al-chlorite (mac for short) and Fe-Al-chlorite (fac for short) are taken as the basis as standard minerals.

They are given here as simplified formulas without taking trivalent Fe into account:

Mg-Chlorite:

Serpentine 6MgO $4SiO_2$ $4H_2O$ in all versions slatenorm and slatecalculation for mc.

Clinochlore 5MgO $1Al_2O_3$ $3SiO_2$ $4H_2O$ in slatenorm version II for mac.

X 4,5MgO $1{,}5Al_2O_3$ $3SiO_2$ $2{,}5H_2O$ in slatenorm version I for mac.

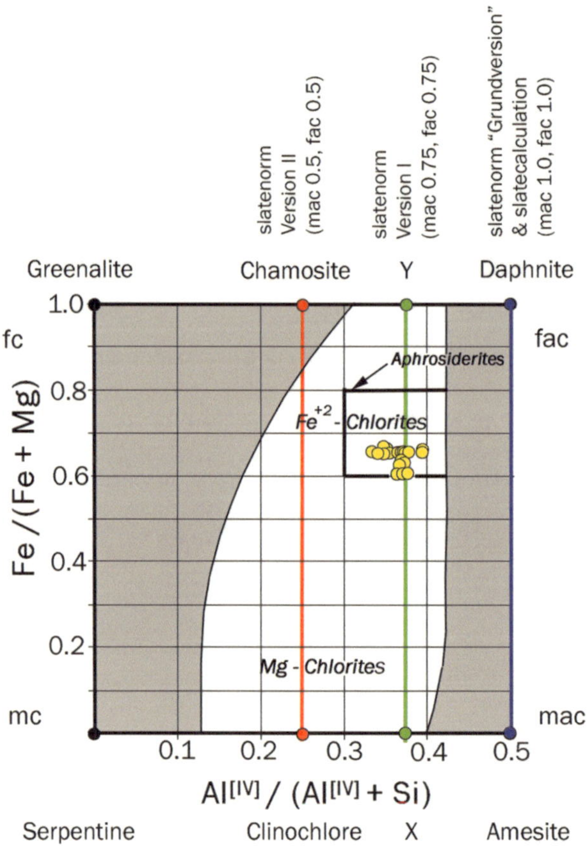

Figure 1.
Composition of chlorite porphyroblasts in Spanish roof slates, light—Field of variation of natural chlorites, further explanations in the text [33].

Amesite 4MgO $2Al_2O_3$ $2SiO_2$ $4H_2O$ in slatenorm "Grundversion" (= GV) and slatecalculation for mac.

Fe-Chlorite:

Greenalite 6FeO $4SiO_2$ $4H_2O$ in all versions slatenorm and slatecalculation for fc.

Chamosite 5FeO $1Al_2O_3$ $3SiO_2$ $4H_2O$ in slatenorm version II for fac.

Y 4,5FeO $1{,}5Al_2O_3$ $3SiO_2$ $2{,}5H_2O$ in slatenorm version I for fac.

Daphnite 4FeO $2Al_2O_3$ $2SiO_2$ $4H_2O$ in slatenorm "Grundversion"(= GV) and slatecalculation for fac.

As an example, full chemical analyses of five different sample (**Tables 3** and **4**) were selected, including a shale, a phyllite, two low-carbonate roofing slates from Germany and Spain, and roofing slate with carbonate (Magog). The biggest difference between slatecalculation and the three versions of slatenorm (I, II, and "Grundversion" = GV) is when hydro-micas (illite and brammallite) are calculated. In this case, more SiO_2 and Al_2O_3 are "consumed" than with micas and thus fewer chlorite minerals are calculated.

In the case of shales (**Table 3**), there are only meaningful results with slatecalculation, i.e., when considering and calculating hydro-micas. The old slatenorm, on the other hand, does not lead to meaningful, sometimes negative results.

	Nigeria	**Germany**		**Spain**		**USA**	**World**
	Shale: Abaka-bili	**Roof slates: Altlay**	**Roof slates: Magog**	**Roof slates: S. Pedro de T.**	**Phyl-lites: Pol Lugo green**	**Pierre shale C870 [7, 35]**	**Ave-rage sedi-men-tary rock after [5, 34]**
	Ø 3		Ø 8	Ø 2	Ø 2		
SiO_2	48.4	56.2	53.1	52.3	58.1	59.7	59.7
TiO_2	1.7	1.0	0.8	1.2	1.0	0.6	0.0
Al_2O_3	31.5	20.6	16.1	22.5	20.0	15.4	14.6
Fe_2O_3	3.5	0.8	1.6	1.4	2.1	4.6	4.8
FeO	0.0	6.6	4.7	7.0	5.0	1.0	
MnO	0.0	0.2	0.1	0.1	0.1	0.2	0.0
MgO	0.6	2.7	3.4	2.9	3.4	2.1	2.6
CaO	0.0	0.7	5.9	0.4	0.4	1.5	4.8
Na_2O	1.0	0.8	1.0	1.3	1.9	1.1	0.9
K_2O	7.8	3.9	3.4	3.6	4.3	2.5	3.2
P_2O_5	0.0	0.1	0.1	0.3	0.1	0.2	0.0
Non-carbonate C	0.3	0.4	0.5	0.7	0.0	0.0	0.0
CO_2	0.0	0.7	4.4	0.2	0.3	0.9	4.7
H_2O+	6.4	4.5	3.9	4.2	3.5	8.5	3.4
S	0.0	0.2	0.4	0.1	0.1	0.6	0.0
Σ	101.2	99.2	99.4	98.1	100.2	98.7	98.7
Fe_2O_3/FeO		0.12	0.33	0.20	0.42	4.75	

Table 3.
Seven exemplary full chemical analyzes [5, 7, 33–35].

	Nigeria				Germany							
	Shales: Abakabili				Roof slates: Altlay				Roof slates: Magog			
	slatenorm			slate-calc.	slatenorm			slate-calc.	slatenorm			slate-calc..
I	II	GV	I		II	GV	I		II	GV		
Quartz	10.4	10.3	10.4	4.2	31.0	31.0	31.3	30.5	27.8	27.8	27.8	27.3
Feldspar	0.2	0.2	0.2	15.1	0.0	0.0	0.0	0.0	0.6	0.6	0.6	3.4
Mica	78.5	76.5	78.5	25.2	42.9	42.9	42.9	38.8	40.9	40.9	40.9	31.6
Hydro-mica	0.0	0.0	0.0	47.4	0.0	0.0	0.0	6.2	0.0	0.0	0.0	7.8
Fe chlorite	5.8	5.7	6.1	0.0	12.3	10.8	13.0	11.2	8.8	8.8	8.8	6.3
Mg chlorite	1.5	1.4	1.5	1.3	6.8	6.9	7.3	7.0	7.9	7.9	7.9	7.9
Chlorite	7.3	7.0	7.6	1.3	19.0	17.8	20.3	18.2	16.6	16.6	16.6	14.2
Carbonates	0.0	0.0	0.0	0.0	1.4	1.4	1.4	1.4	10.1	10.1	10.1	10.1
Pyrrhotite	0.0	0.0	0.0	0.0	0.0	0.0	0.0	0.0	0.0	0.0	0.0	0.0
Pyrite	0.1	0.1	0.1	0.1	0.4	0.4	0.4	0.4	0.7	0.7	0.7	0.7
Sulfides	0.1	0.1	0.1	0.1	0.4	0.4	0.4	0.4	0.7	0.7	0.7	0.7
Ti mineral	1.7	1.7	1.7	6.5	1.9	1.9	1.9	2.7	1.4	1.4	1.4	2.9
Ore minerals	1.7	1.7	1.7	6.5	2.3	2.3	2.3	3.0	2.1	2.1	2.1	3.7
Chloritoid	0.0	00	0.0	0.0	0.0	2.9	0.0	0.0	0.0	0.0	0.0	0.0
Aqua	2.1	2.1	2.1	2.5	0.4	0.3	0.2	0.3	0.2	0.2	0.2	0.4
Σ	100.5	100.2	100.9	102.7	97.7	99.1	99.1	99.2	99.1	99.1	99.1	99.3

	Spain							
	Roof slates: San Pedro de T.				Phyllites: Pol Lugo green			
	slatenorm			slate- calc.	slatenorm			slate- calc.
	I	II	GV		I	II	GV	
Quartz	24.7	24.6	25.0	25.8	23.7	23.7	23.7	24.8
Feldspar	0.0	0.0	0.0	0.0	9.3	9.3	9.3	9.3
Mica	46.7	46.7	46.7	46.7	46.9	46.9	46.9	46.9
Hydro-mica	0.0	0.0	0.0	0.0	0.0	0.0	0.0	0.0
Fe chlorite	13.6	12.2	14.3	11.9	10.3	10.3	10.3	7.05
Mg chlorite	7.6	7.8	8.1	8.4	7.4	7.4	7.4	7.4
Chlorite	21.1	20.0	22.5	20.3	17.7	17.7	17.7	14.5
Carbonates	0.5	0.5	0.5	0.5	0.7	0.7	0.7	0.7
Pyrrhotite	0.0	0.0	0.0	0.1	0.0	0.0	0.0	0.0
Pyrite	0.1	0.1	0.1	0.1	0.2	0.2	0.2	0.2
Sulfides	0.1	0.1	0.1	0.2	0.2	0.2	0.2	0.2
Ti mineral	2.2	2.2	2.2	3.6	1.8	1.8	1.8	3.7
Ore minerals	2.4	2.4	2.4	3.8	2.0	2.0	2.0	4.1

	Spain							
	Roof slates: San Pedro de T.				Phyllites: Pol Lugo green			
	slatenorm			slate- calc.	slatenorm			slate- calc.
	I	II	GV		I	II	GV	
Chloritoid	0.0	2.9	0.0	0.0	0.0	0.0	0.0	0.0
Aqua	−0.3	−0.4	−0.5	0.3	−0.6	−0.6	−0.6	−0.3
Σ	96.5	97.9	97.9	98.0	99.9	99.9	99.9	100.2

Table 4.
Results of a comparison of all versions of slatenorm and slatecalculation (analysis from ***Table*** *3) [33].*

In the German roof slates, too, slatecalculation results in the hydro-micas illite and brammallite, which also lead to lower chlorite values. The appearance of chloritoid in slatenorm version II in the roof slates from Altlay/Germany and San Pedro de T./Spain, which actually do not contain this mineral, shows that this version is incorrect. There are differences in ore minerals including titanium minerals between the results of slatecalculation and all versions of slatenorm. Slatecalculation determines additional minerals, such as pyrrhotite pn and titanomagnetite tm, with the additional consideration of Fe_2O_3. This leads to higher ore mineral values. At the same time, the Fe chlorite values are lower (fc and fac).

Due to the natural variability of chlorites and the use of standard minerals, such as fc and mc, which do not actually occur as end links in nature (**Figure 1**), the results of both standard mineral calculations should only be given as sums (chlorite = mac + mc + fac + fc), possibly divided into Fe-chlorite (= fac + fc) or Mg chlorite (= mac + mc).

The "Grundversion" of slatenorm works better than the other two versions I and II and is therefore preferable. The more complex calculation of slatecalculation leads to better values than slatenorm, especially when hydro-micas are calculated.

8. Applications (slatenorm and slatecalculation)

Slatenorm (later slatecalculation) was originally developed as a method of determining the mineral content to assess the suitability of the rock as a roofing slate. For this purpose, more than 360 analyses were carried out similar to [20] based on practical experience in normal, oxidizing, occasional oxidizing and carbonate slate, and also in high carbon content-slate with carbonate and hard slate divided. There are also shale and phyllite/schist. For reasons of clarity, not all types in [9] are to be treated.

Additional results will be added in ref. [33]. The nine samples of the group "oxidizing slates" are actually only errors in the selection during the extraction or manufacture of "occasional oxidizing slates" (17 samples). As far as all other properties are concerned, it is a normal slate (see **Figure 2**). The 94 samples of the group "high carbon content- slates" are also normal slates, but with a no carbonate carbon content of over 1%. EN 12326 [36] excludes roofing slate with a content of more than 2%.

Cardenes et al. [26] summarized the cleavage and perforability of roofing slate in a diagram of rigid (= standard minerals: qz + an + ab + or = Quartz and Feldspar) and elastic minerals (= standard minerals: mu + pa + ill + br + mac + mc + fac + fc = Mica, Hydro-mica and Chlorite). However, the classes "soft," "medium hard," "hard," and

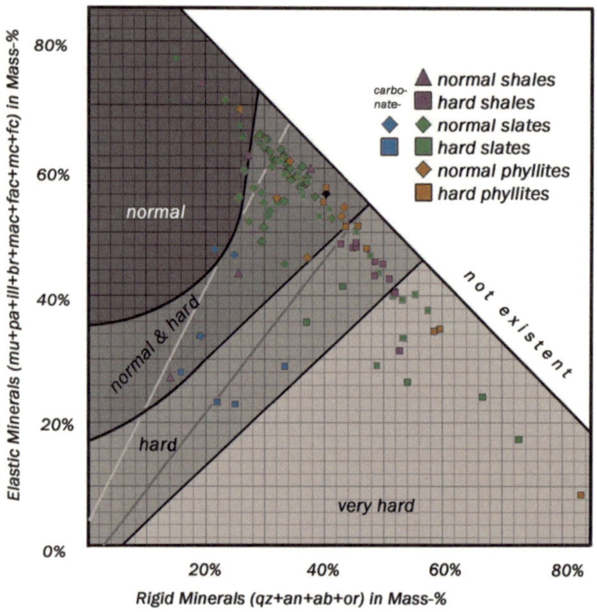

Figure 2.
Diagram showing the ratio of rigid to elastic minerals, calculated with slatecalculation, with changes according to [26].

"very hard" listed there are imprecise. The reasons for these properties are not always related to the rigid mineral content and especially the quartz content. The determination of the quartz content remains an important prerequisite for the evaluation of a roofing slate deposit.

The class boundaries in **Figure 2** (gray lines) provided in ref. [26] should be corrected. Some slates with carbonate that can be processed normally fall into the wrong class there (there is medium-hard). Obviously, the good cleavability of the carbonates or, to a lesser extent, the feldspars or chloritoids have to be included in the assessment. This leads to new class boundaries in **Figure 2** (black lines). There is a narrow field between the classes "normal" and "hard" in which slates of both classes occur. In **Figure 2**, some slates from the groups' carbonate slate, high carbon content-slate, phyllite/schist, and shale have been assigned the properties hard or normal.

The phyllosilicates calculated in slatecalculation (after [9]) show a total mica content from usually above 40% (up to a maximum of 60%) and a chlorite content from more than 10% (up to a maximum of 25%) in normal slates (**Figure 3**). Only for samples with higher carbonate ("carbonate" and "with carbonate") or higher carbon ("high carbon content"), are the proportions lower. The ratio of mica (mu + pa + ill + br) to chlorite (mac + mc + fac + fc) is 3 to 1. In "normal" slates, the Fe-chlorites content (fac + fc) outweighs the Mg-chlorites content (mac + mc). The calculated proportion of hydro-micas (ill and br) could reflect (in addition to the Kübler index or the organic matter reflectance [37, 38], **Figure 4**) the degree of metamorphism in most of the samples.

That is why, phyllites always have hydro-mica values of 0% in the calculation outputs.

The slates of the Iberian Variscides also show very low positive percentages of hydro-micas, while slates from the Central European Variscides (Ardennes and Rhenohercynian zone) have a lower metamorphic grade and show higher values of hydro-micas (**Figures 3** and **4**).

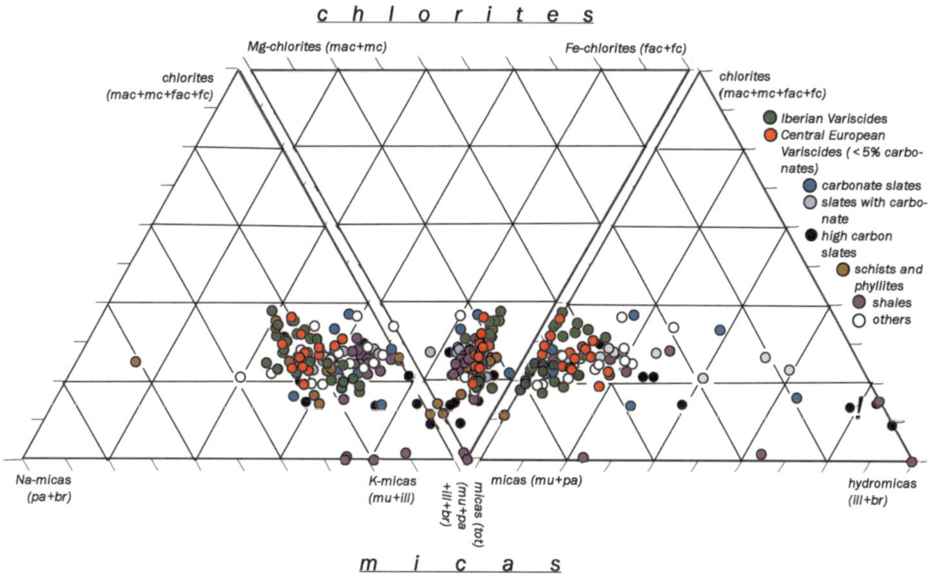

Figure 3.
Phyllosilicates as calculated by slatecalculation: Types of phyllosilicates (after [9]).

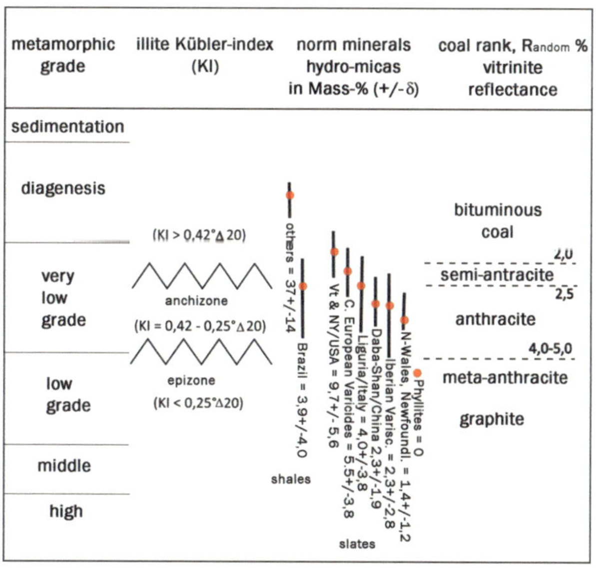

Figure 4.
Methods for determining the grade of metamorphosis (according to [9, 33, 38, 39]).

9. Discussion

Correspondences between the standard methods for sediments (SEDNORM, SEDMIN, PELNORM, subsection 4) on the one hand and slatenorm and slatecalculation, on the other hand, are to be expected. As **Tables 5** and **6** shows, however, these are low. There are rather clear differences.

	Nigeria				Germany							
	Shales: Abakabili				Roof slates: Altlay				Roof slates: Magog			
	Slate-norm	slate-calc.	SED-MIN	SED-MIN **	Slate-norm	slate-calc.	SED-MIN	SED-MIN *	Slate-norm	slate-calc.	SED-MIN	SED-MIN *
Quartz	10.4	4.2	−11.1	0.2	31.3	31.0	27.0	32.7	27.8	27.3	28.4	28.4
Anorthite	0.0	0.0			0.0	0.0			0.5	0.5		
Albite	0.0	5.1	8.6	8.6	0.0	0.0	6.9	6.9	0.1	2.9	8.1	8.1
K-spar	0.0	9.78	32.0	5.8	0.0	0.0	0.0	0.0	0.0	0.0	0.0	0.0
Feldspar	0.2	0.2	0.2	15.1	0.0	0.0	6.9	6.9	0.6	3.4	8.1	8.1
Muskovite	66.0	25.2			32.8	29.6			29.1	25.5		
Paragonite	12.6	0.0			10.1	9.1			15.9	15.9		
Sericite	78.5	25.2	0.0	49.7	42.9	38.8	30.6	30.6	40.9	31.6	24.6	24.6
Illite		39.7	13.2	13.2		4.7	3.8	3.8		5.5	7.6	7.6
Brammallite		7.7				1.5				2.3		
Hydro-mica		47.4	13.2	13.2		6.2	3.8	3.8		7.8	7.6	7.6
Mica+Hm.	78.5	72.7	13.2	13.2	42.9	45.0	34.3	34.3	40.9	39.4	32.2	32.2
Chlorite	7.6	1.3	0.7	0.7	20.3	18.2	9.1	9.1	16.7	14.2	12.2	12.2
Calcite	0.0	0.0	0.9	0.9	0.4	0.4	0.0	0.0	10.1	10.1	8.8	8.8
Dolomite	0.0	0.0	0.0	0.0	1.0	1.0	1.6	1.6	0.0	0.0	0.7	0.7
Ferro-Dolo.	0.0	0.0	0.0	0.0	0.0	0.0	0.0	0.0	0.0	0.0	0.0	0.0
Gypsum			0.0	0.2			1.1	1.1			2.0	2.0
Pyrite/Pn.	0.1	0.1	0.0	0.0	0.4	0.0	0.0	0.0	0.7	0.7	0.0	0.0
Rutile	1.7	0.0	0.8	0.6	0.0	0.0	0.8	1.0	0.0	0.0	0.8	0.8
Titanomagn.		6.5				1.6				2.9		
Ilmenite	0.0	0.0			1.9	0.0			1.4			
Hematite		0.0	2.8	2.6	0.0	0.0	6.5	6.5		0.0	4.5	4.5
Limonite		0.2				0.0				0.1		
Apatite	0.0	0.0	0–0	0.0	0.2	0.2	0.2	0.2	0.3	0.3	0.3	0.3
Smectite			0.0	0.0			0.0	0.0			0.0	0.0
Kaolinite			55.1	18.3			12.5	0.0			0.0	0.0
Phyllosilic,	86.2	74.0	69.0	81.9	63.2	63.2	55.8	43.4	57.6	53.6	44.4	44.4
Aqua	2.1	2.5			0.2	0.3			0.2	0.2		
Σ	100.4	102.3	103	100.8	98.7	97.6	100	93.4	98.6	98.8	98.0	98.0

	Spain							
	Roof slates: S. Pedro de T.				Phyllites: Pol Lugo green			
	Slate-norm	slate-calc.	SED-MIN	SED- MIN *	Slate-norm	slate-calc.	SED-MIN	SED- MIN *
Quartz	25.0	25.8	15.6	26.4	23.7	24.8	20.1	24.4
Anorthite	0.0	0.0			0.0	0.0		
Albite	0.0	0.0	10.9	10.9	9.3	9.3	16.3	16.3
K-spar	0.0	0.0	5.9	0.0	0.0	0.0	7.1	0.0

	Spain							
	Roof slates: S. Pedro de T.				Phyllites: Pol Lugo green			
	Slate-norm	slate-calc.	SED-MIN	SED- MIN *	Slate-norm	slate-calc.	SED-MIN	SED- MIN *
Feldspar	0,0	0.0	16.9	10.9	9.3	9.3	23.4	16.3
Muskovite	30.8	30.8			36.6	36.6		
Paragonite	15.9	15.9			10.2	10.2		
Sericite	46.7	46.7	19.6	28.1	46.9	46.9	22.0	32.2
Illite		0.0	4.5	4.5		0.0	7.4	7.4
Brammallite		0.0				0.0		
Hydro-mica		0.0	4.5	4.5		0.0	7.4	7.4
Mica+Hm.	46.7	46.7	24.1	32.6	46.9	46.9	29.4	39.6
Chlorite	22.5	20.3	10.6	10.6	17.7	14.5	12.6	12.6
Calcite	0.0	0.0	0.0	0.0	0.0	0.0	0.0	0.0
Dolomite	0.2	0.2	0.4	0.4	0.7	0.7	0.2	0.2
Ferro-Dolo.	0.4	0.3	0.0	0.0	0.0	0.0	0.0	0.0
Gypsum			0.4	0.4			0.5	0.5
Pyrite/Pn.	0.1	0.1	0.0	0.0	0.2	0.2	0.0	0.0
Rutile	0.0	0.0	0.8	1.2	0.0	0.0	0.8	1.8
Titanomagn.		2.7				3.7		
Ilmenite	2.2	0.9			1.8	0.0		
Hematite		0.0	7.3	7.3		0.0	5.3	5.3
Limonite		0.0				0.0		
Apatite	0.6	0.6	0.6	0.6	0.3	0.3	0.3	0.3
Smectite			10.6	10.6			0.0	0.0
Kaolinite			23.6	0.0			9.2	0.0
Phyllosilic.	69.2	67.0	68.9	53.7	64.6	61.4	51.2	52.2
Aqua	−0.5	−0.3			−0.6	−0.3		
Σ	97.7	97.1	110.5	100.9	99.9	100	101.8	100.9

Table 5.
*Comparison of the results from slatenorm, slatecalculation and SEDMIN. * = TiO_2 (kaolinite) = 0, ** = TiO_2 (kaolinite) = 1.1. Empty fields = not calculated.*

In **Table 5**, the five examples from **Table 3** are calculated using the specified methods. **Table 6** contains the standard mineral results of two further examples in **Table 3** from the literature given there. Standard minerals remain, which are not calculated and are marked as gaps in **Tables 5** and **6**. In SEDNORM, SEDMIN, and PELNORM, the clay minerals kaolinite ($2SiO_2$ Al_2O_3 $0.05TiO_2$ $2H_2O$, Mol. Wt of mineral: 262.15, according to ref. [6]) and smectite ($4SiO_2$ Al_2O_3 $0.1Na_2O$ $0.1CaO$ 10.9 H_2O: 550.46 of mineral: 550.46 of mineral, according to ref. [6]). In contrast to the other methods, a distinction is made in slatenorm and slatecalculation between Na and K mica or Na and K hydro-mica. In the other methods, on the other hand, only K mica

	USA					World			
	Pierre shale					Average sedimentary rock after [5, 34]			
	Slate-norm	**slate-calc.**	**SED-MIN**	**PEL-NORM M**	**PEL-NORM N**	**Slate-norm**	**slate-calc.**	**SED-MIN**	**SED-NORM ***
Quartz	39.0	34.8	41.9	23.4	26.2	38.1	37.0	39.3	35.3
Anorthite	1.1	1.1				0.6	0.0		
Albite	0.0	0.0	8.6	8.6	0.0	0.6	4.8	7.6	
K-spar	0.0	0.0	0.0			0.0	3.7	0.0	
Feldspar	1.1	1.1	9.2	2.6	5.4	1.2	8.4	7.6	7.8
Muskovite	21.1	0.0				27.1	11.8		
Paragonite	12.6	0.0				27.1	0.0		
Sericite	34.5	0.0	21.1			37.3	11.8	25.3	
Illite		31.3	0.0	30.4	27.8		15.0	3.0	33.0
Brammallite		20.3							
Hydro-mica		51.6		30.4	27.8		21.2	3.0	33.0
Mica+Hm.	34.5	51.6	21.1	30.4	27.8	37.3	33.0	28.3	33.0
Chlorite	14.3	0.0	6.4			11.4	4.0	7.0	
Calcite	2.0	2.0	0.0			6.4	6.4	6.7	8.8
Dolomite	0.0	0.0	2.4			3.9	3.9	3.5	1.8
Ferro-Dolo.	0.0	0.0	0.0			0.0	0.0	0.0	
Carbonate	2.0	2.0	2.4	3.6	2.2	10.4	10.4	10.2	10.6
Gypsum			3.0	2.6	2.2			0.0	
Pyrite/Pn.	1.0	1.0	0.0			0.0	0.0	0.0	
Rutile	0.6	0.0	0.6			0.0	0.0	0.0	
Titanomagn.		2.3				0.0	0.0		
Ilmenite	0.0	0.0				0.0	0.0		
Hematite		0.0	4.6				0.0	3.6	5.0
Limonite		3.7					5.3		
Apatite	0.4	0.3	0–3			0.0	0.0	0.0	0.0
Smectite			0.0	33.0	28.0			0.0	0.0
Kaolinite			0.0	3.0	2.6			0.0	7.2
Clay mineral			0.0	36.0	30.6			0.0	7.2
Phyllosilic,	48.8	51.6	27.4	68.6	59.8	48.7	37.0	35.2	40.2
Aqua	5.3	5.3				0.4	0.6		1.0
Σ	96.1	102.2	89.4	100.8	95.8	98.8	98.7	95.8	99.9

Table 6.
*Comparison of standard calculation results taken from the literature (SEDNORM [5] and PELNORM [7]) with own calculations according to SEDMIN [6], slatenorm, and slatecalculation. M = modal mineral inventory, N = calculated [7], * = 6, 7 in Table 6 in [5]. Analyzed from Table 3.*

(muskovite) or illite is calculated. PELNORM [7] only considers a small proportion of Na_2O in the formula for illite ($0.025Na_2O$ $0.30K_2O$ Al_2O_3 $0.2MgO$ $0.125Fe_2O_3$ $3.40SiO_2$ Mol. Wt of mineral: 374.17). SEDMIN uses the very low TiO_2 content in the above chemical formula (only 0.05 in $2SiO_2$ Al_2O_3 $0.05TiO_2$ $2H_2O$!) to calculate kaolinite. This leads to high kaolinite contents in phyllites and slates, where this mineral is not even present (**Table 5**).

As mentioned in ref. [9], the Abakabili shale (**Table 5**) may contain a small amount of clay minerals. In the literature [40], the main component is illite with 30–38%, smectite/montmorillonite with 20–30%, quartz with 28–30% (instead of 11% in **Table 5**), and kaolinite with only 15–25% (instead of 55% in **Table 5**).

There are often TiO_2 minerals such as rutile (TiO_2), ilmenite (FeO TiO_2 or titanomagnetite FeO Fe_2O_3 TiO_2) in the group of slates, which in slatenorm and slatecalculation are consequently calculated as standard minerals ru, ilm, and/or tm. Using the TiO_2 in the SEDMIN standard calculation of kaolinite leads to the excessively high results in **Table 5**. For this reason, a corrected value of TiO_2 (if used to calculate kaolinite: * = 0 ** = 1.1) was used there as an alternative.

SEDNORM and SEDMIN specify the Fe compounds to a large extent as hematite (Fe_2O_3). However, this mineral is rarely found in shales and slates. Slatenorm and slatecalculation, therefore, count Fe above all to be monovalent Fe sulfides. Hematite is only calculated if, in rare cases, the color of the rock is red and not black or green (see **Figure 1** in [9]). In **Table 6**, the mean value of the worldwide sedimentary rocks is given from the chemical analysis in **Table 3** (after [5, 34]). With SEDNORM, only calculation results are given for which, according to standard mineral calculations, there are no longer any excess MgO, Na_2O, K_2O, or CO_2 (6, 7 in **Table 6** in [5]). As far as the quartz, the carbonates and the sum of the phyllosilicates are concerned, the methods slatenorm, slatecalculation, SEDMIN, and SEDNORM show sufficient agreement.

10. Conclusions

There are several suitable methods for calculating a standard mineral inventory of rock from the full chemical analysis. Among these, the CIPW standard is used, in particular, when the minerals in igneous rocks are particularly fine-grained or not crystallized at all.

The new standard mineral calculations presented here slatenorm and slatecalculation for fine-grained sediments and very low grade and low-grade metamorphic rocks have now been added. The method has already proven itself in assessing roof slates.

In addition to the Kübler index or the reflectance of coal substances with the calculated content of hydro-micas (illite and brammallite), the method also provides information on the degree (grade) of metamorphosis. Further research here seems to be worthwhile in the future.

A detailed comparison of slatenorm and slatecalculation with the standard mineral methods for sedimentary rocks, such as SEDNORM, SEDMIN, and PELNORM, show only a few matches in the results. As far as methodological differences are overcome, the methods for sediments could be appended to it as further calculation steps after the calculation of slatecalculation. The residual water aq should be used as a basis. This requires a new calculation method that has several variants to choose from.

Acknowledgements

The development of the computer program "slatenorm" was funded by Rathscheck Schiefer und Dach-Systeme - ZN der Wilh. Werhahn KG Neuss, D 56727 Mayen-Katzenberg. The publication here was permitted.

Conflict of interest

The author declares no conflict of interest.

DOI: http://dx.doi.org/10.5772/intechopen.102346

Appendix A: flowchart slatecalculation

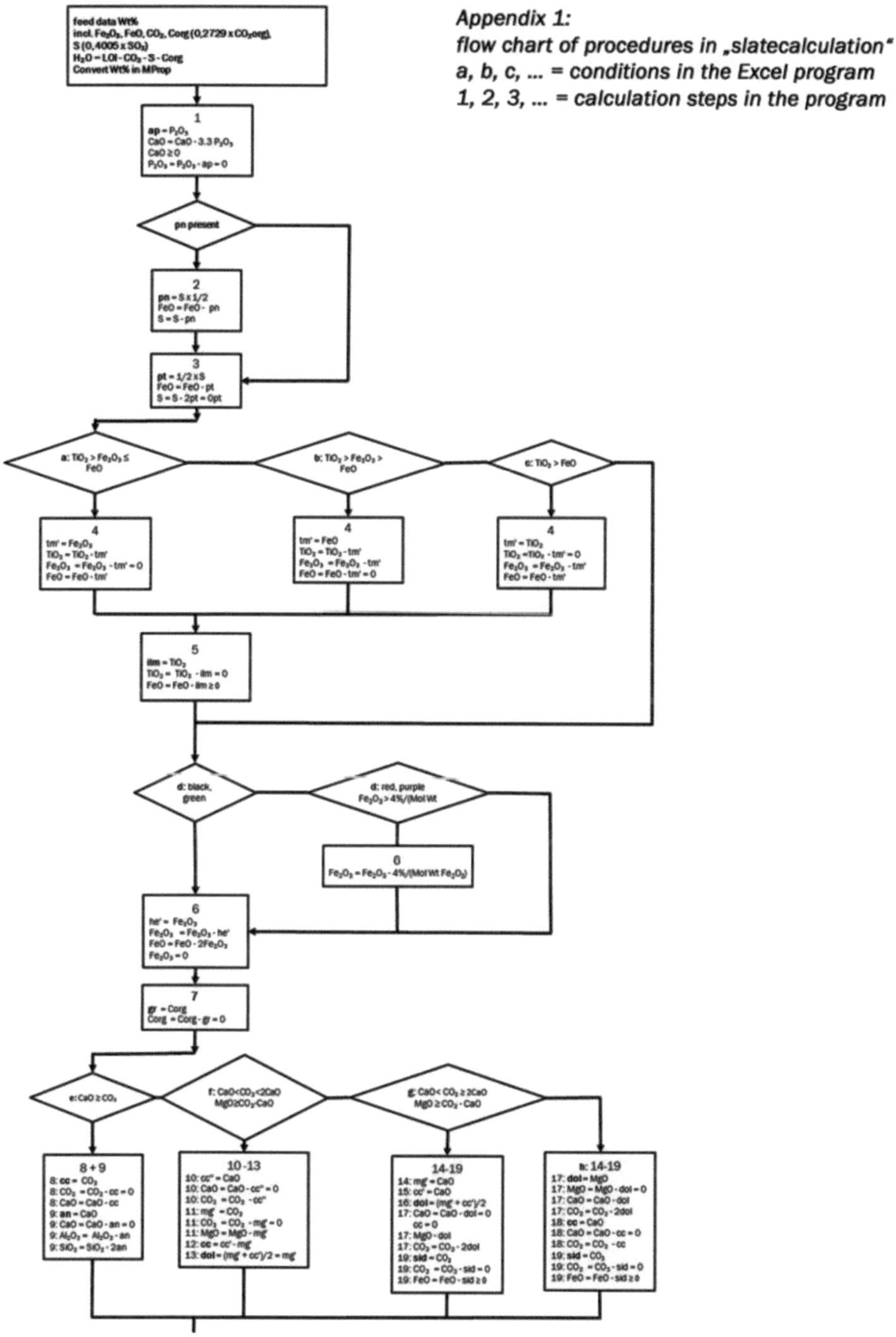

Appendix:
flow chart of procedures in „slatecalculation"
a, b, c, ... = conditions in the Excel program
1, 2, 3, ... = calculation steps in the program
(continued)

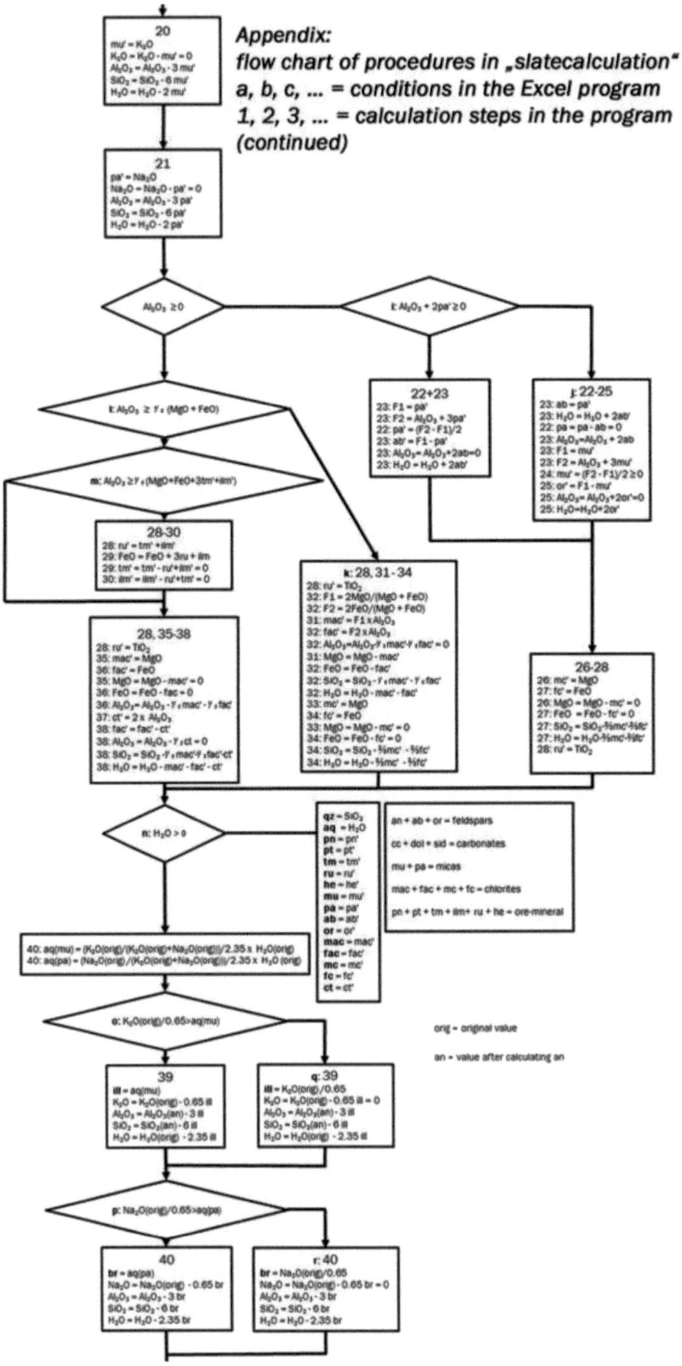

B. Program slatenorm

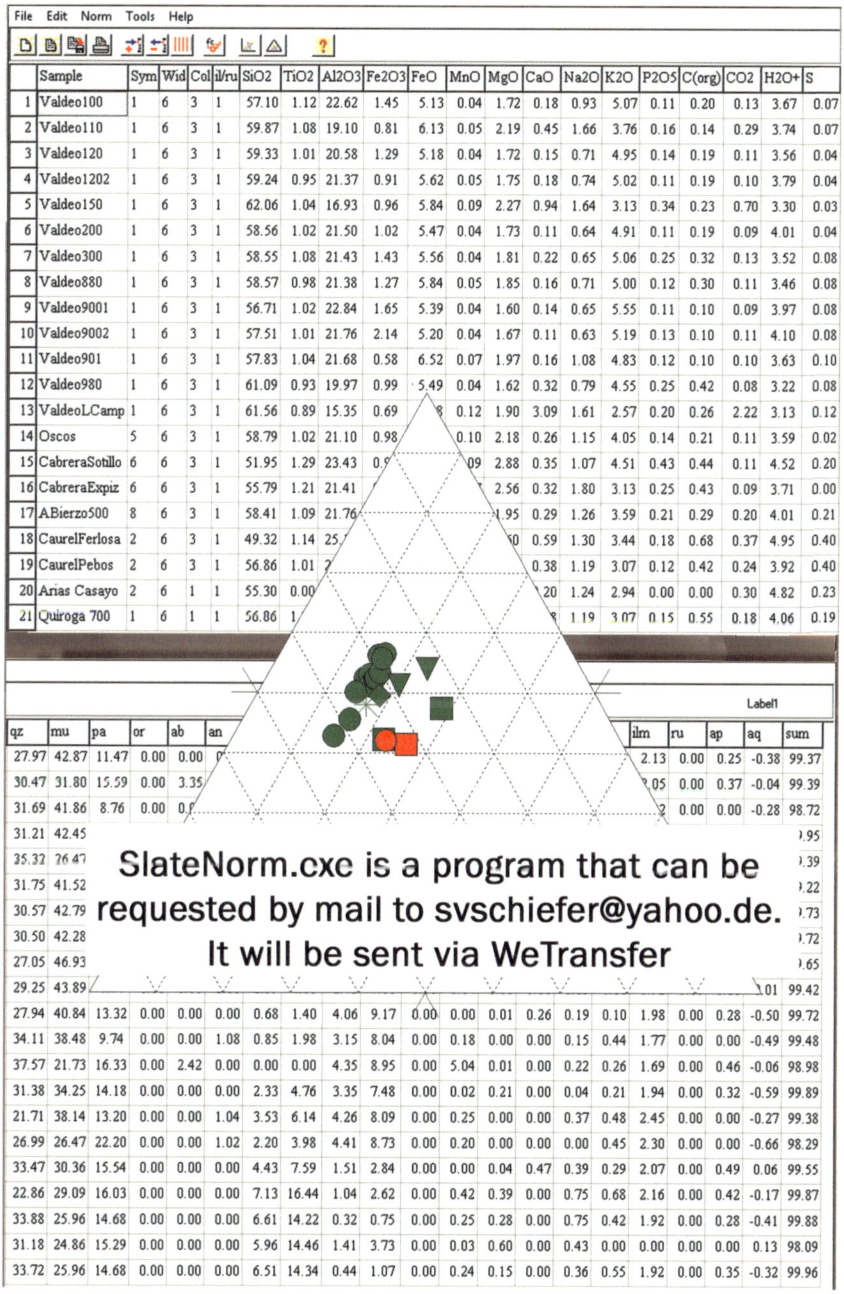

Author details

Hans Wolfgang Wagner
Department of Geology, University of Trier, Mayen, Germany

*Address all correspondence to: wagnerw@uni-trier.de

References

[1] Matthes S. Mineralogie eine Einführung in die spezielle Mineralogie, Petrologie und Lagerstättenkunde. Berlin, Heidelberg, New York, Tokyo: Springer; 1983. p. 417. DOI: 10.1007/978-3-662-8772-5

[2] Martin C, Eiblmaier M. Lexikon der Geowissenschaften. Heidelberg: Spektrum Akademischer Verlag; 2005. p. 2782

[3] Müller G, Braun E. Methoden zur Berechnung von Gesteinsnormen. Vol. 126. Clausthaler tektonische Hefte 15, Clausthal-Zellerfeld: Verlag Eilen Pilgert; 1977. p. 37. DOI: 10.1007/978-3-642-95844-1

[4] Cross W, Iddings JP, Pirsson LV, Washington HS. A quantitative chemico-mineralogical classification and nomenclature of igneous rocks. The Journal of Geology. 1902;**10**:555-690. DOI: 10.1086/621030

[5] Cohan D, Colin CR. SEDNORM - A program to calculate a normative mineralogy for sedimentary rocks based on chemical analyses. Computers & Geosciences. 1991;**17**(9):1235-1253. DOI: 10.1016/0098-3004(91)90026-A

[6] Knackstaetter UR. SEDMIN - Microsoft ExcelTM spreadsheet for calculating fine-grained sedimentary rock mineralogy from bulk geochemical analysis. Central European Journal of Geosciences. 2014; **6**(2):170-181. DOI: 10.2478/s13533-012-0170-3

[7] Merodio JC, Spalletti LA, Bertone LM. A FORTRAN program for the calculation of normative composition of clay minerals and pelitic rocks. Computers & Geosciences. 1992;**18**(1):47-61. DOI: 10.1016/0098-3004(92)90057-x

[8] Podkovyrov VN, Graunov OV, Cullers RL. A linear programming approach to determine the normative composition of sedimentary rocks. Mathematical Geology. 2003;**35**(4): 459-476. DOI: 10.1023/A:1024898024880

[9] Wagner HW, Jung D, Wagner J-F, Wagner MP. Slatecalculation—A practical tool for deriving norm minerals in the lowest-grade metamorphic pelites and roof slates. Minerals. 2020;**10**(395): 349-370. DOI: 10.3390/min10050395

[10] Jung D, Schulz H. Beschreibung von Algol-Programmen zur Berechnung der NIGGLI-Werte und der CIPW-Norm. Neues Jahrbuch für Mineralogie-Abhandlungen. 1965;**103**(3):256-272. DOI: 10.1127/njma/103/1965/256

[11] CIPW. Norm Calculation - Whitman College, Kurt Hollocher's Spreadsheet. New York, USA: Union College; 2004. Available from: https://www.whitman.edu/geology/winter/JDW_PetClass.htm [Accessed: November 18, 2021]

[12] Winter JD. CIPW Norm Excel Spreadsheet. n. s. Available from: http://webspace.pugetsound.edu/facultypages/jtepper/MIN-PET/Norm%20Calculation.XLS [Accessed: February 21, 2020]

[13] Verma S, Torres-Alvarado I, Velasco-Tapia F. A revised CIPW norm. Schweizerische Mineralogische und Petrographische Mitteilungen. 2003;**83**: 197-216. DOI: 10.5169/seals-63145

[14] Verma S, Rivera-Gómez A. Computer programs for the classification and nomenclature of igneous rocks. Episodes. 36. 2013; no. 2: 115-124. DOI: 10.18814/epiiugs/2013/v36i2/005

[15] Niggli P. über Molekularnormen zur Gesteinsberechnung. Schweizerische

Mineralogische und Petrographische Mitteilungen. 1936;**16**:295-317

[16] Barth TFW. Theoretical Petrology. 2nd ed. New York - London: John Wiley & Sons; 1962. p. 416

[17] Barth TFW. A final proposal for calculating the mesonorm of metamorphic rocks. The Journal of Geology. 1962;**70**:497-498. DOI: 10.1086/626841

[18] Stuckless JS. A Program in Hewlett-Packard BASIC for Calculation of Mesonormative Minerals Using HP-Series 80 Computers and VISICALC Electronic Worksheet. Reston VA USA: US Geological Survey Open-File Report 83–923; 1983. pp. 1-24. DOI: 10.3133/OFR83913

[19] Rittmann A. Stable Mineral Assemblages of Igneous Rocks. Berlin - Heidelberg - New York: Springer; 1973. p. 262. DOI: 10.1017/S0016756800045702

[20] Stengelin R, Hewers W. ALGOL-Programs for the Computation of Stable Mineral Assemblages of Igeous Rocks with the Aid of the Rittmann Norm. Tübingen: Fotodruck Präzis, v. Spangenberg KG; 1973

[21] Knackstaetter, UR. SEDMIN. Sedimentary Mineral Calculator.xlsx. Available from: https://www.researchgate.net/publication/293174516_SEDMIN_Sedimentary_Mineral_Calculator/link/56b61f5b08ae3c1b79ad24e0/download [Accessed: December 15, 2021]

[22] Wagner HW, Baumann H, Negendank J, Roschig F. Geological, petrographic, geochemical and petrophysical investigations on roofing slates. Mainzer Geowissenschaftliche Mitteilungen. 1997, 26: 131-184. Available from: https://www.uni-trier.de/fileadmin/fb6/prof/GEO/baumann/pdf_files/slate.pdf [Accessed: February 21, 2020].

[23] Ward C, Gómez-Fernandez F. Quantitative mineralogical analysis of Spanish roofing slates using the Rietveld method and X-ray powder diffraction data. European Journal of Mineralogy. 2003;**15**:1051-1062. DOI: 10.1127/0935-1221/2003/0015-1051

[24] Jung D, Wagner HW. Slatenorm: Calculation of a normative mineral content of slates. Unpublished. 1996-2000

[25] Lorenz W, Gwodsdz W. Roofing slate. In: Sonderhefte Reihe H, Heft SH, editors. Manual on the Geological-Technical Assessment of Mineral Construction Materials. Hannover, Germany: BGR; 2003. pp. 276-290

[26] Cárdenes V, Rubio-Ordóñez A, Wichert J, Cnudde JP, Cnudde V. Petrography of roofing slates. Earth-Science Reviews. 2014;**138**:435-453. DOI: 10.1016/j.earscirev.2014.07.003

[27] Jung D. Zur Frage der Qualitätsbeurteilung von Schiefer. Schriftenreihe des Schiefer-Fachverbandes in Deutschland eV. 2009;**10**:75-107

[28] Wagner HW, Schultheis W. Römischer Dachschiefer—Neue Funde und neue Erkenntnisse. Der Anschnitt. 2011;**63**:202-206

[29] Wagner HW. Geologische Untersuchung und Materialprüfung an Dachschiefer-Altdeckungen. Mainzer Geowissenschaftliche Mitteilungen. 2014;**42**:121-142

[30] Morales-Demarco M, Oyhantçabal P, Stein K-J, Siegesmund S. Dolomitic slates from Uruguay: Petrophysical and petromechanical

characterization and deposit evaluation. Environment and Earth Science. 2013; **69**:1361-1395. DOI: 10.1007/s12665-012-1921-7

[31] Jung D, Geisler-Wierwille T. Program "Slatenorm" Version 1.1. Unpublished. 1999-2000

[32] Wagner HW, Le Bail R, Hacar M, Stanek S. European roofing slates part 1: Remarks to the geology of mineral deposits. Zeitschrift für Angewandte Geologie. 1994;**40**:68-74

[33] Wagner HW. Rohstoffgeologie des Dachschiefers [Habilitationsschrift]. Trier: Department of Geology, University of Trier; 2020. p. 570

[34] Garrels RM, Mackenzie FT. Evolution of Sedimentary Rocks. New York: W. W. Norton; 1971. p. 397. DOI: 10.4319/lo.1972.17.1.0165

[35] Tourtelot HA. Preliminary Investigation of the Geologic Setting and Chemical Composition of the Pierre Shale Great Plains Region. Washington: Geological Survey Professionel Paper 390; 1962. p. 81. Available from: https://pubs.usgs.gov/pp/0390/report.pdf [Accessed: December 14, 2021]

[36] CEN, European Committee for Standardization. EN 12326–1 and –2 Slate and Stone Products for Discontinuous Roofing and Cladding—Part 1: Product Specification. 2014; Part 2: Methods of Test. Berlin, Germany: Beuth; 2010. DOI: 10.31030/2237360, 10.31030/1805528

[37] Cardenes V, Rubio Ordonez A, Lopez Munguira A, Monteroso C. Petrografía y mineralogía de las pizarras para cubiertas de la Península Ibérica en relación con su calidad. Trabajos de Geologia Universidad de Oviedo. 2010; **30**:412-420

[38] Ferreiro Mählmann R, Bozkaya Ö, Potel S, Le Bayon R, Šegvic B, Nieto F. The pioneer work of Bernard Kübler and Martin Frey in very low-grade metamorphic terranes: Paleo-geothermal potential of variation in Kübler-index/organic matter reflectance correlations. A review. Swiss Journal of Geosciences. 2012;**105**:121-152. DOI: 10.1007/s00015-012-0115-3

[39] Arkai P, Sassi F, Desmons J. Very low- to low-grade metamorphic rocks. Recommendations by the IUGS subcommission on the systematics of metamorphic rocks. A Classification and Glossary Terms. 2003;**5**:1-12. Available from: https://www.ugr.es/~agcasco/personal/IUGS/pdf-IUGS/scmr_low_r2_verylowtolowgrademetamorphicrocks.pdf [Accessed: November 29, 2021]

[40] Nweke OM. Evaluating the suitability of clays from Abakaliki Area, Southeastern Nigeria for oil industrial application using geotechnical and rheological properties. Science Innovation. 2015;**3**(2):22-31. DOI: 10.11648/j.si.20150302.11

Chapter 6

Minerals as Prebiotic Catalysts for Chemical Evolution towards the Origin of Life

Yamei Li

Abstract

A transition from geochemistry to biochemistry has been considered as a necessary step towards the emergence of primordial life. Nevertheless, how did this transition occur is still elusive. The chemistry underlying this transition is likely not a single event, but involves many levels of creation and reconstruction, finally reaching the molecular, structural, and functional buildup of complexity. Among them, one apparent question is: how the biochemical catalytic system emerged from the mineral-based geochemical system? Inspired by the metal–ligand structures in metalloenzymes, many researchers have proposed that transition metal sulfide minerals could have served as structural analogs of metalloenzymes for catalyzing prebiotic redox conversions. This assumption has been tested and verified to some extent by several studies, which focused on using Earth-abundant transition metal sulfides as catalysts for multi-electron C and N conversions. The progress in this field will be introduced, with a focus on the CO_2 fixation and ammonia synthesis from nitrate/nitrite reduction and N_2 reduction. Recently developed methods for screening effective mineral catalysts were also reviewed.

Keywords: Origin of life, mineral catalysis, CO_2 fixation, ammonia synthesis, chemical evolution

1. Introduction

Origins of life remain a mystery for our humankind. The concept of "chemical evolution" describes a general evolutionary route from the abiotic to the biotic world through a variety of chemical and physical processes. Kitadai. et al. summarized the reactions explored in the lab for the chemical evolution (**Figure 1**). Starting from geologically abundant molecules (e.g., N_2, CO_2, H_2, PO_4, NO_3^-, HCN, etc.), high energy input drives the synthesis of small organic molecules as precursors. These small organic molecules react with each other to form life's building blocks (e.g., amino acids, nucleobases, sugars, aliphatic acids, etc.). Subsequently, these monomers polymerize into functional polymers which assemble into the so-called protocell.

IntechOpen

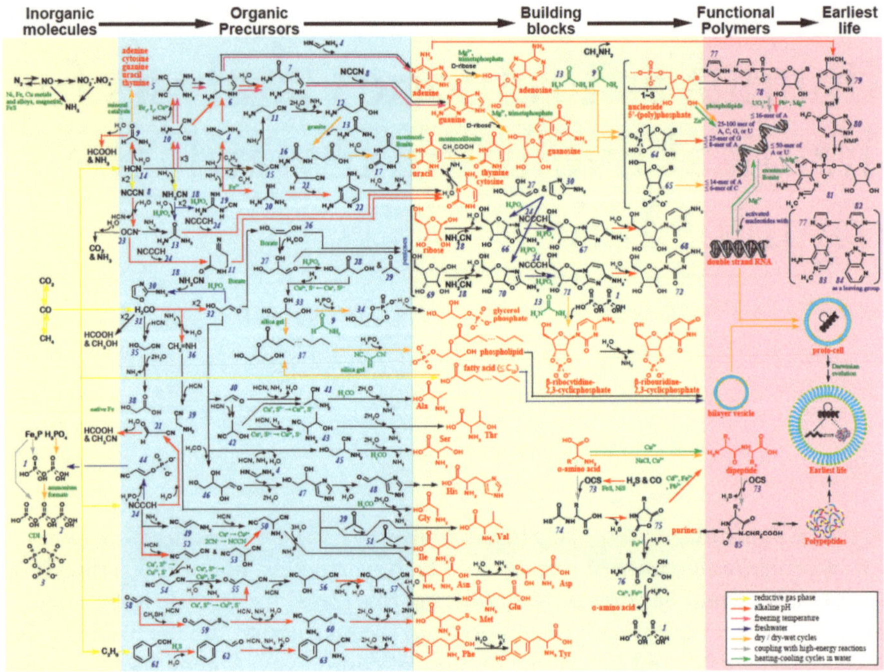

Figure 1.
Overview of the chemical evolution of life, adapted from ref. [1].

As many life's building blocks are not stable at temperatures higher than 100°C [2], a geological setting with moderate temperature is considered to be more favorable for life's emergence. In addition, a moderate temperature can render the chemical system a kinetic control that will otherwise only generate the most stable products following thermodynamics under a high-temperature regime. Kinetic control is required to form metastable products. There are many challenges in chemical evolution, two of which are caused by reaction kinetics. First, in the beginning, how geologically abundant inorganic molecules were activated and converted into small organic molecules? Second, how the chemical reactions are directed towards a high molecular complexity and product diversity for selectively generating life's building blocks? In this regard, catalysis is at the center of chemical evolution. A catalyst lowers the energetic barrier and enhances the reaction rate of activation of inert molecules. Different catalysts with tuned surface property and electronic characteristics can regulate the reaction pathways by adjusting the transition states of the intermediates.

Geological molecules, as the feedstocks of prebiotic synthesis, are typically chemically inert despite their high abundances. The activation of small geological molecules, such as H_2, CO_2, etc., requires redox processes. H_2 needs to be oxidized to release the chemical energy while CO_2 needs to be reduced to synthesize organics which usually show intermediate valence states of carbon (from +3 to −3) [3]. Similarly, methane (CH_4), which was considered to be abundant on the early Earth in some scenarios [4], needs to be oxidized to synthesize useful organics. Therefore, in general, redox processes play an important role in chemical evolution.

The importance of redox processes for energy conservation is also reflected in modern biology, where modern living organisms are relying on enzymes for catalyzing biochemical reactions and maintaining homeostasis. In particular, redox enzymes are important for organisms to harvest energy from geologically available molecules in their surrounding environments. For example, methanogens convert CO_2 and H_2 into methane, with the generation of proton gradient for ATP synthesis [5, 6]. Nitrogen-fixing organisms use redox enzyme nitrogenase to reduce N_2 into ammonia for N assimilation, which is a $6e^-/6H^+$ reaction ($N_2 + 6e^- + 6H^+ = 2NH_3$) [7, 8]. These metabolic reaction pathways were considered to be very ancient based on phylogenetic analysis, which could have appeared in the last universal common ancestor (LUCA) [9–11]. Redox enzymes accounting for these reactions highly rely on earth-abundant transition metals (e.g., Fe, Ni, Mo, etc.) due to the electron-shuttling property of these metal sites and the relatively high affinity of the *d*-orbital electrons with the small molecules. Meanwhile, the interaction between amino acid residuals from the surrounding peptides and the metallic center, and the ligands in the first coordination sphere, also plays an important role in the catalytic processes [12, 13]. This includes mediating proton transfers, stabilizing the intermediates through electronic interaction, etc.

However, before life emerged, it has long been considered that enzymes are too complex to be readily available. What are the geo-catalysts responsible for activating small molecules (including C-, N-, and S-related compounds)? Earth owns more than 5700 known species of minerals, with new species being identified every year (e.g., https://rruff.info/ima/). Both the variety and relative abundances of minerals have changed dramatically over the Earth's history, through various chemical, physical, and biological processes [14]. To understand the role of minerals in the origin of life, determining the first place to spawn the first life is an essential question. There are two dominating and contrasting scenarios of origins of life: those predicting that life emerged in the submarine, alkaline hydrothermal vent systems where the redox, pH, and T gradients keep the system far from equilibrium and serve as energy sources for prebiotic synthesis, as pioneered by Russell, et al. [4, 9, 15–18]; and those predicting that life emerged within subaerial environments with prebiotic synthesis driven by UV photolysis pioneered by Sutherland et al. [3, 19–23]. Both of these two scenarios implicitly emphasize the importance of redox processes for activating inert molecules. The former scenario proposed minerals (such as sulfides and hydroxides) as key players, while the latter relies on radicals and solvated electrons for redox conversions. Recently, an alternative scenario of origins of life in volcanic hot-spring water or the so-called "land-based pool" scenario was proposed by Damer and Deamer [24–26], to solve the self-assembly problem for membrane formation in the salty ocean while allowing condensation/polymerization through wet-dry cycle provided by the fluctuating boundary conditions. This scenario has been testified with self-assembly experiments simulating the hot-spring conditions [25]. Since the role of minerals hasn't been explicitly considered in the scenarios by Sutherland et al. and Deamer et al., only the scenarios involving mineral catalysis (e.g., alkaline hydrothermal vent (AHV) theory, iron–sulfur world theory by Wächtershäuser [27–31]) will be discussed in this chapter.

AHV theory was proposed based on the notion that the far-from-equilibrium condition in alkaline hydrothermal vent systems resembles biochemistry in the following aspects: (1) the large chemical disequilibrium is akin to the conditions the biology tends to live on and stably maintained through Earth's geological time; (2) the pH gradient sustained by the chimney rock wall resembles the chemiosmotic energy conservation shared by all life forms; (3) the transition metal-bearing mineral walls are rich in sulfides, which share the similar metal center and sulfur ligands with the modern Ni-, and Fe-bearing redox enzymes (e.g., carbon monoxide dehydrogenases, hydrogenase, ferredoxin, etc.) [9], thus could have catalyzed similar chemical conversions; (4) many chemoautotrophic microorganisms were discovered in the deep-sea hydrothermal vents and their metabolism is suggested to be phylogenetically old and energetically fueled by the chemicals in the vents; (5) different from the acidic type, high temperature hydrothermal vents, the low temperature (<120°C), alkaline, lost-city type hydrothermal vents renders kinetic control and could stabilize biomolecules formed in-situ.

Regardless of the scenarios, minerals have shown special functions in different types of prebiotic synthesis. Here in this chapter, a special focus will be posted on the redox catalysis mediated by minerals for the prebiotic synthesis, involving C, N, S, which are the fundamental elements of life and involved in a variety of redox conversions.

2. Why a catalytic system is important for prebiotic chemistry?

Before reviewing the state-of-art of mineral-catalyzed organic synthesis, larger questions here are: (1) Why catalysis is required for chemical evolution? (2) At which evolutionary period did catalysis begin to play an important role? The emergence of first life and the subsequent evolution from prokaryotes to eukaryotes all require well-regulated chemical conversion for efficient energy harvesting, sustainable supply of building blocks, and maintaining intracellular homeostasis. Eukaryotes developed more complex energy harvesting organelles that rely on respiration electron transfer chain and photosynthesis to metabolize with a higher transformation efficiency of energy and mass [32] (**Figure 2**). This is essential for maintaining their high cellular complexity in terms of both structure and functionality by balancing the enthalpy and entropy [36]. Notably, the enzymes responsible for these chemical conversions are catalytic, namely, the enzyme catalysts do not change chemically after one cycle or turnover of reaction, although enzymes indeed need replacement after the

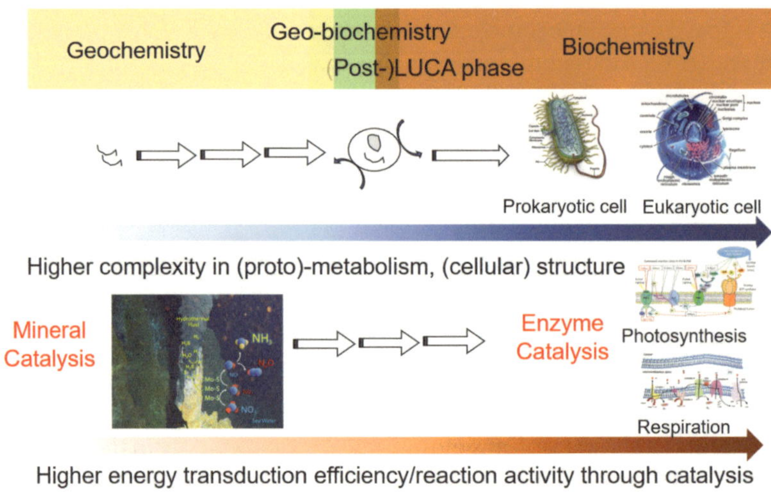

Figure 2.
The scenario of co-evolved catalytic system and life. During the continuous evolution from geochemistry to biochemistry, and the evolution of eukaryotes from prokaryotes, the gradually evolving catalytic systems serve as the physicochemical and energetic basis for promoting an increased energy transduction efficiency and reaction activity for supporting the higher complexity of (proto-)metabolism or (cellular) structures. Schemes of prokaryotic and eukaryotic cellular structures are adapted from Ref. [33]. Schemes for the electron transfer chains in photosynthesis and respiration are adapted from Ref. [34] and Ref. [35], respectively.

expiration of their lifetime. This catalytic feature is essential for boosting the reaction kinetics, saving energy for re-synthesizing enzymes, adapting to different substrate conditions, reversibly promoting both the two directions of the reaction, and so on [37]. As a comparison, in a non-catalytic, stoichiometric reaction, the active species that reacts with the geochemical substrates to target organics end up with a change in their chemical structures in an irreversible manner. After the complete consumption of the active species, the reaction can no longer proceed. From a top-down point of view, the prebiotic chemistry probably needs to evolve towards a catalytic, sustainable type of reaction network, to solve the problem of the shortage of supply of the building blocks/precursors, promote the reaction kinetics, and finally become self-independent when being encapsulated in a protocell. However, it should be noted that, at the initial stage of prebiotic synthesis, both catalytic and stoichiometric reactions are important for the synthesis of organic molecules to accumulate these organic precursors for subsequent conversion. As will be shown later, a large portion of prebiotic syntheses to date have been focusing on a stochiometric type reaction, therefore, relying on active agents. However, for some reason, the term "catalyst" has been used occasionally and misleading. In the following session, special care will be paid to differentiate the "catalytic" and "stoichiometric" types of reaction.

3. Minerals promote the prebiotic synthesis

3.1 Carbon fixation and C-C bond formation

Miller-Urey experiments open up the whole field of prebiotic chemistry [38]. At that time, the early Earth atmosphere was considered to be reducing and mainly composed of reducing species, such as methane, H_2, CO, NH_3, etc. However, this has been questioned, and the current consensus is that the early Earth atmosphere was oxidizing, with CO_2, N_2, and H_2O as the major components, with a trace amount of H_2 [39]. Based on this, a chemoautotrophic scenario pioneered by Wächtershäuser was proposed [27–31]. The chemoautotrophic origin of life scenario relies on primordial carbon fixation within a sulfide-rich hydrothermal vent. Driven by the reducing energy and activation ability of carbon monoxide (CO), many types of reactions were demonstrated. For example, at 100°C, C-C bond formation with the generation of acetate proceeds on FeS and NiS from a mixture of CO and CH_3SH, or from a mixture of CO and H_2S alone with the addition of Se [27]. This reaction resembles the reductive acetyl-coenzyme A (acetyl-CoA) pathway, where the key enzyme, acetyl-CoA synthase, contains a Ni-Fe-S active center and forms acetyl-CoA from coenzyme A, CO, and a methyl group. CO was also demonstrated to promote the reductive polymerization of HCN to α-hydroxy acids and α-amino acids at 80 ~ 120°C on FeS or NiS precipitates [29], where glycine and alanine were formed accompanied by glycolic and lactic acid. The polymerization of amino acids into short peptides was also demonstrated by CO activation [28]. In these experiments, high-pressure CO gas was used (1 ~ 75 bar) [29].

Based on the modeling of the atmosphere in the late Hadean period [39], the most abundant abiotic carbon feedstock on the early Earth is carbon dioxide (CO_2), with a trace amount of CO. Therefore, for the autotrophic origin of life scenarios, CO_2 was a more preferable and primary carbon source for primordial biosynthesis. CO_2 dissolved in the ocean and resulted in a mildly acidic ocean (pH ~ 5.5) [40]. Compared to CO, CO_2 is a chemically inert molecule that requires high activation energy. The acetyl-CoA (AcCoA, or Wood-Ljungdahl) pathway is considered to be the most ancient autotrophic CO_2 fixation pathway in nature [41]. To answer the question of how CO_2 reduction occurs before the evolution of proteins, CO_2 reduction has been demonstrated by Varma et al. [42], that native transition metals (Fe^0, Ni^0, and Co^0) can reduce CO_2 to acetate and pyruvate in millimolar concentrations. Moreover, in the AHV theory proposed by Russell and colleagues [9, 15, 18], the carbon fixation was driven by the direct redox coupling of CO_2 and H_2 on metal sulfides or oxyhydroxides [16]. Later, Lane and Martin [43, 44] also discussed the plausible relevance of the pH gradients in membrane-separated alkaline hydrothermal vent systems with the H^+ gradients across the cell membranes that drive ubiquitous chemiosmotic coupling in all life forms. This scenario was approved in recent work by Sojo et al., that in a microfluidic system with a freshly precipitated thin Fe-, or Ni-sulfide mineral membrane, CO_2 in simulated sea water side was reduced to formate at several micromolar yield [45]. The reaction was likely promoted by pH gradient as evidenced by the boosted yield with increased pH gradient. This shows that CO_2 was reduced by Fe or Ni sulfide, probably through an electrochemical process coupled with the oxidation of H_2.

Inspired by Russell's AHV origin of life theory, in the past decade, an alternative scenario "geo-electrochemical driven carbon fixation" has been explored by Nakamura, Yamamoto, Kitadai, and their colleagues [46–55]. This theory is based on the pH, redox, and thermal gradients between the alkaline hydrothermal vent and seawater. Those gradients thermodynamically drive the redox conversions by coupling H_2 oxidation in the hydrothermal fluid/mineral interface with CO_2 reduction at the seawater/mineral interface. The experimental results show that CO_2 was effectively reduced to CO on certain types of sulfide, such as Ag_2S and CdS [53], with much higher efficiency than FeS and NiS despite their higher geological abundances [48, 53]. The product selectivity highly depends on the identity of the metal in the sulfide minerals. Using the CO gas generated by electrochemical reduction, many reactions that were reported in Wächtershäuser's experiments were confirmed [53]. Since the disequilibrium and gradients in the deep-sea hydrothermal vent system can be maintained throughout the Earth's early history, the geo-electrochemical CO_2 reduction provided a stable and sustainable source of CO which could have fueled the prebiotic synthesis. By simulating the geo-electrochemical conditions of alkaline hydrothermal vents, other researchers also reported CO_2 reduction to a variety of products, including formate [49, 53], acetate [49], methane [48, 52], pyruvate [49], C_2H_6 [52], methanol [49] on Fe- or Ni-containing sulfides.

Regarding the reaction mechanisms, relevance with biological enzymes has been suggested. In biology, the enzyme catalyzing the reduction of CO_2 to CO is carbon monoxide dehydrogenase (CODH) with a $[NiFe_4S_4]$ cluster [56]. The reaction is considered to be the oldest pathway of biological carbon fixation and therefore may have been involved in the origin of life [9, 10]. Yamaguchi et al. first studied two metal sulfides greigite (Fe_3S_4) and violarite ($FeNi_2S_4$) and found that Ni-bearing sulfides show higher efficiency in reducing CO_2 [48]. Further, Lee et al. reported the in-situ FTIR spectroscopic analyses of the surface intermediate during electrochemical CO_2 reduction on these two minerals [52]. Intermediate species assignable to surface-bound CO_2 and formyl groups were found to be stabilized in the presence of Ni, lending insight into its role in enhancing the multistep CO_2 reduction process. These researches suggested an evolutionary link between mineral-catalyzed prebiotic reaction and enzyme-catalyzed biochemical reaction.

3.2 Nitrogen fixation

The most geologically abundant N source on the early Earth is dinitrogen (N_2). This molecule is chemically inert because of the stable N ≡ N triple bond. Under high-temperature conditions, N_2 can be reduced hydrothermally to ammonia, where reductants were considered to be abundant H_2S [57] or sulfide minerals. The yield of ammonia using H_2S as the reductant at low temperature (120°C) is relatively low even with iron monosulfide as the catalyst and is considered to be insufficient for providing ammonia for prebiotic synthesis. On the other hand, there have been accumulating reports on electrochemical reduction of N_2 on Fe_2O_3 [58–60], or FeS [61], CuS [62–64], Mo sulfides [65, 66] at ambient temperature and pressure. These types of reactions could contribute to the prebiotic synthesis of ammonia, following the geo-electrochemistry-driven prebiotic synthesis scenario.

Another chemically more active form of inorganic N species on the early Earth is nitrogen oxyanions including nitrate (NO_3^-) and nitrite (NO_2^-). These compounds were formed by lightning and photochemical processes of atmospheric N_2 and CO_2 with subsequent hydration during rainfall. This could lead to the accumulation of these nitrogen oxyanions in the early ocean with a concentration of micromolar level that is expected to be sufficient for serving as high-potential electron acceptors for the emergence of life in the oceanic environment [67]. NO_3^- and NO_2^- are high-potential electron acceptors ($E^0(NO_3^-/NO_2^-)$ = 0.835 V vs. NHE (normal hydrogen electrode), $E^0(NO_2^-/NO)$ = 1.202 V vs. NHE) [68]. These electron acceptors are invoked to participate in redox coupling with the oxidation of reducing species, such as methane [17], for the synthesis of active methyl-bearing species such as Acetyl-CoA-like molecules.

Despite their relatively higher reactivity, the reduction of NO_3^- is still kinetically demanding due to the low chemical affinity and low complexation ability with metal sites. Therefore, industrial reduction of NO_3^- typically requires relatively harsh conditions, such as very acidic pH, UV-photolysis, or high temperature [69]. Fe-based species have been extensively studied for ammonia synthesis via reduction of NO_3^- and NO_2^-. These include mackinawite (FeS) [70, 71], Fe^{2+} [72, 73], pyrite (FeS_2) [74, 75], and green rust ($Fe^{II}_4Fe^{III}_2(OH)_{12}SO_4{\cdot}yH_2O$) [76]. Fe(II) ions and green rust can reduce nitrite to ammonia at neutral to alkaline pH (pH ≥ 7) [73, 77]. Although Fe(II) cannot reduce nitrate, the addition of a trace amount of Cu^{2+} enables the generation of ammonia at pH 8 [77]. Green rust and pyrite can also reduce nitrate into ammonia at pH ~8, which however requires an anion-free environment due to the strong competing adsorption effect from many types of anions [74, 76]. Moreover, the reduction ability of Fe(II) and green rust decrease upon decreasing the solution pH to acidic pH [73, 76]. Therefore, this reaction could have consumed ferrous ions in the ocean. In high temperature (300°C), high pressure (50 ~ 500 MPa) hydrothermal setting, a variety of minerals (Fe-, Ni-, Cu-sulfides, and magnetite) can reduce nitrate into ammonia [78] and the ammonia is maintained stably in contact with these minerals. Accordingly, it was argued that the hydrothermal vent system could have supplied sufficient ammonia for the prebiotic synthesis of biomolecules, such as amino acids. However, at the same time, these mineral-promoted reactions are stoichiometric and strongly affected by the presence of other ions and low pH [73, 76].

Metals other than iron have rarely been considered to account for the geochemical reduction of nitrogen oxyanions. Nevertheless, biological nitrate reduction is catalyzed by nitrate reductase enzyme. All types of nitrate reductases exclusively utilize mononuclear molybdenum as the active center which is bounded by one or two dithiolene groups (-S-C-C-S-) ligated to a pterin group and other ligands (oxo, water, sulfur, etc.) [79, 80]. Inspired by the enzyme structures, the bio-inspired mineral catalysts provide another approach to tackle the kinetic problem. An oxo-bearing molybdenum sulfide as a structural analog of nitrate reductase was synthesized using the hydrothermal method. Notably, this mineral catalyzes both nitrate and nitrite reductions at a wide range of pH, with the generation of a variety of products, including NO, N_2O, NH_4^+, and N_2 [51, 69]. The reaction mechanism of nitrate reduction resembles that of the enzyme, relying on a redox-active, pentavalent $[(Mo^V{=}O)S_4]$ species as the active intermediate. This species was likely generated by a concerted proton-coupled electron transfer step, as evidenced by the near Nernstian behavior revealed by the pH dependence [69]. During nitrite reduction, this mineral show ability to decouple the proton transfer with electron transfer, facilitating a pH-regulated reaction selectivity towards the N-N coupling process [81, 82]. Therefore, this study shows that minerals can not only catalyze a similar reaction with the enzyme but also share a similar reaction mechanism, therefore reinforcing the evolutionary link between geo- and biochemistry.

3.3 Prebiotic S chemistry

Thioesters (R-(C=O)-SR') are organics with high energy (C=O)-S bonds and was invoked to act as an alternative and prebiotic version of phosphoesters (such as ATP) to drive endergonic reactions coupled with the hydrolysis of (C=O)-S bond. The reasons are as follows as described in a recent paper [83]: (1) both thioester and phosphoester bonds have similar standard free energies of hydrolysis; (2) the thioester synthesis precedes the synthesis of phosphoesters in metabolism, such as in glycolysis and the Wood-Ljungdahl pathway; (3) based on computational studies employing network extension algorithms to a phosphate-free core metabolism, thioesters can promote similar reactions like phosphoesters through energetic coupling, for example, reductive TCA cycle and biosynthesis of amino acid [84]. In addition, using thioester as a prebiotic energy coupling agent can partially solve the problem of the scarcity of phosphate in the ferruginous Archean ocean [85] despite the presence of other reactive P sources (e.g., phosphite) [86]. Researchers have reported the synthesis of thioester by abiotic processes. Driven by active CO, at 100°C, the Wächtershäuser group reported C-C bond formation with the generation of CH_3-CO-SCH_3 thioesters proceeds on FeS and NiS from a mixture of CO and CH_3SH [27]. Sandan et al. reported the generation of thioester by reacting of thioacetate and thiols in presence of Fe^{3+} at 70°C in water [83]. In addition, by further adding ferredoxin–heme maquettes, [4Fe-4S] cluster was formed based on the characteristic UV–Vis absorption band at 384 and 447 nm. This process also generates iron sulfide minerals. Recently, Kitadai et al. demonstrated the synthesis of S-methyl thioacetate (MTA) synthesis from CO and methanethiol on NiS at room temperature at neutral pH [54]. NiS was partially reduced to Ni under simulated geo-electrochemical conditions. This partially reduced Ni^0/NiS mixture catalyst is important to accumulate surface-bound CO by electro-reduction of CO_2. CO accumulation process on Ni^0/NiS surface was crucial for subsequent thioester formation in early ocean hydrothermal systems. The pH and temperature conditions are mild and geologically plausible.

4. Prebiotic catalyst screening

4.1 Availability of metals in the Hadean ocean

It is not surprising that many bio-essential metal elements are also Earth-abundant, considering the high reliance of the biosphere on the geosphere. These include mainly d-block elements (vanadium (V), chromium (Cr), manganese (Mn), iron (Fe), cobalt (Co), nickel (Ni), copper (Cu), molybdenum (Mo), tungsten (W), and zinc (Zn)) and exist in numerous oxidation states and be bonded by various ligands (O, S, etc.) with varied crystal structures and stoichiometries in minerals. The variety and relative abundances of minerals evolve with the Earth's history and also depend on the geological type of the locality [14]. In the ocean, changes in elemental abundances on geological time scales are intimately linked to evolutionary processes [87]. The availability of soluble transition metals changes progressively with time, with the greatest change in the redox-sensitive elements. However, this has provided environments enriched with an immobilized form of minerals, which would have provided the active surface for promoting prebiotic organic synthesis. The redox state of the environment evolved through at least three stages (adapted from ref. [87]), with major oxygenation events occurring ~2.4 billion to 1.8 billion years ago during the first of these stages, the ocean was largely devoid of dissolved O_2, and iron was abundant in the form of dissolved Fe^{2+} complexes. Much of the sulfur at that time was in the form of insoluble sulfide minerals locked in the continental crust. Besides Fe, the ocean abundances of transition metals such as manganese, cobalt, nickel, copper, zinc, and molybdenum are sensitive to environmental redox conditions, and also precipitate as sulfide minerals.

The scarcity of many bio-essential transition metals due to precipitation as insoluble sulfides has been considered to limit the size and shape the metabolism of the primordial biosphere [87, 88]. However, in terms of prebiotic chemistry, both the soluble and precipitated forms of transition metals could contribute to promoting the reactions, as have discussed in Section 3. The immobilized form of metal sulfides could have provided active surfaces with enormous potentials for catalytic functions. Therefore, for future prebiotic synthesis studies, more work using non-iron elements should be conducted to screen optimal geological catalysts.

4.2 Chemical diversity of metal sulfides

Since transition metals mainly existed in sulfides during Hadean and early Archean eon, these sulfides have been studied for prebiotic synthesis, as described in Section 3. However, in most cases, the activities of these minerals are low compared with their enzymatic counterparts and their contribution to prebiotic synthesis has rarely been quantitatively constrained based on their kinetics with some exceptions [73]. A possible reason accounting for the low reactivity is that prebiotic synthesis researches have been heavily focused on the most earth-abundant minerals (e.g., FeS, and NiS). To explore the chemical diversity of sulfide minerals, Li et al. evaluated the chemical diversity of metal sulfides of Co, Cu, Fe, Mn, Mo, Ni, V, and/or W with 135,434 species-locality pairs recorded in the mineralogy database (http://rruff.info/ima/) [88]. The diversity and distribution of these metal sulfides were analyzed in terms of the following aspects: locality frequency, multiple metal composition, crystal structure, and valence state of dominating elements. It was found that natural

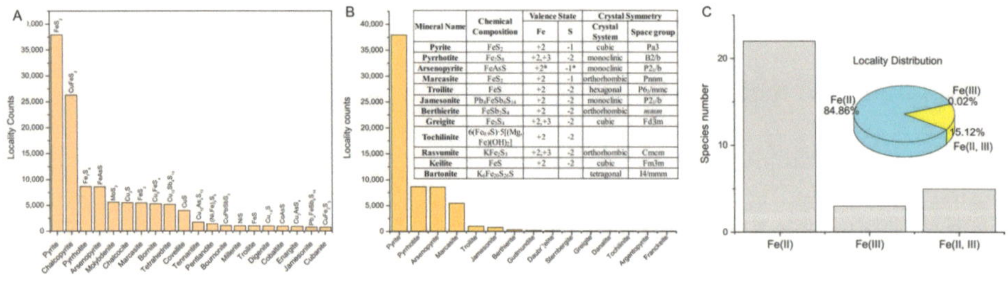

Figure 3.
(A) Metal sulfide distribution in natural environments. The 20 most frequently observed species are ranked in order of locality counts. The chemical composition of each species is shown. (B) Distribution and chemical diversity (chemical composition, Fe/S valence states, and crystal symmetry) of Fe single-metal sulfides. (C) Relative abundances and locality distribution of Fe^{2+}-, Fe^{3+}-, and Fe^{2+} plus Fe^{3+}-sulfides. Adapted from ref. [88].

metal sulfides show marked variations in chemical composition, crystal structure, and metal/sulfur valence states, suggesting a large chemical space associating with chemical variations of sulfide minerals still waits for exploration (**Figure 3**). For Fe sulfides, unexpectedly, mackinawite (FeS) is not among the top-ten mostly frequently observed species. This suggest that it may be problematic to use this mineral as a dominant target for prebiotic synthesis. Rather, pyrite is the most frequently observed species. The observation of the S_2^{2-} state suggests that not only metals can mediate redox change, sulfur ligands can also participate in the redox reaction with the valence change (S_2^{2-}/S^{2-}).

Each metal also has the dominant valent states. In Fe-S species, more than 84.86% of localities feature only Fe^{2+} species, followed by those containing both Fe^{2+} and Fe^{3+} (15.12%), and by three sulfides that only contain Fe^{3+} (0.02%). Generally, metals with low valence states predominate the library of metal sulfides, except for Co, allowing these minerals to act as an electron source or a catalytic center for charge accumulation during redox conversion. The minerals with mixed-valence states could exhibit unique functions due to special electron transfer and surface adsorption properties [89, 90]. Moreover, binary metal compositions are ubiquitous in natural sulfides. For example, Ni-Fe sulfides have ten species, and seven of them contain Ni and Fe as substitutional cations with a wide Ni/Fe ratio range (0 ~ 35 at%), with pentlandite $(Ni,Fe)_9S_8$ being the most prevalent form. The other three species contain fixed Fe/Ni ratios. The capability of Ni-Fe binary sulfides to have both fixed and varied Ni/Fe ratios in their structures is a unique characteristic different from that of Cu-Fe binary sulfides, in which Cu and Fe tend to form specific structures with fixed stoichiometries. The great chemical diversity provides a wide variety of catalytic functions and suggests that there is still a large chemical space of minerals for the exploration of unknown reactivities.

4.3 Enzyme analog minerals

Based on the discussion above, screening suitable geological catalysts is challenging and requires rational approaches. In this regard, machine learning or big-data mining could provide promising solutions.

Given an envisioned evolution from geochemistry to biochemistry, the mineral-and enzyme-based catalytic systems could be compared to understand the evolutionary link between them. In 2014, Michael Russell and his colleagues proposed that minerals sharing similar metal sites and ligands with enzymes could serve as a prebiotic catalyst for activating small geological molecules [4] (**Figure 4**) because the similar structure

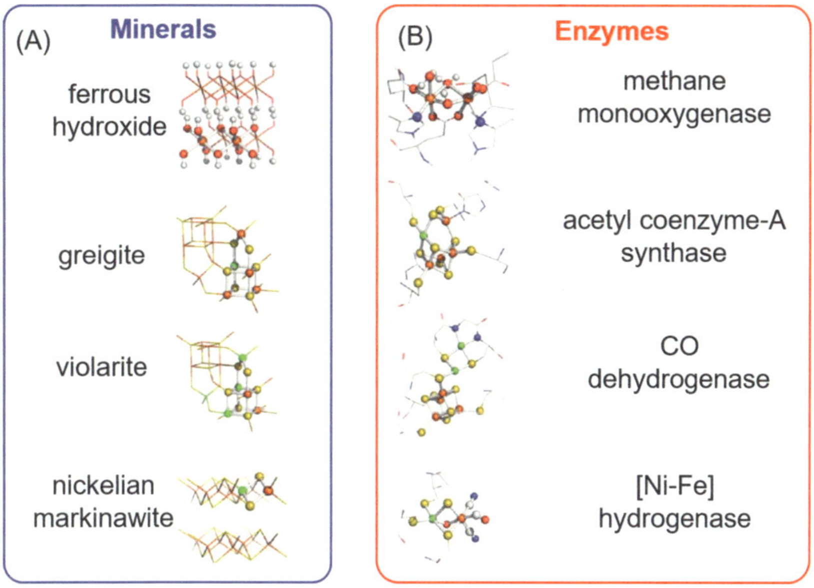

Figure 4.
Structural comparison between minerals and enzyme active centers. Adapted from ref. [4].

could exhibit similar chemical affinity towards the same substrate. Many prebiotic syntheses are influenced by similar perceptions and pursue the prebiotic carbon and nitrogen fixation using enzyme mimetic mineral catalysts [27, 45, 48, 49, 51, 52, 54, 69]. This idea could narrow down the candidates of geo-catalysts, however, an inherent difficulty in studying the property of minerals is the wide range of data and parameters to consider when searching for an appropriate catalyst for a specific enzymatic reaction [91]. Especially, since the structure (particularly the first coordination structure) alone doesn't dictate the overall catalytic property, the structural resemblance between minerals and enzymes doesn't ensure a definite functional similarity.

To solve this problem, a computational approach has been employed to systematically compare the metal–ligand structure of minerals and enzymes, as reported in recent work by Zhao et al. [91]. They compared the metalloenzyme cluster structure recorded in the protein database and the mineral structural data in the mineralogy database, using molecular similarity metrics. Therefore, this is probably the first attempt to quantify the structural similarity between biological machineries and minerals. In this study, iron–sulfur and nickel–iron–sulfur ligands were analyzed. Except for greigite and mackinawite, other iron sulfide species (marcasite and troilite) that were less studied previously were also predicted to have high structural similarity with iron–sulfur clusters in biology. Therefore, these results highlight the predictability of the modeling method for searching less studied minerals that hold potential as early prebiotic catalysts.

5. Conclusions

Minerals have been considered as the key player for prebiotic synthesis, and up-to-date researches have verified the catalytic property of many prebiotic mineral catalysts towards the carbon and nitrogen fixation reactions. Based on these discussions, mineral-mediated processes are probably critical for the evolution of protometabolism towards autotrophic origins of life. However, the rare demonstration of the formation of other types of life's building blocks, such as sugars, nucleobases, etc., accompanied by the difficulty in polymerization in water, pose several obvious challenges for many related origins of life scenarios [92]. To solve this problem, extensive screening of prebiotic catalysts based on the mineralogy database combined with the numerical prediction of structure–function relations should be helpful. Alternatively, other types of membrane and replicating systems have been proposed [16]. Additionally, the evolution of the mineral catalytic system could be further explored, by considering the hybridization of organic ligands with minerals. These organic ligands could be short peptides that can be formed under geologically plausible conditions [93, 94], or other types of polymers (e.g., polyesters [95, 96]). The hybridization of these organic ligands introduces stereochemical and electronic control on the whole reaction systems, which could help with overcoming the kinetic problems for certain reactions.

Acknowledgements

This work is funded by the Japan Society for the Promotion of Science KAKENHI grants No. 19 K15671 and No. 20H04608.

Appendices and nomenclature

Place appendix and nomenclature before Reference list.

Author details

Yamei Li
Earth-Life Science Institute, Tokyo Institute of Technology, Tokyo, Japan

*Address all correspondence to: yamei.li@elsi.jp

References

[1] Kitadai N, Maruyama S. Origins of building blocks of life: A review. Geoscience Frontiers. 2018;**9**:1117-1153

[2] Larralde R, Robertson MP, Miller SL. Rates of decomposition of ribose and other sugars: implications for chemical evolution. Proceedings of the National Academy of Sciences. 1995;**92**:8158

[3] Sasselov DD, Grotzinger JP, Sutherland JD. The origin of life as a planetary phenomenon. Science Advances. 2020;**6**:eaax3419

[4] Russell MJ, Barge LM, Bhartia R, Bocanegra D, Bracher PJ, Branscomb E, et al. The Drive to Life on Wet and Icy Worlds. Astrobiology. 2014;**14**:308-343

[5] Thauer RK, Shima S. Methane and microbes. Nature. 2006;**440**:878-879

[6] Kristjansson JK, Schönheit P, Thauer RK. Different Ks values for hydrogen of methanogenic bacteria and sulfate reducing bacteria: An explanation for the apparent inhibition of methanogenesis by sulfate. Archives of Microbiology. 1982;**131**:278-282

[7] Burgess BK, Lowe DJ. Mechanism of Molybdenum Nitrogenase. Chemical Reviews. 1996;**96**:2983-3012

[8] Schneider K, Müller A. In: Smith BE, Richards RL, Newton WE, editors. Catalysts for Nitrogen Fixation: Nitrogenases, Relevant Chemical Models and Commercial Processes. Netherlands, Dordrecht: Springer; 2004. pp. 281-307

[9] Russell MJ, Martin W. The rocky roots of the acetyl-CoA pathway. Trends in Biochemical Sciences. 2004;**29**:358-363

[10] Preiner M, Igarashi K, Muchowska KB, Yu M, Varma SJ, Kleinermanns K, et al. A hydrogen-dependent geochemical analogue of primordial carbon and energy metabolism. Nature Ecology & Evolution. 2020;**4**:534-542

[11] Weiss MC, Sousa FL, Mrnjavac N, Neukirchen S, Roettger M, Nelson-Sathi S, et al. The physiology and habitat of the last universal common ancestor. Nature Microbiology. 2016;**1**:16116

[12] Wodrich MD, Hu X. Natural inspirations for metal–ligand cooperative catalysis. Nature Reviews Chemistry. 2017;**2**:0099

[13] Kawai K, Nagata N. Metal–ligand interactions: An analysis of zinc binding groups using the Protein Data Bank. European Journal of Medicinal Chemistry. 2012;**51**:271-276

[14] Hazen RM, Papineau D, Bleeker W, Downs RT, Ferry JM, McCoy TJ, et al. Mineral evolution. American Mineralogist. 2008;**93**:1693-1720

[15] Russell MJ, Hall AJ, Martin W. Serpentinization as a source of energy at the origin of life. Geobiology. 2010;**8**:355-371

[16] Russell MJ. The "Water Problem"(sic), the Illusory Pond and Life's Submarine Emergence-A Review. Life. 2021;**11**:1-28

[17] Russell MJ, Nitschke W. Methane: Fuel or Exhaust at the Emergence of Life? Astrobiology. 2017;**17**:1053-1066

[18] Russell MJ, Hall AJ. The emergence of life from iron monosulphide bubbles at a submarine hydrothermal redox and pH front. Journal of the Geological Society. 1997;**154**:377

[19] Green NJ, Xu J, Sutherland JD. Illuminating Life's Origins: UV Photochemistry in Abiotic Synthesis of Biomolecules. Journal of the American Chemical Society. 2021;**143**:7219-7236

[20] Liu Z, Wu LF, Bond AD, Sutherland JD. Photoredox chemistry in the synthesis of 2-aminoazoles implicated in prebiotic nucleic acid synthesis. Chemical Communications. 2020;**56**:13563-13566

[21] Liu Z, Wu LF, Xu J, Bonfio C, Russell DA, Sutherland JD. Harnessing chemical energy for the activation and joining of prebiotic building blocks. Nature Chemistry. 2020;**12**:1023-1028

[22] Rimmer PB, Thompson SJ, Xu J, Russell DA, Green NJ, Ritson DJ, et al. Timescales for Prebiotic Photochemistry Under Realistic Surface Ultraviolet Conditions. Astrobiology. 2021;**21**: 1099-1120

[23] Xu J, Green NJ, Russell DA, Liu Z, Sutherland JD. Prebiotic Photochemical Coproduction of Purine Ribo- and Deoxyribonucleosides. Journal of the American Chemical Society. 2021;**143**:14482-14486

[24] Damer B, Deamer D. The Hot Spring Hypothesis for an Origin of Life. Astrobiology. 2019;**20**:429-452

[25] Milshteyn D, Damer B, Havig J, Deamer D. Amphiphilic Compounds Assemble into Membranous Vesicles in Hydrothermal Hot Spring Water but Not in Seawater. Life. 2018;**8**:1-15

[26] Deamer D, Damer B, Kompanichenko V. Hydrothermal Chemistry and the Origin of Cellular Life. Astrobiology. 2019;**19**:1523-1537

[27] Huber C, Wächtershäuser G. Activated Acetic Acid by Carbon Fixation on (Fe, Ni) S Under Primordial Conditions. Science. 1997;**276**:245-247

[28] Huber C, Wächtershäuser G. Peptides by Activation of Amino Acids with CO on (Ni,Fe) S Surfaces: Implications for the Origin of Life. Science. 1998;**281**:670-672

[29] Huber C, Wächtershäuser G. α-Hydroxy and α-amino acids under possible hadean, volcanic origin-of-life conditions. Science. 2006;**314**:630-632

[30] Wächtershäuser G. Before enzymes and templates: theory of surface metabolism. Microbiological Reviews. 1988;**52**:452-484

[31] Wächtershäuser G. Pyrite formation, the first energy source for life: A hypothesis. Systematic and Applied Microbiology. 1988;**10**:207-210

[32] Lane N. The Vital Question: Energy, Evolution, and the Origins of Complex Life. New York: W. W. Norton & Company; 2015

[33] Viva, 10 Major Difference Between Prokaryotes And Eukaryotes Cells (With Examples & Pictures). Available from: https://vivadifferences.com/understanding-prokaryotes-vs-eukaryotes-cells/

[34] Kanechi M. In: Lizárraga JCGC a GLL, editor. Photosynthesis-From Its Evolution to Future Improvements in Photosynthetic Efficiency Using Nanomaterials. Rijeka: IntechOpen; 2018

[35] Agrawal A, Mabalirajan U. Rejuvenating cellular respiration for optimizing respiratory function: Targeting mitochondria. American Journal of Physiology. Lung Cellular and Molecular Physiology. 2015;**310**:L103-L113

[36] Dragicevic V, Sredojevic S. In: Moreno-Piraján JC, editor.

Thermodynamics-Systems in Equilibrium and Non-Equilibrium. Rijeka: IntechOpen; 2011

[37] Fourmond V, Plumeré N, Léger C. Reversible catalysis. Nature Reviews Chemistry. 2021;**5**:348-360

[38] Miller SL. A production of amino acids under possible primitive earth conditions. Science. 1953;**117**:528-529

[39] Kasting James F. Earth's early atmosphere. Science. 1993;**259**:920-926

[40] Macleod G, McKeown C, Hall AJ, Russell MJ. Hydrothermal and oceanic pH conditions of possible relevance to the origin of life. Origins of Life and Evolution of the Biosphere. 1994;**24**:19-41

[41] Fuchs G. Alternative pathways of carbon dioxide fixation: Insights into the early evolution of life? Annual Review of Microbiology. 2011;**65**:631-658

[42] Varma SJ, Muchowska KB, Chatelain P, Moran J. Native iron reduces CO2 to intermediates and end-products of the acetyl-CoA pathway. Nature Ecology & Evolution. 2018;**2**:1019-1024

[43] Lane N. Bioenergetic constraints on the evolution of complex life. Cold Spring Harbor Perspectives in Biology. 2014;**6**:1-18

[44] Lane N, Martin WF. The origin of membrane bioenergetics. Cell. 2012;**151**:1406-1416

[45] Hudson R, de Graaf R, Strandoo Rodin M, Ohno A, Lane N, McGlynn SE, et al. CO2 reduction driven by a pH gradient. Proceedings of the National Academy of Sciences. 2020;**117**:22873

[46] Yamamoto M, Nakamura R, Oguri K, Kawagucci S, Suzuki K, Hashimoto K, et al. Generation of electricity and illumination by an environmental fuel cell in deep-sea hydrothermal vents. Angewandte Chemie International Edition. 2013;**52**:10758-10761

[47] Yamamoto M, Nakamura R, Kasaya T, Kumagai H, Suzuki K, Takai K. Spontaneous and widespread electricity generation in natural deep-sea hydrothermal fields. Angewandte Chemie (International Ed. in English). 2017;**56**:5725-5728

[48] Yamaguchi A, Yamamoto M, Takai K, Ishii T, Hashimoto K, Nakamura R. Electrochemical CO2 reduction by Ni-containing iron sulfides: How is CO2 electrochemically reduced at bisulfide-bearing deep-sea hydrothermal precipitates? Electrochimica Acta. 2014;**141**:311-318

[49] Roldan A, Hollingsworth N, Roffey A, Islam HU, Goodall JB, Catlow CR, et al. Bio-inspired CO2 conversion by iron sulfide catalysts under sustainable conditions. Chemical Communications. 2015;**51**:7501-7504

[50] Nakamura R, Takashima T, Kato S, Takai K, Yamamoto M, Hashimoto K. Electrical current generation across a black smoker chimney. Angewandte Chemie (International Ed. in English). 2010;**49**:7692-7694

[51] Li Y, Yamaguchi A, Yamamoto M, Takai K, Nakamura R. Molybdenum sulfide: A bioinspired electrocatalyst for dissimilatory ammonia synthesis with geoelectrical current. The Journal of Physical Chemistry C. 2017;**121**:2154-2164

[52] Lee JE, Yamaguchi A, Ooka H, Kazami T, Miyauchi M, Kitadai N, et al. In situ FTIR study of CO2 reduction on inorganic analogues of carbon monoxide dehydrogenase. Chemical Communications. 2021;**57**:3267-3270

[53] Kitadai N, Nakamura R, Yamamoto M, Takai K, Li Y, Yamaguchi A, et al. Geoelectrochemical CO production: Implications for the autotrophic origin of life. Science Advances. 2018;**4**:eaao7265

[54] Kitadai N, Nakamura R, Yamamoto M, Okada S, Takahagi W, Nakano Y, et al. Thioester synthesis through geoelectrochemical CO2 fixation on Ni sulfides. Communications Chemistry. 2021;**4**:37

[55] Ang R, Khan AU, Tsujii N, Takai K, Nakamura R, Mori T. Thermoelectricity generation and electron-magnon scattering in a natural chalcopyrite mineral from a deep-sea hydrothermal vent. Angewandte Chemie (International Ed. in English). 2015;**54**:12909-12913

[56] Dobbek H, Svetlitchnyi V, Gremer L, Huber R, Meyer O. Crystal structure of a carbon monoxide dehydrogenase reveals a [Ni-4Fe-5S] cluster. Science. 2001;**293**:1281-1285

[57] Schoonen MAA, Xu Y. Nitrogen reduction under hydrothermal vent conditions: Implications for the prebiotic synthesis of C-H-O-N compounds. Astrobiology. 2001;**1**:133-142

[58] Cui X, Tang C, Liu X-M, Wang C, Ma W, Zhang Q. Highly selective electrochemical reduction of dinitrogen to ammonia at ambient temperature and pressure over iron oxide catalysts. Chemistry – A European Journal. 2018;**24**:18494-18501

[59] Wang Z, Zheng K, Liu S, Dai Z, Xu Y, Li X, et al. Electrocatalytic nitrogen reduction to ammonia by Fe2O3 nanorod array on carbon cloth. ACS Sustainable Chemistry & Engineering. 2019;**7**:11754-11759

[60] Xiang X, Wang Z, Shi X, Fan M, Sun X. Ammonia synthesis from electrocatalytic N2 reduction under ambient conditions by Fe2O3 nanorods. ChemCatChem. 2018;**10**:4530-4535

[61] Xiong W, Guo Z, Zhao S, Wang Q, Xu Q, Wang X. Facile, cost-effective plasma synthesis of self-supportive FeSx on Fe foam for efficient electrochemical reduction of N2 under ambient conditions. Journal of Materials Chemistry A. 2019;**7**:19977-19983

[62] Kong J, Kim M-S, Akbar R, Park HY, Jang JH, Kim H, et al. Electrochemical nitrogen reduction kinetics on a copper sulfide catalyst for NH3 synthesis at low temperature and atmospheric pressure. ACS Applied Materials and Interfaces. 2021;**13**:24593-24603

[63] Kim HS, Choi J, Kong J, Kim H, Yoo SJ, Park HS. Regenerative electrocatalytic redox cycle of copper sulfide for sustainable NH3 production under ambient conditions. ACS Catalysis. 2021;**11**:435-445

[64] Kim M-C, Nam H, Choi J, Kim HS, Lee HW, Kim D, et al. Hydrogen bonding-mediated enhancement of bioinspired electrochemical nitrogen reduction on Cu2–xS catalysts. ACS Catalysis. 2020;**10**:10577-10584

[65] Xu X, Tian X, Sun B, Liang Z, Cui H, Tian J, et al. 1 T-phase molybdenum sulfide nanodots enable efficient electrocatalytic nitrogen fixation under ambient conditions. Applied Catalysis B: Environmental. 2020;**272**:118984

[66] Suryanto BHR, Wang D, Azofra LM, Harb M, Cavallo L, Jalili R, et al. MoS2 polymorphic engineering enhances selectivity in the electrochemical reduction of nitrogen to ammonia. ACS Energy Letters. 2019;**4**:430-435

[67] Wong ML, Charnay BD, Gao P, Yung YL, Russell MJ. Nitrogen Oxides

in Early Earth's Atmosphere as Electron Acceptors for Life's Emergence. Astrobiology. 2017;**17**:975-983

[68] Rosca V, Duca M, de Groot MT, Koper MTM. Nitrogen cycle electrocatalysis. Chemical Reviews. 2009;**109**:2209-2244

[69] Li Y, Go YK, Ooka H, He D, Jin F, Kim SH, et al. Enzyme mimetic active intermediates for nitrate reduction in neutral aqueous media. Angewandte Chemie (International Ed. in English). 2020;**59**:9744-9750

[70] Gordon AD, Smirnov A, Shumlas SL, Singireddy S, DeCesare M, Schoonen MAA, et al. Reduction of nitrite and nitrate on nano-dimensioned FeS. Origins of Life and Evolution of Biospheres. 2013;**43**:305-322

[71] Summers DP. Ammonia formation By the reduction of nitrite/nitrate by Fes: Ammonia formation under acidic conditions. Origins of Life and Evolution of Biospheres. 2005;**35**:299-312

[72] Buresh RJ, Moraghan JT. Chemical reduction of nitrate by ferrous iron. Journal of Environmental Quality. 1976;**5**:320-325

[73] Summers DP, Chang S. Prebiotic ammonia from reduction of nitrite by iron (II) on the early Earth. Nature. 1993;**365**:630-633

[74] Singireddy S, Gordon AD, Smirnov A, Vance MA, Schoonen MAA, Szilagyi RK, et al. Reduction of nitrite and nitrate to ammonium on pyrite. Origins of Life and Evolution of Biospheres. 2012;**42**:275-294

[75] Stirling A, Rozgonyi T, Krack M, Bernasconi M. Prebiotic NH3 formation: Insights from simulations. Inorganic Chemistry. 2016;**55**:1934-1939

[76] Hansen HCB, Koch CB, Nancke-Krogh H, Borggaard OK, Sørensen J. Abiotic nitrate reduction to ammonium: Key role of green rust. Environmental Science & Technology. 1996;**30**:2053-2056

[77] Ottley CJ, Davison W, Edmunds WM. Chemical catalysis of nitrate reduction by iron (II). Geochimica et Cosmochimica Acta. 1997;**61**:1819-1828

[78] Brandes JA, Hazen RM, Yoder HS. Inorganic nitrogen reduction and stability under simulated hydrothermal conditions. Astrobiology. 2008;**8**:1113-1126

[79] Enemark JH, Cooney JJA, Wang J-J, Holm RH. Synthetic analogues and reaction systems relevant to the molybdenum and tungsten oxotransferases. Chemical Reviews. 2004;**104**:1175-1200

[80] Hille R, Hall J, Basu P. The mononuclear molybdenum enzymes. Chemical Reviews. 2014;**114**:3963-4038

[81] He D, Li Y, Ooka H, Go YK, Jin F, Kim SH, et al. Selective electrocatalytic reduction of nitrite to dinitrogen based on decoupled proton–electron transfer. Journal of the American Chemical Society. 2018;**140**:2012-2015

[82] He D, Ooka H, Kim Y, Li Y, Jin F, Kim SH, et al. Atomic-scale evidence for highly selective electrocatalytic N–N coupling on metallic MoS₂. Proceedings of the National Academy of Sciences. 2020;**117**:31631

[83] Sanden SA, Yi R, Hara M, McGlynn SE. Simultaneous synthesis of thioesters and iron-sulfur clusters in water: Two universal components of energy metabolism. Chemical Communications. 2020;**56**:11989-11992

[84] Goldford JE, Hartman H, Smith TF, Segrè D. Remnants of an ancient metabolism without phosphate. Cell. 2017;**168**:1126-1134.e1129

[85] Pasek MA. Rethinking early Earth phosphorus geochemistry. Proceedings of the National Academy of Sciences. 2008;**105**:853

[86] Pasek MA, Harnmeijer JP, Buick R, Gull M, Atlas Z. Evidence for reactive reduced phosphorus species in the early Archean ocean. Proceedings of the National Academy of Sciences of the United States of America. 2013;**110**:10089-10094

[87] Anbar Ariel D. Elements and evolution. Science. 2008;**322**:1481-1483

[88] Li Y, Kitadai N, Nakamura R. Chemical diversity of metal sulfide minerals and its implications for the origin of life. Life. 2018;**8**:1-26

[89] Santos-Carballal D, Roldan A, Dzade NY, de Leeuw NH. Reactivity of CO2 on the surfaces of magnetite (Fe3O4), greigite (Fe3S4) and mackinawite (FeS). Philosophical Transactions of the Royal Society A: Mathematical, Physical and Engineering Sciences. 2018;**376**:20170065

[90] Roldan A, de Leeuw NH. Catalytic water dissociation by greigite Fe3S4 surfaces: density functional theory study. Proceedings of the Royal Society A: Mathematical, Physical and Engineering Sciences. 2016;**472**:20160080

[91] Zhao D, Bartlett S, Yung YL. Quantifying mineral-ligand structural similarities: Bridging the geological world of minerals with the biological world of enzymes. Life. 2020;**10**:1-9

[92] Marshall M. How the first life on Earth survived its biggest threat — water. Nature. 2020;**588**:4

[93] Parker ET, Zhou M, Burton AS, Glavin DP, Dworkin JP, Krishnamurthy R, et al. A plausible simultaneous synthesis of amino acids and simple peptides on the primordial earth. Angewandte Chemie International Edition. 2014;**53**:8132-8136

[94] Forsythe JG, Yu S-S, Mamajanov I, Grover MA, Krishnamurthy R, Fernández FM, et al. Ester-mediated amide bond formation driven by wet–dry cycles: A possible path to polypeptides on the prebiotic earth. Angewandte Chemie International Edition. 2015;**54**:9871-9875

[95] Chandru K, Mamajanov I, Cleaves HJ, Jia TZ. Polyesters as a model system for building primitive biologies from non-biological prebiotic chemistry. Life. 2020;**10**:1-16

[96] Chandru K, Guttenberg N, Giri C, Hongo Y, Butch C, Mamajanov I, et al. Simple prebiotic synthesis of high diversity dynamic combinatorial polyester libraries. Communications Chemistry. 2018;**1**:30

Chapter 7

Ionic Conductivity of Strontium Fluoroapatites Co-doped with Lanthanides

Khouloud Kthiri, Mohammed Mehnaoui, Samira Jebahi, Khaled Boughzala and Mustapha Hidouri

Abstract

Britholites derivatives of apatite's that contain lanthanium and neodymium in the serial compounds $Sr_8La_{2-x}Nd_x(PO_4)_4(SiO_4)_2F_2$ with $0 \leq x \leq 2$ were subject of the present investigation. The solid state reaction was the route of preparing these materials. Several techniques were employed for the analysis and characterization of the synthesized powders. The chemical analysis results indicated that molar ratio $\frac{Sr+La+Nd}{P+Si}$ was of about 1.67 value of a stoichiometric powder. The X-ray diffraction data showed single-phase apatites crystallizing in hexagonal structure with $P_{63/m}$ space group were successively obtained. Moreover, the substitution of lanthanium by neodymium in strontium phosphosilicated fluorapatite was total. This was confirmed by the ***a*** and ***c*** lattice parameters contraction when (x) varies coherently to the sizes of the two cations. The infrared spectroscopy and the ^{31}P NMR (MAS) exhibited the characteristic bands of phosphosilicated fluorapatite. The pressureless sintering of the material achieved a maximum of 89% relative density. The sintered specimens indicated that the Nd content as well as the heating temperature affected the ionic conduction of the materials and the maximum was 1.73×10^{-6} S cm^{-1} obtained at 1052 K for x = 2.

Keywords: fluorobritholites, lanthanium-neodymum substitution, sintering, ionic conductivity

1. Introduction

The phosphosilicate apatites containing a coupled substitution of the divalent cation by a trivalent lanthanide or a tetravalent actinide ion and the trivalent groupment PO_4 by a tetravalent SiO_4 groupment in the general formula $Me(XO_4)6Y_2$ (Me: divalent cation; XO_4: anionic groupment and Y: monovalent anion) allow to obtain materials called britholite [1–4]. Such materials were found in the natural nuclear reactors Alko of Gabon which demonstrated that they are storing some radionuclides such as uranium U, thorium Th, plutonium Pu and minor actinides like neptinium Np, americium Am and curium Cm [5–8]. Moreover, silicate based apatite samples were found to contain up to 50 wt% of lanthanides (La, Ce, Nd) and actinides

(U, Th) in Ouzzal site of Algeria [9]. Hence, britholites were considered as natural nuclear waste disposal and allowing the confinement of radionuclides and some fission byproducts produced by the nuclear industry [10–12]. In fact, many studies indicated that britholites are able to confine radionuclides with continuous irradiation for millions of years with conserved structure and thermal and chemical stability [13, 14]. Indeed, due to the stability and flexibility of their structure, apatites offer many possibilities for substitutions. Moreover, britholite materials favored many cationic and anionic substitutions in their crystallographic structure. These later might be in a total or limited range [15–17]. Therefore, these substitutions are governed by the ionic sizes, the valence, the electronegativity and the polarizability [18]. In this context, several processes have been developed for the preparation of these materials containing various elements such as actinides and lanthanides via solid state reaction or mechanical synthesis [19–27].

On the other hand, many investigations have revealed that britholites might be a good ionic conductor for their use in fuel cells. The conductivity was proved as a thermal process at intermediate temperature range 400–900°C [28–32]. Therefore, the electrical properties allow using the materials as a solid electrolyte in solid oxide fuel cells (SOFCs) [33, 34].

Like-apatite, phosphosilicate apatites have a hexagonal structure and a space group P63/m [15, 35]. Their framework is built on the sixth XO_4 groups and the Me is divided between two crystallographic sites: four are located in the site Me(1), coordination 9, and six other are located in the site Me(2), coordination 7. Hence, in order to highlight the capacity of these materials to store non radioactive elements similar to radionuclides as well as their potentialities as ionic conductors, the sintered materials series $Sr_8La_{2-x}Nd_x(PO_4)_4(SiO_4)_2F_2$ with $0 \leq x \leq 2$ were investigated.

2. Materials and methods

2.1 Powder synthesis

A solid state method was adopted to prepare strontium fluorobritholites compounds $Sr_8La_{2-x}Nd_x(PO_4)_4(SiO_4)_2F_2$ with $0 \leq x \leq 2$ [36]. The starting reagents: strontium fluoride SrF_2 (99.99%. Merck)), strontium carbonate $SrCO_3$ (≥99.00% Fluka), silica SiO_2 (Prolabo), lanthanum and neodymium oxide ($La_2O_3{\cdot}Nd_2O_3$) (99.99% Merck) and strontium diphosphate ($Sr_2P_2O_7$) were used. The reaction equation (1) is the following:

$$\begin{aligned} &3SrCO_3 + 2Sr_2P_2O_7 + SrF_2 + 2SiO_2 + \frac{2-x}{2}La_2O_3 + \frac{x}{2}Nd_2O_3 \\ &\rightarrow \mathbf{Sr_8La_{2-x}Nd_x(PO_4)_4(SiO_4)_2F_2} + 3CO_2\uparrow \text{with } 0 \leq x \leq 2. \end{aligned} \quad (1)$$

$Sr_2P_2O_7$ was synthesized by the following reaction at 900°C:

$$2SrCO_3 + 2(NH_4)_2HPO_4 \rightarrow Sr_2P_2O_7 + 2CO_2 + 3H_2O + 4NH_3 \quad (2)$$

$SrCO_3$ (>96% Riedel de Haen), Gd_2O_3 (>99.5% Prolabo), Nd_2O_3 (>99.5% Prolabo) SiO_2 (>99.5% Alfa), SrF_2 (>99.5% Prolabo) and $(NH_4)_2HPO_4$ (>99% Acros Organics) were used as raw materials. For each composition the molar ratio (Sr + La + Nd)/(P + Si) and the obtained quantity of each composition should be respectively 1.67 and 1.5×10^{-3} moles. Before synthesis, each quantity of lanthanum and neodymium

Reactants	SrLa$_2$F	SrLa$_{1.5}$Nd$_{0.5}$F	SrLa$_1$Nd$_1$F	SrLa$_{0.5}$Nd$_{1.5}$F	SrNd$_2$F
La_2O_3	0.4887	0.3665	0.2443	0.1221	—
Nd_2O_3	—	0.1261	0.2523	0.3785	0.5047

Table 1.
Masses in grams of lanthanum and neodymium oxides used in the synthesis of $Sr_8La_{2-x}Nd_x(PO_4)_4(SiO_4)_2F_2$ with $0 \leq x \leq 2$.

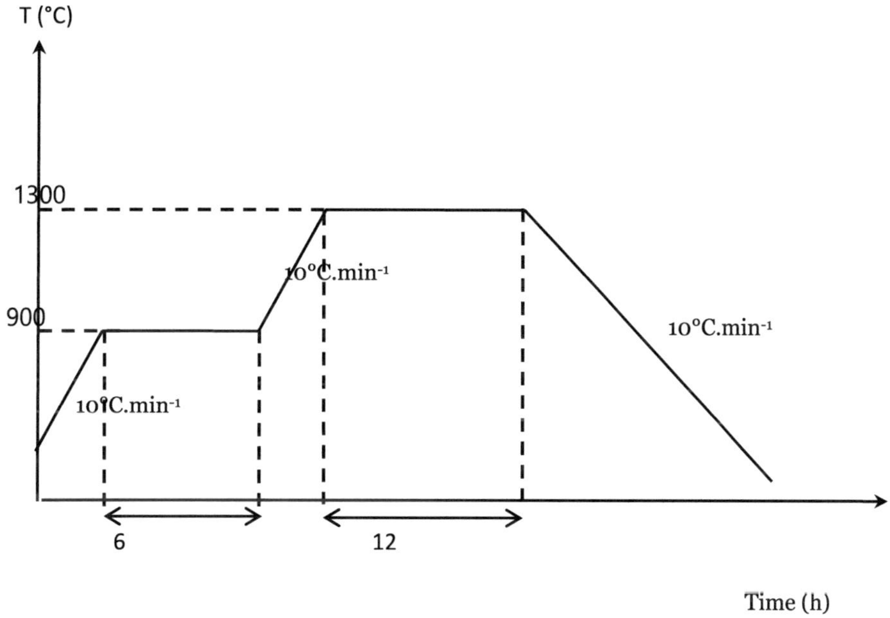

Figure 1.
Thermal cycle used for strontium fluorobritholite sintering.

oxides given in **Table 1** was furnaced at 1000°C for 12 h to avoid the formation of Ln-hydroxide. Then, the solid mixture was milled and homogenized in an agate mortar for about 30 min, and then cold pressed under 100 MPa into pellets (30 and 3 mm). During sintering, the pellets were sintered in the temperature range 1250–1450°C in a carbolyte type furnace with controlled argon atmosphere. The temperature varied with 50°C for each value of x. The sintering cycle is shown in **Figure 1**. The heating and cooling rate was of 10°C min^{-1}. In the following sections, the samples will be named SrLa$_{2-x}$Nd$_x$F where x is the substituted Nd rate.

2.2 Analysis and characterization

A PANalytical X'pert Pro diffractometer with a KαCu anode (λ = 1.54 Å) operating 40 kV and 40 mA was the apparatus used for the XRD patterns recording. The scans range was between 10 and 70° (2θ) with a step size of 0.02°. The crystallite size of the powder D_{hkl} was calculated using the (300) and (002) reflections following Debye Sheerer equation [37]:

$$D = \frac{K\lambda}{\beta^{1/2}\cos\theta} \tag{3}$$

Needs to remember that λ is the X-ray wavelength of the monochromatic X-ray beam. For the apatitic crystallites K is a constant equal to 0.9. $\beta^{1/2}$ is the full width at half maximum of the selected reflection and θ is the Bragg's diffraction angle.

The Fourier transformed infrared (FTIR)-attenuated total reflection (ATR) spectra were performed at room temperature on a Perkin Elmer spectrometer in the spectral range 4000–400 cm^{-1}.

The chemical analysis of Sr., P, Si, La and Nd ions in the synthesized samples was determined via an inductively coupled plasma atomic emission spectroscopy (ICP-AES) (JY-Horiba Ultima-C spectrometer). The samples were thus previously mixed with 99.9% lithium metaborate, fused at 1000°C for 25 min and dissolved in HCl (0.6 M). The fluoride content in the synthesized samples was measured by a specific ion-selective electrode.

The complex impedance measurements were performed on pellets sintered at between 1250 and 1450°C for 24 h. Their densities varied from 72 to 83% of the theoretical density as a function the sintering temperature. The two faces of the pellets were coated with a silver paint and then two platinum wires electrodes linked them to a Hewlett-Packard 4192-A impedance analyzer. The measurements were recorded with the temperatures variation from 450 to 780°C and frequencies from 10 Hz to 13 MHz.

3. Results and discussion

3.1 Chemical content

The samples' quantitative chemical analyses are shown in **Table 2**. As observed there is a satisfactory agreement between the elements amount determined from the analyses and those introduced in the starting. As a consequence, the experimental formula was close to the theoretical ones. The $\frac{Sr+La+Nd}{P+Si}$ molar ratios are very close to the theoretical value of 1.667 for stoichiometric apatite.

3.2 X-Ray diffraction analysis

Figure 2 showed the XRD patterns of all compositions. It is evident that the samples were single apatite phase. By comparaison to the JCPDS 17-0609 file data for the strontium fluorapatite, the samples are characteristic of the hexagonal symmetry and the P63/m space group. No additional diffraction lines relative to supplementary phases were detected in any of the patterns. However, the presence of very small quantities of impurities was not excluded. The XRD patterns of the **Figure 3** indicated that when the substitution level increased, the peaks slightly shift towards the high 2θ angles

Samples	Sr	La	Nd	P	Si	F	Molar ratio $\frac{Sr+La+Nd}{P+Si}$
$SrLa_2F$	7.96	1.97	—	3.98	1.97	1.99	1.668
$SrLa_{1.5}Nd_{0.5}F$	7.96	1.47	0.47	3.98	1.96	1.98	1.666
$SrLa_1Nd_1F$	7.97	0.98	0.97	3.99	1.96	1.96	1.667
$SrLa_{0.5}Nd_{1.5}F$	7.99	0.47	1.48	3.99	1.98	1.97	1.664
$SrNd_2F$	7.98	—	1.97	3.98	1.98	1.97	1.669

Table 2.
Number of atoms per unit cell of $Sr_8La_{2-x}Nd_x(PO_4)_4(SiO_4)_2F_2$ $(0 \leq x \leq 2)$.

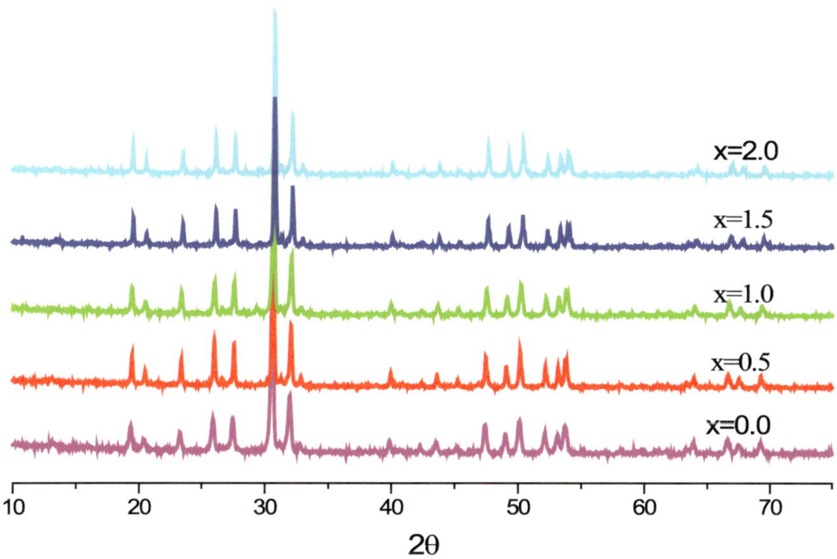

Figure 2.
DRX spectra of strontium fluorbritholites $Sr_8La_2Nd_x(PO_4)_4(SiO_4)_2F_2$ with ($0 \leq x \leq 2$).

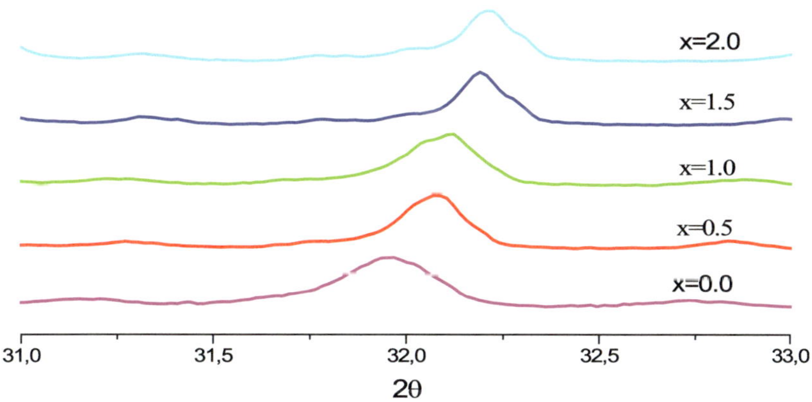

Figure 3.
Radiation (300) of fluorbritholites $Sr_8La_{2-x}Nd_x(PO_4)_4(SiO_4)_2F_2$ ($0 \leq x \leq 2$).

indicating a contraction of the unit cell. This contraction, which agrees with the Nd^{3+} radius (${}^{VI}\mathbf{r}_{Nd}{}^{3+}$ = 0.983 Å) [38], that is smaller than that of La^{3+} (${}^{VI}\mathbf{r}_{La}{}^{3+}$ = 1.032 Å) [39] confirms the incorporation of Nd^{3+} into the apatite structure.

As shown in **Figure 4** and **Table 3**, the cristallographic parameters calculated using the Fullprof program without any structural refinement of the all compositions depended on the substitution level. In fact, if Nd content rose, ***a*** and ***c*** decreased. The calculated cristallographic parameters were similar to that existing in the literature [40, 41]. Moreover the evolution of the lattice parameters was linear in accordance with the Vegard's law:

$$\boldsymbol{a} = -0.0092x + 9.7346\ \text{Å}; \sigma(a) = 2.5 \times 10^{-3}\text{Å}\ (\sigma : \text{standard deviation})$$
$$\boldsymbol{c} = -0.009x + 7.2806\ \text{Å}; \sigma(c) = 2 \times 10^{-3}\ \text{Å} \quad (4)$$

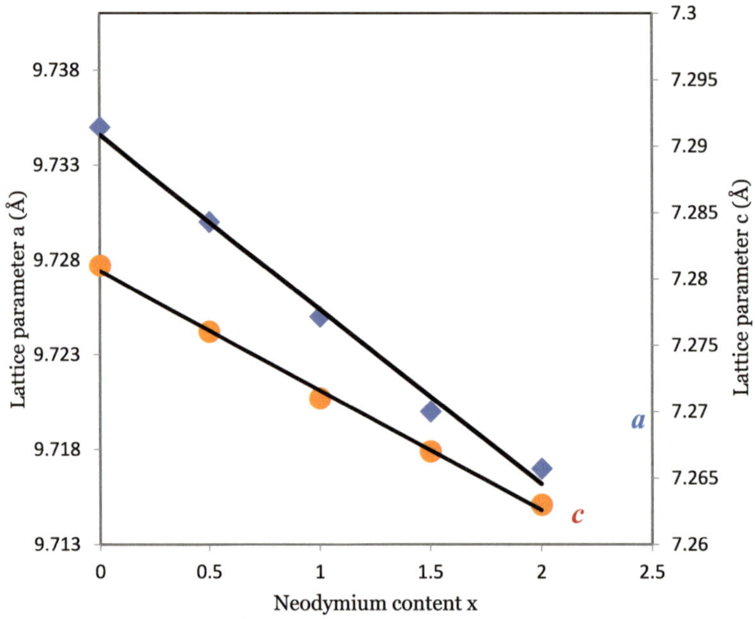

Figure 4.
Lattice parameters as a function neodymium level in the $Sr_8La_{2-x}Nd_x(PO_4)_4(SiO_4)_2F_2$ *with* $(0 \leq x \leq 2)$.

Sample	a (Å)	c (Å)	V (Å^3)	D_{300} (Å)	D_{300} (Å)
$SrLa_2F$	9.735(2)	7.281(2)	597.55(2)	304(3)	387(4)
$SrLa_{1.5}Nd_{0.5}F$	9.730(3)	7.278(3)	596.69(2)	278(2)	362(3)
$SrLa_1Nd_1F$	9.725(2)	7.271(2)	595.43(2)	254(3)	347(2)
$SrLa_{0.5}Nd_{1.5}F$	9.720(3)	7.267(2)	594.57(3)	237(3)	326(3)
$SrNd_2F$	9.717(3)	7.263(2)	593.87(3)	223(4)	307(4)

Table 3.
Crystallographic parameters of strontium fluorbritholites $Sr_8La_{2-x}Nd_x(PO_4)_4(SiO_4)_2F_2$ $(0 \leq x \leq 2)$.

indicates the existence of a continuous solid solution in the explored substitution domain.

The rational parameters that govern the site occupation are the nature, the electronegativities, the valences and the polarizabilities of the ions. The bibliography studies' results indicated that like those observed in natural phosphosilicate apatites, the substituted cations in the apatite structure had preferential occupation for Me(2) sites [42–45]. Thus, it could be concluded that La^{3+} and Nd^{3+} ions substituting Sr^{2+} with $0 \leq x \leq 2$ in our studied samples were subsequently preferentially localized in Me(2) sites.

3.3 FTIR spectroscopy

The FTIR spectra of the samples were given in **Figure 5**. The identification of all the bands was done by comparison with un- and substituted strontium fluorapatite the previously reported in the literature [40, 41]. The characteristic absorption bands of SiO_4 and PO_4 were observed [41].

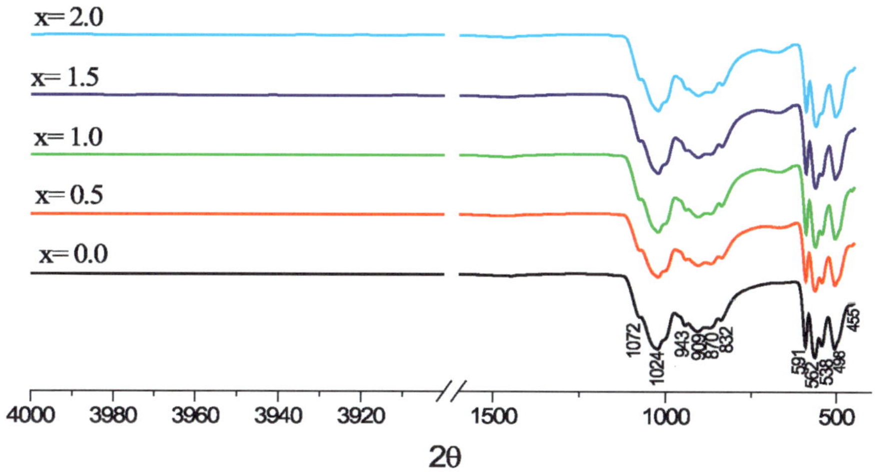

Figure 5.
Infrared spectra of $Sr_8La_{2-x}Nd_x(PO_4)_4(SiO_4)_2F_2$ ($0 \leq x \leq 2$).

The PO_4characteristic bands observed at 1072–1024 cm $^{-1}$ coincide with to the asymmetric stretching mode (υ_3). The band at 943 cm^{-1} is attributed to the symmetrical stretching mode (υ_1). The asymmetric bending mode (υ_4) bands appeared in the 538–562 cm^{-1} range and the symmetric one (υ_2) are shown at 455 cm^{-1}. The bands observed at 909–943 (υ_3), 832–870 (υ_1), about 538 (υ_4) and 460–498 cm^{-1} (υ_2) were assigned to SiO_4. Moreover as the neodymium amount in the samples increased a shift of the PO_4 and SiO_4 absorption bands towards the low wave numbers was detected. This was attributed to a reduction in size of the lattice inducing an increase in anion-anion repulsion (PO_4 vs. SiO_4) [45]. This observation was in good agreement with those obtained by diffraction of X-rays and confirms that the neodymium substitution reduced the lattice size.

3.4 ^{31}P NMR spectroscopy

In the **Figure 6** are represented the ^{31}P NMR-MAS spectra. A single isotropic signal was observed for all the spectra. It indicated also that a unique crystallographic site for the PO_4 tetrahedron in the apatite structure was present. However a slight chemical shift towards the lower values was observed as well as a broadening of the peaks was attributed to the Nd substitution. This fact was related to a disorder induced in the apatite network caused by the substitution of La by Nd. This was previously seen with doped with rare earth apatite's [46–48].

3.5 Materials sintering

Materials densification optimization has been performed by sintering the synthesized samples in the temperatures range 1250–1500°C with a fixed holding time of 6 h. Relative density d_{re} was calculated using the equation:

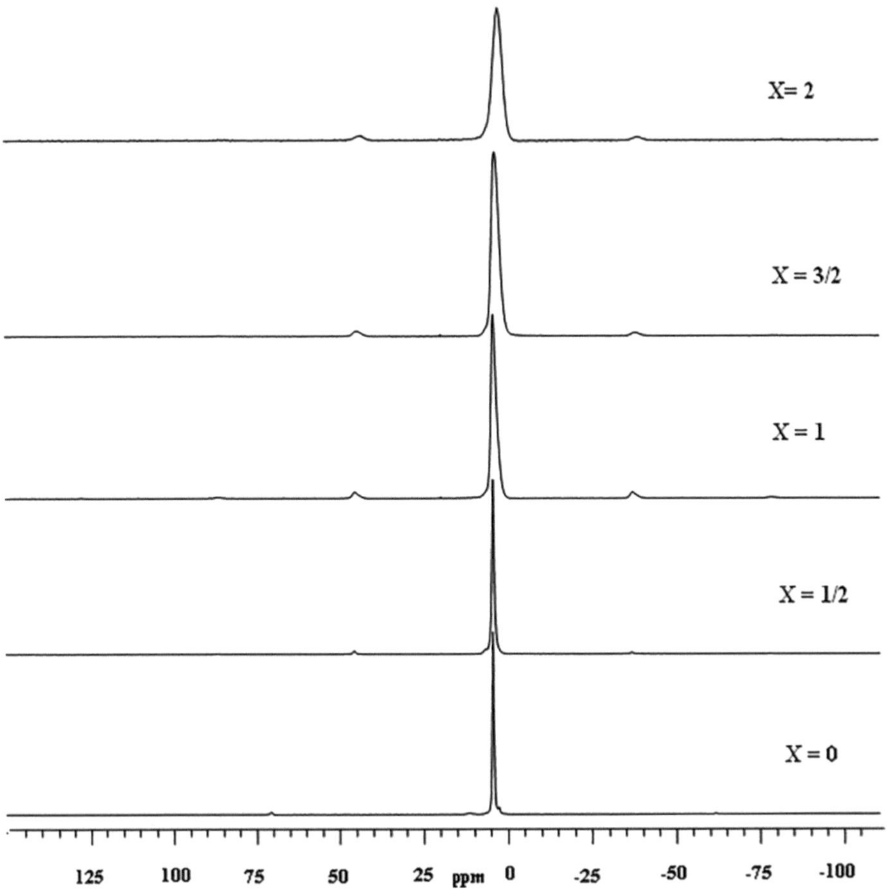

Figure 6.
^{31}P NMR-MAS spectra of fluorbritholites $Sr_8La_{2-x}Nd_x(PO_4)_4(SiO_4)_2F_2$ with $(0 \leq x \leq 2)$.

$$d_{re} = \frac{\rho_{the}}{\rho_{exp}} \times 100 \quad (5)$$

where the theoretical density ρ_{the} was calculated using the equation:

$$\rho_{the} = \frac{ZM}{NaV} \quad (6)$$

(Z: number of molecules/unit cell, M: molecular weight Na: Avogadro number and V: volume of the unit cell) and experimental density determined from the mass and the dimension of sintered pellets by means of the equation

$$\rho_{the} = \frac{m}{\pi h r^2} \quad (7)$$

Figure 7 shows that relative density of the sintered samples strictly depends on sintering temperatures as well as on Nd content. An irregular trend was noted and the highest relative density 89% was obtained with x = 2 Nd content when sintered only at 1250°C. The remaining samples presents lower than densifications ratios obtained at

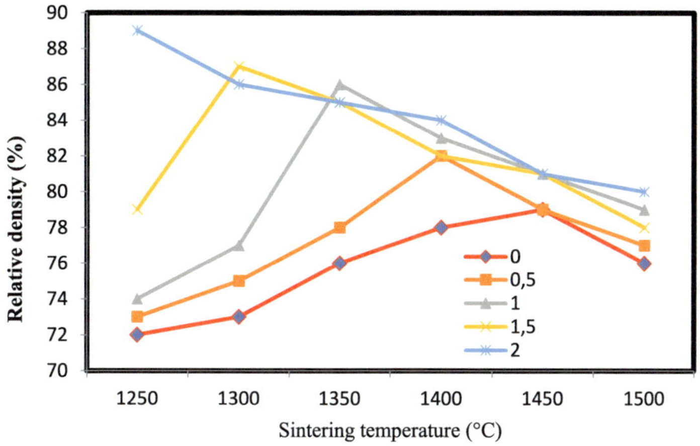

Figure 7.
Relative density versus sintering temperature of $Sr_8La_{2-x}Nd_x(PO_4)_4(SiO_4)_2F_2$ *with* $(0 \leq x \leq 2)$.

higher temperatures. From these data, it can be deduced that the grains morphology and size modification strongly depends on Nd content and sintering temperature. The Nd doping should improve the materials densification by reducing the porosity. This was confirmed by the percentage porosity of the higher densified samples calculated by the following equation:

$$p = (1 - dr) \times 100 \tag{8}$$

As plotted on **Figure 8**, the porosity of the samples decreased as Nd content increased. This result muched the evolution of the relative density suggested to

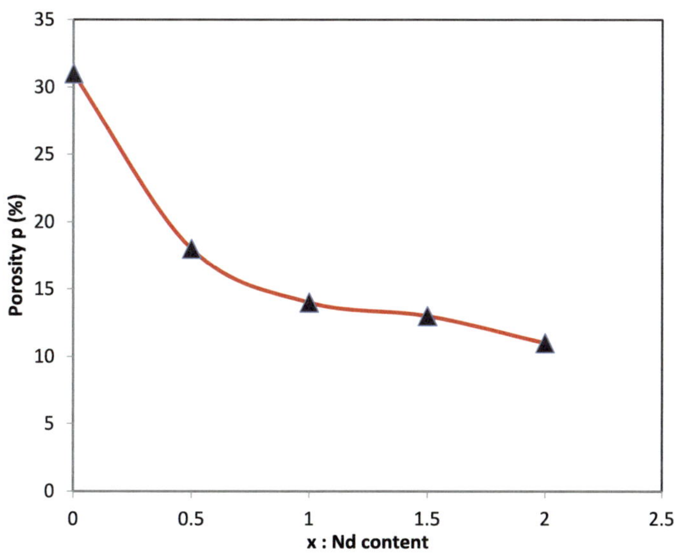

Figure 8.
Porosity versus Nd content of maximum densified samples.

increase when crystallite size is reduced (i.e. grain size). This should promotes the materials densifications by eliminating the intergranular porosity.

The microstructure of the samples given on **Figure 9** is closely coherent with the densification rates as well as porosity. Indeed, the micrographs show a progressive removal of the porosity when the Nd rate rises. Thus with x = 0 the microstructure is of intergranular aspect revealing the presence of abundant porosity. With x = 0.5, although some pores persist on the surface the porosity was reduced,. When x = 1 the open porosity has almost disappeared and only the closed porosity remains, reflecting the 89% densification.

$$Sr_8La_{2-x}Nd_x(PO_4)_4(SiO_4)_2F_2 \text{ with } (0 \leq x \leq 2). \tag{9}$$

3.6 Impedance spectroscopy

The ionic conductivity of the samples was determined between 400 and 800°C with a step of 20°C by complex impedance plots. Thus, for each sample, 20 complex impedance plots (plane, Z″ vs Z′) were plotted. The intercept of the semicircular arcs with the real axis allow obtaining the bulk resistance R. The ionic conductivity of the sintered samples was calculated from the equation:

$$\sigma = \frac{e}{SR} \tag{10}$$

The thickness and the area of the sample were e and S, respectively. **Figure 10** reprinted the ionic conductivity σ versus the neodymium substitution. The first deduction is that σ depends on this substitution and particularly at higher temperatures. The curves obtained at 604 (877 K) and 482°C (755 K) indcated that the measured conductivity was about 4.4×10^{-7} S cm^{-1}. By contrary with the increase of Nd content, σ rose up to 1.73×10^{-6} S cm^{-1} at 779°C (1052 K). Hence, the electric conductivity of the samples depend onthe Nd substituted level.

The total activation energy of the samples was obtained from the Arrhenius equation:

$$\sigma T = Ae^{-\frac{Ea}{KT}} \tag{11}$$

The parameters to define are the pre-exponential factor A, activation energy Ea, Boltzmann constant k and absolute temperature T, respectively. **Figure 11** shows an Arrhenius-type plot indicating that the electrical conduction of the materials is activated by heating. The σ values were slightly different from those found in the

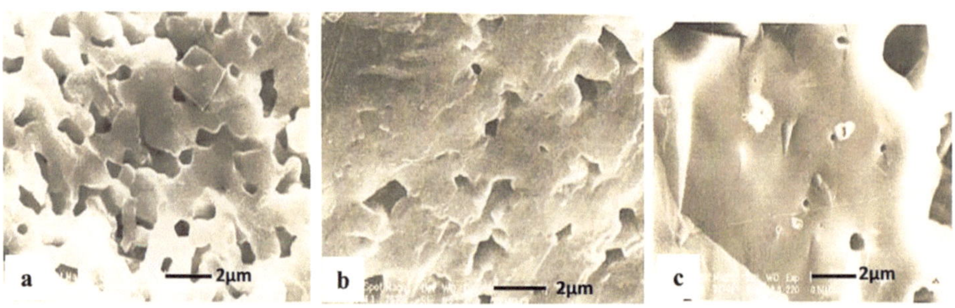

Figure 9.
Micrographs of sintered samples $Sr_8La_{2-x}Nd_x(PO_4)_4(SiO_4)_2F_2$ (a) x = 0.0; (b) x = 0.5; (c) x = 1.0.

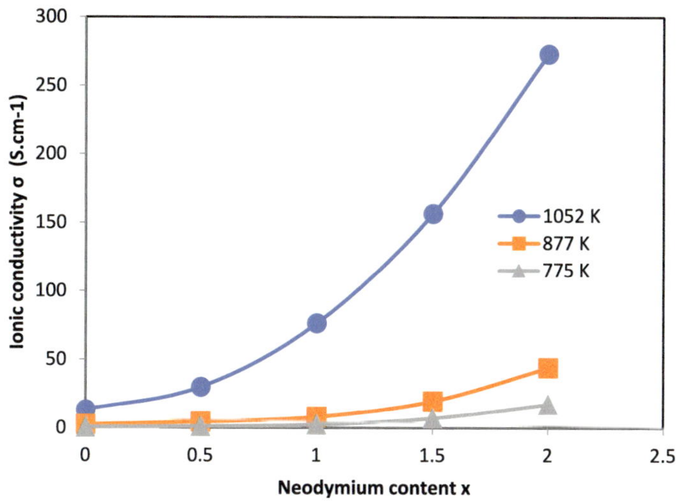

Figure 10.
Ionic conductivity versus neodymium content.

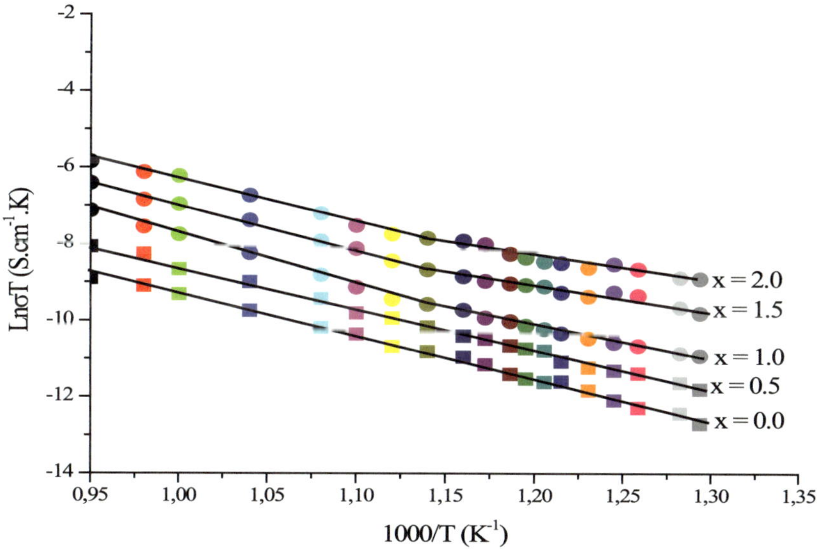

Figure 11.
Plots of LnσT versus 1000/T of fluorbritholites $Sr_8La_{2-x}Nd_x(PO_4)_4(SiO_4)_2F_2$ with $(0 \leq x \leq 2)$.

literature [49, 50]. The difference might have resulted from the preparation and sintering methods reflected by the difference in densification ratios (range 72–83%). The slope in the Arrhenius plots versus temperatures gives the activation energy. This later parameter increased when Nd level rose reaching a maximum of 1.1 eV when x = 1 then decreased to 0.91 eV (**Table 4**). Moreover a slight break in slope for $x \geq 1$ was detected in the Arrhenius plots. This was related to the Sr/Nd—F bond likely to the work of Njema and al [49]. In fact, in $Sr_8La_{2-x}Nd_x(PO_4)_4(SiO_4)_2F_2$ with $(0 \leq x \leq 2)$ samples, the mobility of F^- along the ***c*** axis ensure the charge motion. Thus as the Nd-doping increased the F^- mobility was enhanced and the conductivity was improved. The fluoride ions motion along the structure should be related to the

x	0	0.5	1	1.5	2
Ea (eV)	0.87	0.95	1.1	1.03	0.91

Table 4.
Activation energy of $Sr_8La_{2-x}Nd_x(PO_4)_4(SiO_4)_2F_2$ *with* $(0 \leq x \leq 2)$.

neodymium polarizability slightly higher than its of lanthanium. Herein, the polarizabilities of lanthanum and neodymium were 4.82, 5.01 $Å^3$, respectively [51, 52]. Laghzizil and *al* emphasized the improved fluoride mobility in the presence of polarizable cations localized in Me(2) site [53, 54].

4. Conclusion

Strontium fluorbritholites $Sr_8La_{2-x}Nd_x(PO_4)_4(SiO_4)_2F_2$ with $(0 \leq x \leq 2)$ were successfully prepared by reaction in the solid state. Characterization by several techniques revealed that all the powders were composed of a single apatite phase. The lattice parameters *a* and *c* were inversely proportional to Nd content. The La-Nd substitution was totally coherent with the ionic size of lanthanum and neodymium ions. The chemical analysis showed that for all the compositions the ratio $\frac{Sr+La+Nd}{P+Si}$ was close to the stoichiometric value (1.667) indicating the stoichiometry of the powders. The Infrared absorption apectroscopy study confirmed the presence of the absorption bands relative to PO_4 and SiO_4 groups. The ^{31}P NMR (MAS) showed the presence of a unique isotropic signal confirming the existence of a single crystallographic site for the phosphorus nucleus. The materials' ionic conductivity measured via impedance spectroscopy was found to be heating-dependent. The maximum of 1.73×10^{-6} S cm^{-1} is obtained with higher Nd level and at a temperature of around 1052 K.

Conflict of interest

The authors declare no conflict of interest.

Author details

Khouloud Kthiri[1], Mohammed Mehnaoui[2], Samira Jebahi[3], Khaled Boughzala[4] and Mustapha Hidouri[1]*

1 High Institute of Applied Sciences and Technology, Gabes University, Gabes, Tunisia

2 Faculty of Sciences of Bizerte, Laboratory of Applied Chemistry to Natural Resources and Substances and to the Environment, University of Carthage, Bizerte, Tunisia

3 RU Analysis and Applied Process for Environment, High Institute for Applied Sciences and Technology, Mahdia, Tunisia

4 Energy and Matter Research Laboratory, National Center for Nuclear Sciences and Technology, Sidi Thabet, Tunisia

*Address all correspondence to: mustapha.hidouri@laposte.net

References

[1] Oberti R, Ottolini L, Della Ventura G, Parodi GC. On the symmetry and crystal chemistry of britholite: New structural and microanalytical data. American Mineralogist. 2001;**86**:1066-1075

[2] Njema H, Debbichi M, Boughzala K, Said M, Bouzouita K. Structural electronic and thermodynamic properties of britholites $Ca_{10-x}La_x(PO_4)_{6-x}(SiO_4)_xF_2$ $(0 \leq x \leq 6)$: Experiment and theory. Materials Research Bulletin. 2014;**51**:210-216

[3] Noe DC, Hughes JM, Mariano AN, Drexler JW, Kato A. The crystal structure of monoclinic britholite-(Ce) and britholite-(Y). Zeitschrift für Kristallographie. 1993;**206**:233-246

[4] Arden KM, Halden NM. Crystallization and alteration history of britholite in rare-earth-element-enriched pegmatitic segregations associated with the Eden Lake Complex, Manitoba, Canada (ISO4). 1999;**37**: 1239-1253

[5] Bros R, Carpena J, Sère V, Beltritti A. Occurence of Pu and fissiogenic RRE in hydrothermal apatite from the fossil nuclear reactor 16 of Oklo (Gabon). Radiochimica Acta. 1996;**74**: 277-282

[6] Sère V. Geochemistry of neoformed minerals in Oklo (Gabon). Geological history of the Oklo Basin: Contribution for geological disposal studies for radioactive waste [Doctoral thesis]. France: Paris University; 1996. p. 7

[7] Bodu R, Bouzigues H, Morin N, Pfiffelmann JP. On the existence of isotopic anomalies encountered in uranium from Gabon. Comptes Rendus de l'Académie des Sciences - Series IIC. 1972;**275**:1731-1734

[8] Trocellier P. Immobilization of radionuclides in single-phase crystalline waste forms: A review on their intrinsic properties and long term behavior. Annales de Chimie Science des Materiaux. 2000;**25**:321

[9] Carpena J, Kienast R, Ouzegane JK, Jehanno C. Evidence of the contrasted fission-track clock behavior of the apatites from In Ouzzal carbonatites (northwest Hoggar): The low-temperature thermal history of an Archean basement. Geological Society of America Bulletin. 1988;**100**:1237-1243

[10] Gauthier-Lafaye F. 2 billion year old natural analogs for nuclear waste disposal: The natural nuclear fission reactors in Gabon (Africa). Comptes Rendus Physique. 2002;**3**:839-849

[11] Bros R, Carpena J, Sere V, Beltritti A. Occurrence of Pu and fissiongenic REE in hydrothermal apatites from the fossil nuclear reactor 16 at Oklo (Gabon). Radiochimica Acta. 1996;**74**:277

[12] Bhattacharyya DK. Issues on the disposal of waste containing NORM. Applied Radiation and Isotopes. 1998;**49**: 215-226

[13] Chaumont J, Soulet S, Krupa JC, Carpena J. Competition between disorder creation and annealing in fluoroapatite nuclear waste forms. Journal of Nuclear Materials. 2002;**301**:122-128

[14] Soulet S, Carpena J, Chaumont J, Kaitasov O, Ruault MO, Krupa JC. Simulation of the α-annealing effect in apatitic structures by He-ion irradiation: Influence of the silicate/phosphate ratio and of the OH^-/F^- substitution. Nuclear Instruments & Methods in Physics Research, Section B (Beam Interactions with Materials-and Atoms). 2001;**184**:383

[15] Elliot JC. Structure and Chemistry of the Apatites and Other Calcium Orthophosphates. Vol. 5. Amsterdam: Elsevier; 1994

[16] LeGeros RZ. Biological and synthesized apatites. In: Brown PW, Constantz B, editors. Hydroxyapatite and Related Materials. Boca Raton: CRC Press; 1994. pp. 3-28

[17] Pan Y, Fleet ME. Compositions of the apatite-group minerals: Substitution mechanisms and controlling factors. Reviews in Mineralogy and Geochemistry. 2002;**48**:13-49

[18] Zhua K, Yanagisawa K, Shimanouchi R, Ondaa A, Kajiyoshi K. Preferential occupancy of metal ions in the hydroxyapatite solid solutions synthesized by hydrothermal method. Journal of the American Ceramic Society. 2006;**26**:509-513

[19] Gmati N, Boughzala K, Abdellaoui M, Bouzouita K. Mechanochemical synthesis of strontium britholites: Reaction mechanism. Comptes Rendus Chimie. 2011,**14**: 896-903

[20] Njema H, Boughzala K, Boughzala H. Structural analysis by Rietveld refinement of calcium and lanthanum phosphosilicate apatites. Journal of Rare Earths. 2013;**31**:897-904

[21] Romanchuk AY, Gracheva NN, Bryukhanova KI, Seregina IF, Trigub AL, Gavrichev KS, et al. Sorption behavior and speciation of AmIII in orthophosphates of rare-earth elements. Mendeleev Communications. 2018;**28**: 303-305

[22] Philippe K, Raison E, Philippe M, Martin L, Solari PL, Rudy D, et al. Structural investigations of PuIII phosphate by X-ray diffraction. MAS-NMR and XANES spectroscopy. Journal of Solid State Chemistry. 2015; **230**:169-174

[23] Clavier N, Dacheux N, Martinez P, Du Fou de Kerdaniel E, Aranda L, Podor R. Sintering of β-thorium–uranium (IV) phosphate–diphosphate solid solutions from low-temperature precursors. Chemistry of Materials. 2004; **16**:3357-3366

[24] Terra O, Clavier N, Dacheux N, Podor R. Preparation and characterization of lanthanum–gadolinium monazites as ceramics for radioactive waste storage. New Journal of Chemistry. 2003;**27**:957-967

[25] Benard P, Brandel V, Dacheux N, Jaulmes S, Launay S, Lindecker C. $Th_4(PO_4)_4P_2O_7$. A new thorium phosphate: Synthesis. characterization and structure determination. Chemistry of Materials. 1996;**8**:181-188

[26] Terra O, Dacheux N, Audubert F, Podor R. Immobilization of tetravalent actinides in phosphate ceramics. Journal of Nuclear Materials. 2006;**352**: 224-232

[27] Dacheux N, Clavier N, Robisson AC, Terra O, Audubert F, Lartigue JÉ. Immobilisation of actinides in phosphate matrices. Comptes Rendus Chimie. 2004;**7**:1141-1152

[28] Nakayama S, Kageyama T, Aono H, Sadaoka H. Ionic conductivity of lanthanoid silicates, $Ln_{10}(SiO_4)_6O_3$(Ln = La, Nd, Sm, Gd, Dy, Y, Ho, Er and Yb). Journal of Materials Chemistry. 1995;**5**: 1801-1805

[29] Nakayama S, Aono H, Sadaoka Y. Ionic conductivity of $Ln_{10}(SiO_4)_6O_3$ (Ln = La, Nd, Sm, Gd and Dy). Chemistry Letters. 1995;**24**:431

[30] Nakayama S, Sakamoto M. Electrical properties of new type high oxide ionic conductor $RE_{10}Si_6O_{27}$ (RE = La, Pr, Nd, Sm, Gd, Dy). Journal of the European Ceramic Society. 1998;**18**:1413-1418

[31] Wenhui Y, Yaping G. Synthesis and ionic conduction of cation-deficient apatite $La_{9.332x/3}M_xSi_6O_{26}$ doped with Mg, Ca, Sr. Chinese Journal of Chemical Engineering. 2008;**16**:488-491

[32] Arikawa H, Nishiguchi H, Ishihara T, Takita Y. Oxide ion conductivity in Sr-doped $La_{10}Ge_6O_{27}$ apatite oxide. Solid State Ionics. 2000;**136-137**:31-37

[33] Panteix PJ, Julien I, Bernache-Assollant D, Abélard P. Synthesis and characterization of oxide ions conductors with the apatite structure for intermediate temperature SOFC. Materials Chemistry and Physics. 2006; **95**:313-320

[34] Célérier S, Robert CL. Synthesis by sol-gel route of oxyapatite powders for dense ceramics: Applications as electrolytes for solid oxide fuel cells. Journal of the European Ceramic Society. 2005;**25**:2665-2668

[35] Elliott JC, Wilson RM, Dowker SEP. JCPDS-International Centre for Diffraction Data. Advances in X-ray Analysis. 2002;**45**:172-181

[36] Boughzala K, Hidouri M, Ben Salem E, BenChrifa A, Bouzouita K. Insertion du cesium dans des britholites au strontium. Comptes Rendus Chimie. 2007;**10**:1137-1146

[37] Hata M, Okada K, Iwai S. Cadmium hydroxyapatite. Acta Crystallographica. 1978;**34**:3062-3064

[38] Shannon RD. Revised effective ionic radii and systematic studies of interatomic distances in halides and chalcogenides. Acta Crystallographica. 1976;**32**:751-767

[39] Schroeder LW, Mathew M. Cation ordering in $[Ca_2][La_8][([SiO_4])_6][O_2]$. Journal of Solid State Chemistry. 1978; **26**:383

[40] Boughzala K, Ben Salem E, Ben Chrifa A, Gaudin E, Bouzouita K. Synthesis and characterization of strontium-lanthanum apatites. Materials Research Bulletin. 2007;**42**:1221

[41] Hassine A, Jaba N, Panczer G, Bouzouita K. Synthesis and characterization of strontium-neodymium-fluorbritholites. Comptes Rendus Chimie. 2010;**13**:1460-1467

[42] Fleet ME, Pan Y. Site preference of Nd in fluorapatite $[Ca_{10}(PO_4)_6F_2]$. Journal of Solid State Chemistry. 1994;**111**:78

[43] Carpena J, Boyer L, Fialin M, Kienast JR, Lacout JL. Ca^{2+}, $PO_4^{3-} \leftrightarrow Ln^{3+}$, SiO_4^{4-} coupled substitution in the apatitic structure: Stability of the mono-silicated fluor-britholite. Comptes rendus de l'Académie des sciences. Series IIA. 2001;**333**:373

[44] Boughzala K, Ben Salem E, Kooli F, Gravereau P, Bouzouita K. Spectroscopic studies and Rietveld refinement of strontium-britholites. Journal of Rare Earths. 2008;**26**:483-489

[45] Fowler BO. Infrared studies of apatites. I. Vibrational assignments for calcium, strontium, and barium hydroxyapatites utilizing isotopic substitution. Inorganic Chemistry. 1974; **13**:194-207

[46] Hidouri M, Albeladi N. Lanthanum-neodymium-co-substituted calcium fluorobritholites. Journal of Chemical Engineering and Materials Science. 2018; **6**:151-162

[47] Hidouri M. Sintering and ionic conduction of neodymium-bearing fluorobritholites. Materials Chemistry and Physics. 2019;**228**:254-262

[48] Fleet ME, Liu X, Pan Y. Site preference of rare earth elements in hydroxyapatite [$Ca_{10}(PO_4)_6(OH)_2$]. Journal of Solid State Chemistry. 2000; **149**:391

[49] Njema H, Boughzala K, Chaabène A, Bouzouita K. Study of the ionic conductivity of $Ca_6La_4(PO_4)_2(SiO_4)_4F_2$ and $Ca_4La_6(SiO_4)_6F_2$. Journal of Rare Earths. 2013;**31**:897

[50] Laghzizil A, Elherch N, Bouhaouss A, Lorente G, Coradin T, Livage J. Electrical behavior of hydroxyapatites $M_{10}(PO_4)_6(OH)_2$ (M = Ca, Pb, Ba). Materials Research Bulletin. 2001;**36**:953-962

[51] Shannon RD. Dielectric polarizabilities of ions in oxides and fluorides. Journal of Applied Physics. 1993;**73**:348

[52] Vineis C, Davies PK, Negas T, Bell S. Microwave dielectric properties of hexagonal perovskites. Materials Research Bulletin. 1996;**31**:431-437

[53] Laghzizil A, Bouhaouss A, Ferhat M, Barboux P, Morineau R, Livage J. Anionic conductivity in fluorapatites: Correlation between structure and electrical properties. Advanced Materials Research. 1994;**1**:479-488

[54] Laghzizil A, El Herch N, Bouhaouss A, Lorente G, Macquete J. Comparison of electrical properties between fluoroapatite and hydroxyapatite materials. Journal of Solid State Chemistry. 2001;**156**:57-60

Section 2

Minerals from Magmatic Rocks

Chapter 8

Mg-Ilmenite from Kimberlites, Its Origin

Sergey I. Kostrovitsky

Abstract

The main regularities of the saturation of kimberlite rocks with the accessory mineral Mg-ilmenite (Ilm), the peculiarities of the distribution of Ilm compositions in individual pipes, in different clusters of pipes, in diamondiferous kimberlite fields, are considered as the example of studies carried out within the Yakutian kimberlite province (Siberian Craton). Interpretation of different crystallization trends in MgO-Cr_2O_3 coordinates (conventionally named "Haggerty's parabola", "Steplike", "Hockey stick", as well as the peculiarities of heterogeneity of individual zonal and polygranular Ilm macrocrysts made it possible to propose a three-stage model of crystallization Ilm: (1) Mg-Cr poor ilmenite crystallizing from a primitive asthenospheric melt; (2) Continuing crystallization in the lithospheric contaminated melt by MgO and Cr_2O_3; (3) Ilmenite subsequently underwent sub-solidus recrystallization in the presence of an evolved kimberlite melt under increasing oxygen fugacity (fO_2) conditions.

Keywords: magnesian ilmenite, kimberlite, kimberlite field, kimberlite cluster, macrocrysts, mantle xenoliths, asthenosphere, lithosphere

1. Introduction

Ilmenite proper, corresponding to the chemical formula $FeTiO_3$, often forms a series of solid solutions with isostructural minerals - heikilite $MgTiO_3$, pyrophanite - MnTiO3, hematite - Fe_2O_3. Along with Mg^{2+}, Mn^{2+}, Fe^{3+}, ilmenite can contain isomorphic impurities of Al, Cr, Nb, V, etc. There is a geochemical specialization of impurity elements in ilmenite, depending on the type of rocks. For example, ilmenite from basic rocks is characterized by the presence of V, Cr, Co, Ni. In ilmenites from kimberlites, there is an increased content of Cr, Al, Nb, Zr. A typical impurity for ilmenites is Mg, while the MgO content can reach up to 20 wt.% [1]. Ilm with MgO content >6 wt% is commonly referred to as Mg-ilmenite. The existence of a continuous series $FeTiO_3$-$MgTiO_3$ is assumed. Isomorphic impurity Fe_2O_3 occurs according to the scheme of heterovalent isomorphism $Fe^{3+} + Ti^{4+} \leftrightarrow 2F^{3+}$.

Mg-ilmenite (Ilm) is an important kimberlite indicator mineral, which is widely used in diamond exploration to identify primary deposits. In kimberlites, Ilm forms discrete monomineralic grains (i.e., megacrysts, macrocrysts, and micro-phenocrysts), whose content varies widely (from 0.1 to 2–3 wt.%]. Less frequently, Ilm occurs in mantle xenoliths [2–4] and in Ilm-Prx intergrowths [5, 6]. Due to the fact that syngenetic mineral inclusions of olivine, clinopyroxene, and garnet in Ilm macrocrysts are extremely rare, it is difficult to elucidate the genesis of Ilm. That is why the issues of the occurrence of ilmenite, its mantle sources, and its genetic connection with kimberlite melt continue to be discussed. Potential origins of Ilm macrocrysts and megacrysts include: (I) the disaggregation of Ilm-bearing lithospheric mantle lithologies [1, 7, 8]; (II) crystallization within the asthenosphere [9–12]; (III) crystallization from an asthenospheric melt within the lithosphere associated with kimberlite magmatism [4, 13–17]; a modern take on this previous model is (IV) formation in a "metasomatic aureole" surrounding the (proto-) kimberlite melt and or previous pulses of failed (proto-)kimberlite melt, alongside other megacryst suite minerals and sheared xenoliths [18–22].

This section of the book is a compilation of two published articles [23, 24], written on the basis of a study of the representative collections of Ilm collected by the author. Before proceeding to the presentation of our model of the origin of Mg-ilmenite [24], let us consider different trends of crystallization Ilm in MgO-Cr_2O_3 coordinates, which are characteristic of individual pipes, pipe clusters, and diamond-bearing fields, as well as the peculiarities of the heterogeneity of the composition of individual zonal and polygranular Ilm megacrysts.

2. The composition of Mg-ilmenite

Mineralogical assessment of most kimberlite pipes in the four diamond fields of the Yakutian kimberlite province (YaKP) (**Figure 1**) provided the author with an opportunity to study the compositions of Ilm macrocrysts. A representative number of both the studied pipes (94) and the Ilm macrocrysts (11,003) were studied. Microprobe analyzes were carried out at the Central Analytical Laboratory of the Botuobinskaya Geological Survey of ALROSA on a Superprobe JXA 8800R. Repeated attempts by researchers [1, 25, 26] to reveal the compositional features of Ilm from individual fields were unsuccessful. The reason for the failure lies in the fact that the researchers focused on the comparison of statistical parameters of the distribution of the composition. Consideration of the trends in the variability of the Ilm composition is much more informative. Here we predominantly focus on bivariate plots MgO-Cr_2O_3, since these coordinates are the most informative for demonstrating differences between Ilm from various fields, clusters, and pipes [4, 7, 27]. It is important to note that the average composition of Ilm and its MgO-Cr_2O_3 distribution does not vary with sampling depth, or with the textural type (i.e., unit) of kimberlite within a single pipe. Therefore, the composition of Ilm is an invariant characteristic unique to a given kimberlite [22]. A comparison of the Ilm compositions from different fields indicates that their common feature is (**Table 1**) their fairly consistent homogeneous composition. With wide variations in the content of the main Ilm oxides from different fields, with the exception of the Mirninsky field, they are characterized by a very similar average composition. The Mirninsky field kimberlites contain higher-Fe Ilm, with a higher content of the hematite (Fe_2O_3) component.

Despite the closeness of the average oxide contents (in three fields), Ilm from each field shows completely different distributions of the composition points in the MgO-Cr_2O_3 plots (**Figure 2**). Below we provide a brief description of them.

Mirninsky field, despite the small number of pipes (only 9), is one of the most productive - 5 pipes (Mir, Internatsionalnaya, 23 KPSS, Dachnaya, Taezhna) belong to the diamond deposits. The distribution of the composition points of the Ilm composition on the MgO-Cr_2O_3 graph (**Figure 2a**) resembles the type of distribution, which is

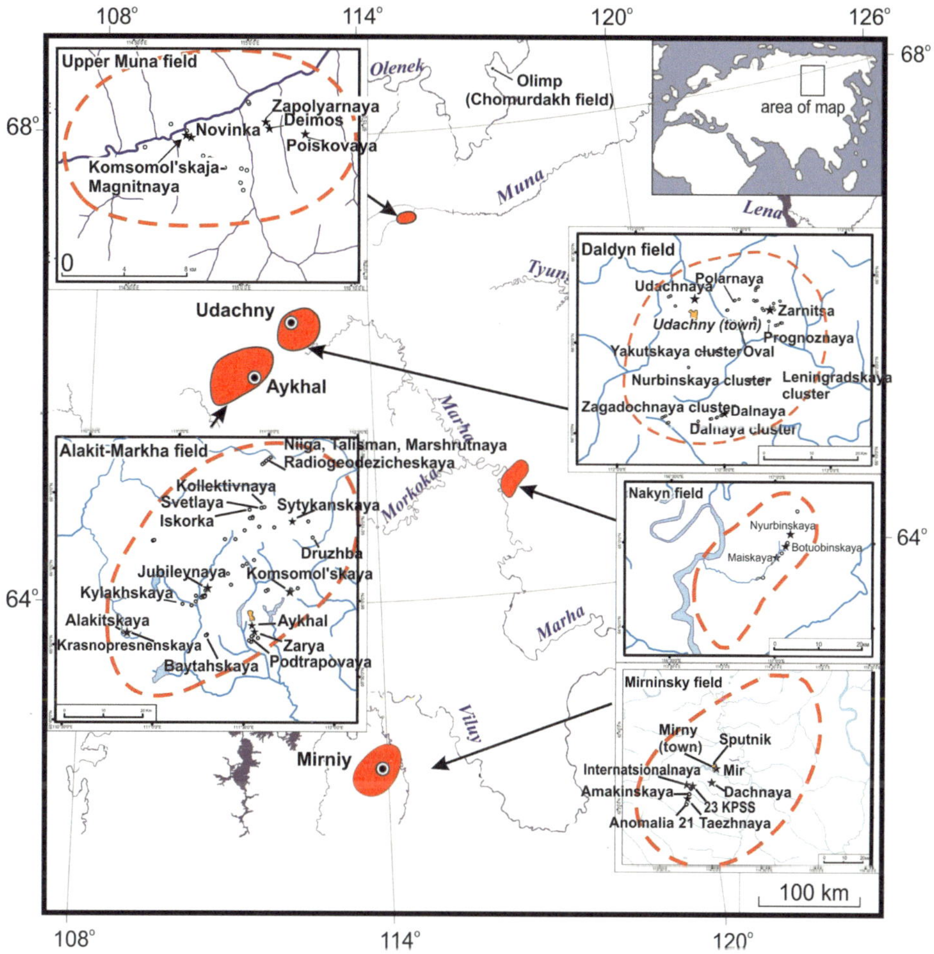

Figure 1.
Map showing the location of diamondiferous kimberlite fields in the Yakutian province.

conventionally named "Haggerty's parabola " after the name of the researcher who first discovered it [8]. The clearest and most numerous group of composition points belongs to low-Cr Ilm (<0.5 wt% Cr_2O_3) with a variable MgO content, covering almost the entire range of its variation. Other groups of points of composition, corresponding to low-Mg and high-Mg Ilm with variable content of Cr_2O_3, form two branches of the parabola on the graph and generally demonstrate a scattered type of distribution.

A feature of the MgO-Cr_2O_3 plot for the Daldynsky field (**Figure 2b**) is the presence of three distinct groups of composition points that show no or weak correlation between oxides. The presence of three Ilm groups in terms of Cr_2O_3 content is found for most of the Daldynsky field pipes, but not for all. For example, Ilm from pipes of the Dalnya cluster are characterized by a unimodal distribution of Cr_2O_3 content.

The Alakit-Marhinsky field also consists of more than 60 pipe and dike bodies. But the Ilm compositions (707 analyzes) were studied from only 12 pipes, which is due to the higher-Mg composition of kimberlites in this field and, accordingly, the limited number of pipes containing Ilm. The overall plot for the entire field in MgO-Cr_2O_3 coordinates (**Figure 2c**) reflects the overlap of different distribution types, which are

	Mirninsky (1600)	Daldynsky (4171)	Alakit-Marhinsky (4634)	Verhnemunsky (598)
TiO_2	45.8 28.5–56.5	48.0 36.4–55.2	48.3 36.7–60.3	47.4 37.7–59.5
Al_2O_3	0.6 0–4	0.53 0–1.9	0.4 0–1.4	0.5 0–3.8
Cr_2O_3	1.0 0.1–9	1.0 0.2–14.5	1.4 0–7	1.4 0.1–12.6
FeO	18.8 0–43.4	14.4 0.1–44	6.7 0–24	15.5 0–46.3
Fe_2O_3	24.7 8–44.5	25.8 0.2–45.9	12.9 0–30.7	26.2 8.9–40.3
MnO	0.16 0.1–2.4	0.25 0.1–2.9	0.3 0.1–0.9	0.2 0.1–1
MgO	8.8 0–15.6	9.6 0.1–16.2	9.9 6–19.7	9.1 0.9–18.1

Table 1.
Average, the interval of variation in the composition of Mg-ilmenite from different diamondiferous fields of the Yakutian province (in parentheses - the number of analyses).

demonstrated by the plots for different pipes (**Figure 3a–e**). In general, the MgO-Cr_2O_3 plot for Alakit-Marhinsky field is peculiar in the form of individual clusters of points of composition and certainly differs from the corresponding graphs from other diamondiferous fields.

There are 16 known kimberlite pipes in the Verhnemunsky field. The database characterizes the compositions of Ilm from most of the pipes in this field and includes 513 analyzes. A distinctive feature of the field (**Figure 2d**) is the presence in each pipe of the low-Mg group Ilm (6.5–8 wt% MgO), which demonstrates the change in the MgO content at constant Cr_2O_3. Ilm with a MgO content of more than 8 wt% are characterized by a scattered type of distribution in the coordinates MgO-Cr_2O_3, reflecting wide variations in the composition of the mineral.

The Daldynsky field, in which about 60 kimberlite pipes were discovered, was studied for most of the pipes, and therefore the author classified it as an etalon field, namely, thanks to the studies of this field, the most important conclusions about the origin of Ilm were made. **Table 2** reports the most representative average compositions of Ilm grains, in terms of the oxides TiO_2, MgO, Cr_2O_3, and FeOtotal, from pipes in the Daldynsky field (4171 analyses of Ilm). It is evident that Ilm from different pipes of the same cluster yield similar values, while Ilm from different clusters have a significantly different composition. For example, Ilm from the pipes of the Malyutka and Zarnitsa clusters have closely similar MgO abundances, though they differ markedly in Cr_2O_3 content. Ilmenite from the Dalnya, Leningradska, and Dolgozhdana clusters display similar MgO contents, but are different in Cr_2O_3 and so on. Some clusters of pipes demonstrate local heterogeneities in Ilm composition. For example, in the Yakutska cluster, closely located pipes (Akademicheska and Aeros'emochna on the one hand, and Yakutska and Ilmenitova, on the other hand), exhibit very similar Ilm compositions. **Figure 3** presents a plot of the average contents of MgO and Cr_2O_3, showing the proximity of compositions of Ilm from different pipes of the cluster and the differences in Ilm compositions between different clusters. As a rule, the points of average Ilm composition from pipes of one cluster are grouped near each other.

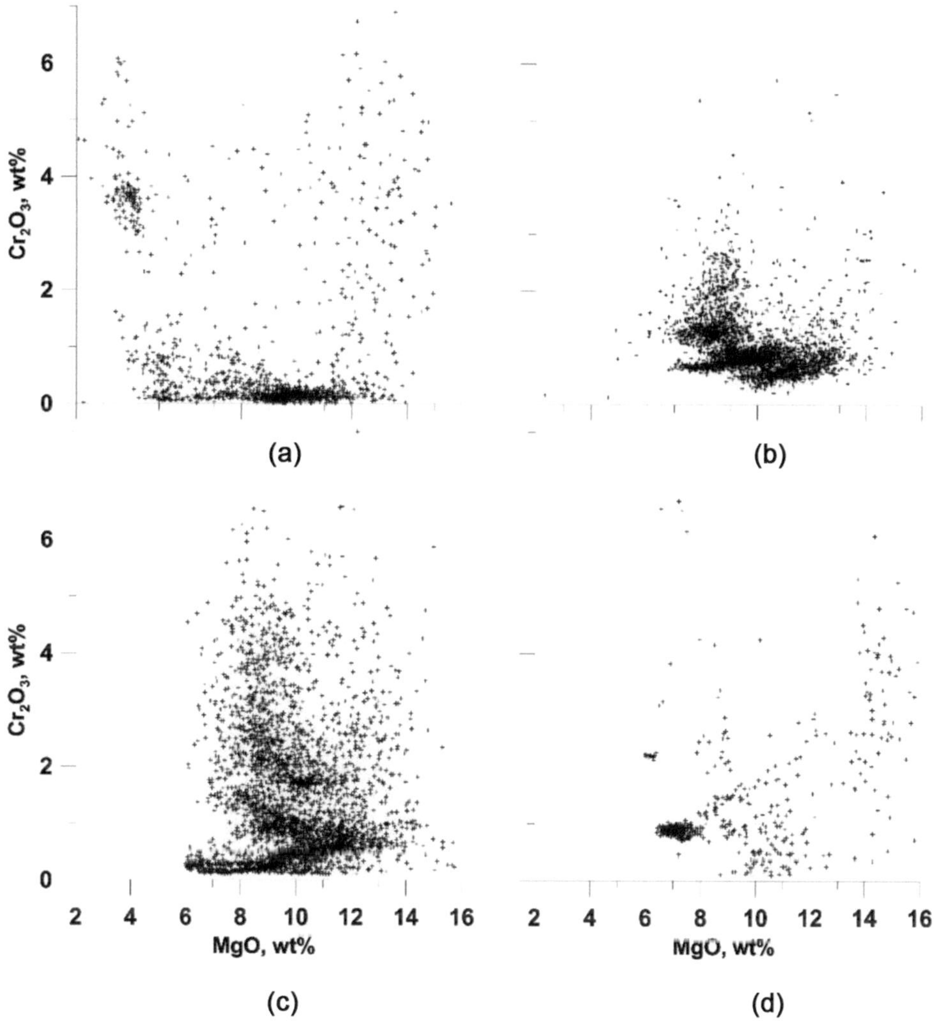

Figure 2.
MgO vs. Cr2O3 bivariate plots showing the distribution of ilmenite compositions from different diamond-bearing kimberlite fields of the Yakutian province: (a) Mirninsky; (b) Daldynsky; (c) Alakit-Marhinsky; (d) Verhnemunsky.

It is evident that the Daldynsky field is characterized by regional heterogeneity along with a clustered distribution of Ilm compositions. The highest Mg content and low Cr_2O_3 content are found in Ilm from pipes in the southern part of the Daldyn field (Dalnya, Leningradska, Yakutska clusters, **Figure 4**), while the northern part of the field predominantly contains clusters of pipes (Zarnitsa, Letnya, and Malyutka) with low MgO and high Cr_2O_3 Ilm (**Table 2**). By combining the MgO–Cr_2O_3 plots with histograms of Cr_2O_3 content (**Figures 5** and **6**), we can clearly identify significant differences in the distribution of Ilm compositions between different pipes. The histograms of Ilm composition in the Daldynsky field show different types of distribution: (1) unimodal, e.g. pipes of the Dalnya (**Figure 6**) and Leningradska clusters; (2) bimodal, e.g. pipes of the Yakutska and Rot-Front clusters, according to MgO content; (3) tri-modal, with distinct minima dividing the analyzed Ilm grains into three separate groups, e.g. pipes of the Zarnitsa (**Figure 5**) and Malyutka clusters.

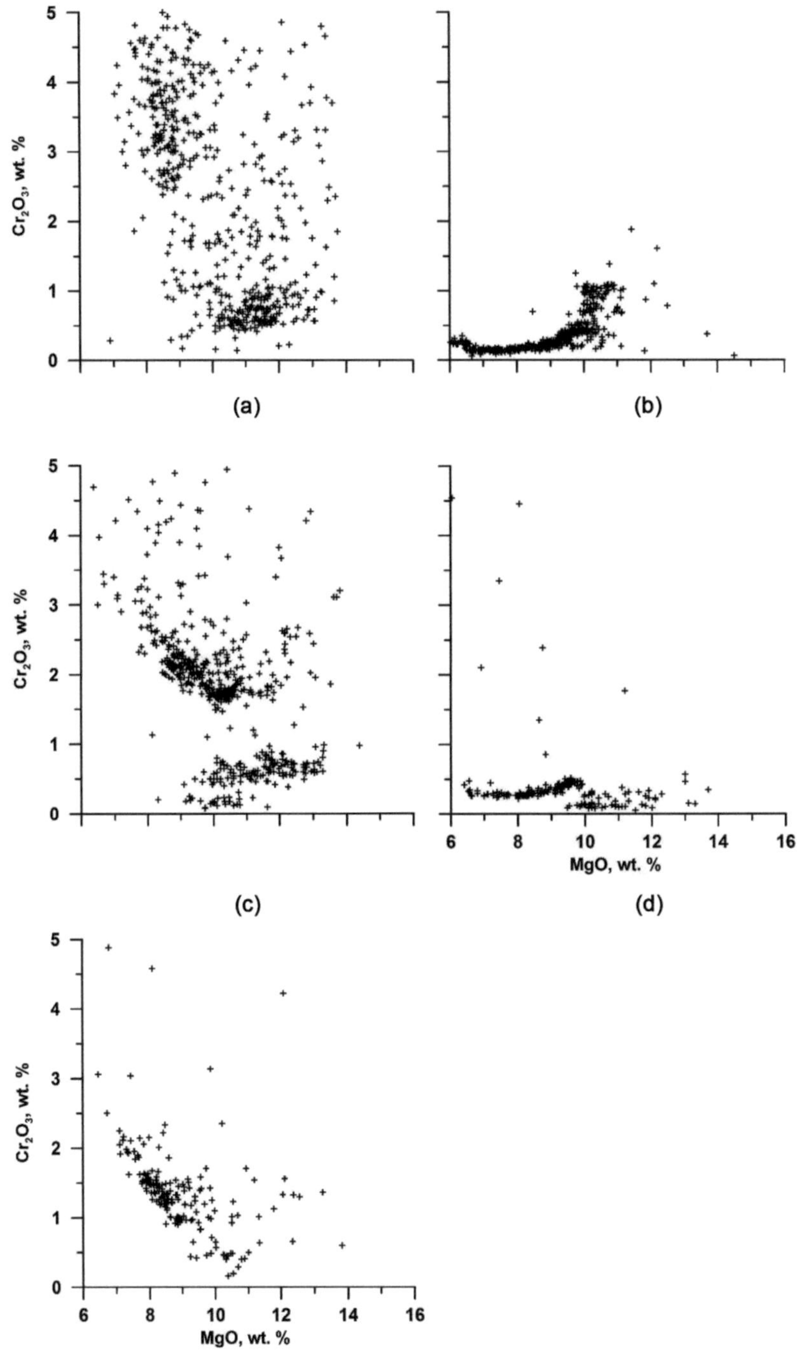

Figure 3.
Different types of distribution of Ilm composition from different pipes from the Alakit-Markhinsky field: (a) "Haggerty parabola" (pipe Yubileynaya); (b) "hockey stick-like" (3 pipes of Iskorka cluster: Iskorka, Svetlaya, Kollektivnaya); (c) "step-scattered" (pipes Komsomolskaya, Sitikanskaya); (d) "stepped" (pipe Druzhba); (e) "inclined", with an inverse correlation between MgO and Cr_2O_3 *(3 pipes of NIIGA cluster, NIIGA, Marshrutnaya, Talisman).*

Cluster of pipes	N	Pipe	Number analyzes	Average composition (in wt.%)			
				Cr_2O_3	MgO	FeOtot	Mg#
Udachnaya	4	Udachnaya-west	88	1.2	9.07	40.28	18.42
	5	Udachnaya-east	82	1.01	9.6	39.92	19.46
Polyarna	6	Lyubimaya	124	0.84	9.3	39.56	19.05
	7	Volzhanka	97	0.82	9.29	40.41	18.79
	8	Studecheska	91	0.76	9.58	39.71	19.48
	9	Polyarna	92	0.79	9.54	39.45	19.52
Bukovinska	11	Jila-75	lmenites are absent in all high-Mg kimberlites pipes of the Bukovinska cluster				
	12	Gornyztska					
	13	Bukovinska					
	14	Popova					
Malyutka	15	Daykova	93	1.74	8.7	39.66	18.04
	16	Sosednya	89	1.45	9.05	41.08	18.07
	17	Malyutka	92	1.71	8.82	41.16	17.67
Zarnitsa		Nevidimka	95	0.85	8.99	39.92	18.27
	20	Zarnitsa	114	0.94	8.96	40.66	18.16
	21	Popugaeva	116	0.93	8.84	40.92	17.95
	22	Cheburashka	30	1.16	8.35	41.82	16.66
	23	Elektra	105	0.97	9.2	40.52	18.63
	24	Argys	100	1.01	8.98	40.35	18.34
	25	Iksovaya	123	0.93	9.22	40.64	18.04
Letnya	27	Osennya	64	1.81	8.48	40.20	17.51
	28	Zimnya	85	1.55	8.21	40.52	17.11
	29	Letnya	87	1.56	8.67	40.45	17.74
Rot-Front	30	Rot-Front	83	1.2	10.3	37.92	21.43
	31	Oval	127	1.18	11.1	35.46	23.78
Yakutska	37	Aeroc'emochna	111	0.86	9.91	38.43	21.04
	38	Academicheska	122	0.9	9.79	37.44	20.72
	39	Ilmenitova	79	1.06	9.33	38.33	19.65
	40	Yakutska	85	0.74	9.15	38.21	19.40
Nurbinska	42	Dolgozhdanna	105	1.1	10.8	36.51	22.90
	43	Nyurbinska	111	0.93	10.3	37.11	21.76
Leningradska	44	Geophysicheska	37	0.74	12	34.09	25.98
	45	Leningradska	90	0.77	11.1	34.73	24.41
	46	Molodezhna	80	0.73	10.8	35.62	23.37
Dalnya	50	Mambo	110	0.84	10.7	36.79	22.63
	51	Nu-Pogodi	78	0.8	11	36.94	23.06
	52	Ugadayka	79	0.85	10.9	36.79	23.03

Cluster of pipes	N	Pipe	Number analyzes	Average composition (in wt.%)			
				Cr_2O_3	MgO	FeOtot	Mg#
	53	Saratovska	101	0.77	11.3	36.38	23.71
	54	Jila-70	71	0.84	11.6	35.46	24.73
	55	Dalnya	82	0.8	10.9	36.41	23.05
	56	Jila-74	66	0.86	11.2	35.54	23.95

Table 2.
Average composition of Mg-imenite in pipes of the Daldynsky field.

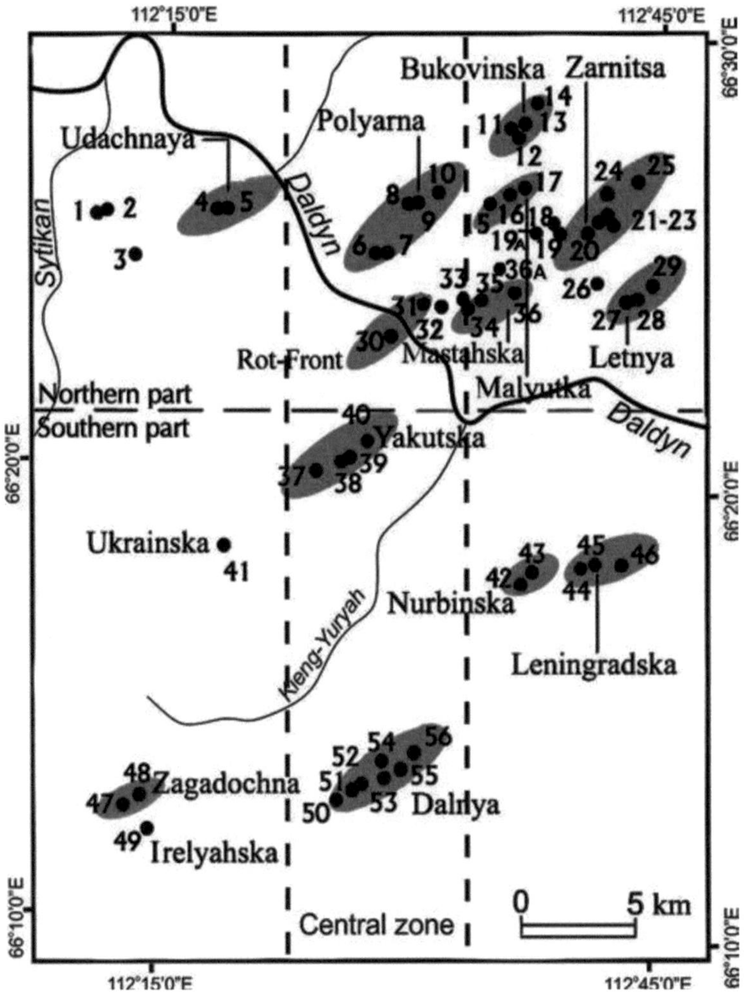

Figure 4.
Sketch map showing the location of kimberlite pipes in the Daldyn field. The names of the pipes are given in ***Table 2*** *(the numbers for pipes on the map and in the table are the same). The clusters of pipes are shown on the map as solid ovals, which group together pipes containing Ilm of similar composition.*

While the first type of distribution is dominant for Ilm from the southern part of the Daldyn field, the third type is essentially characteristic of pipes from the northern part of the field.

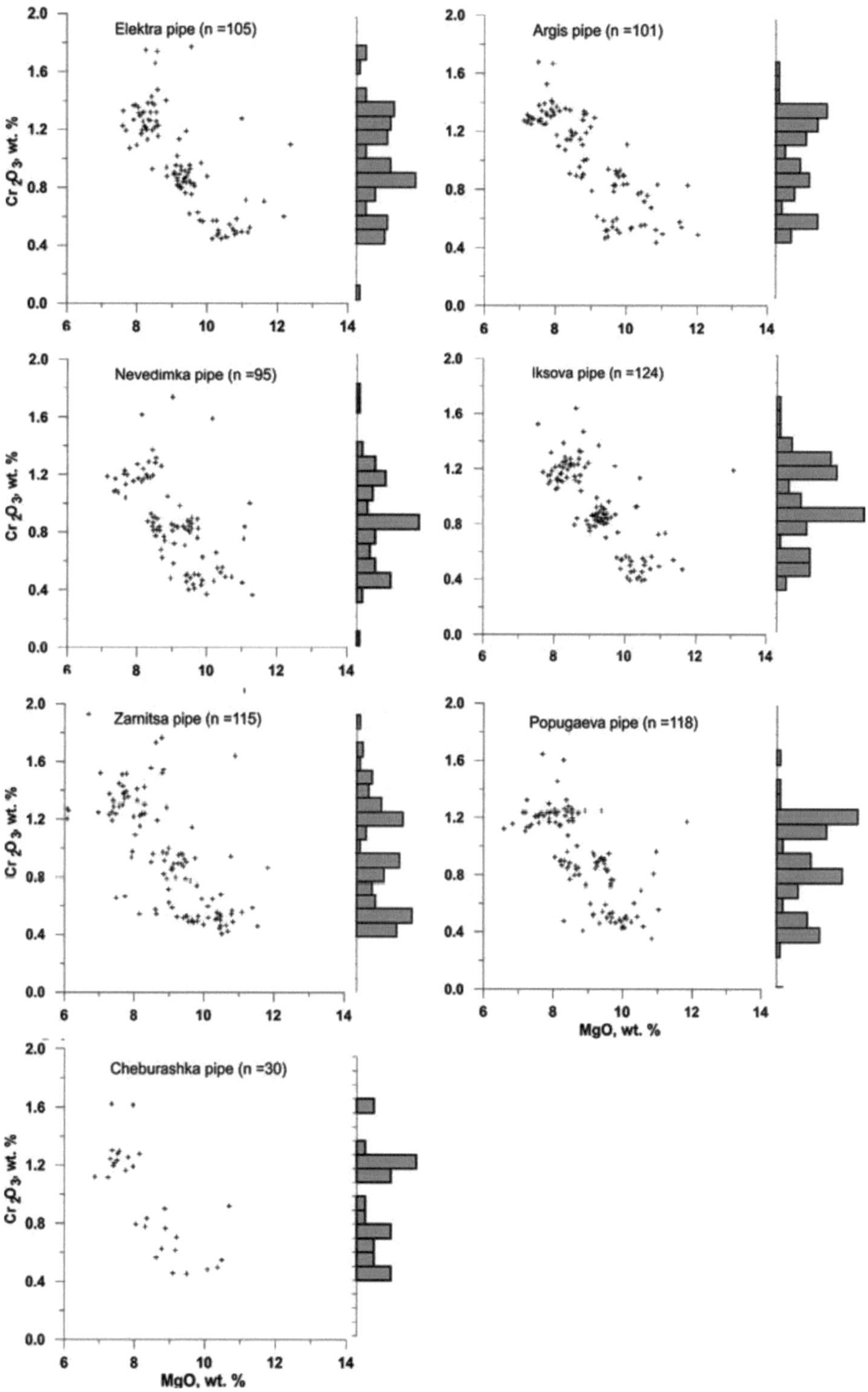

Figure 5.
MgO–Cr_2O_3 correlation diagrams combined with histograms of Cr_2O_3 distribution in Ilm from kimberlites in the Zarnitsa cluster.

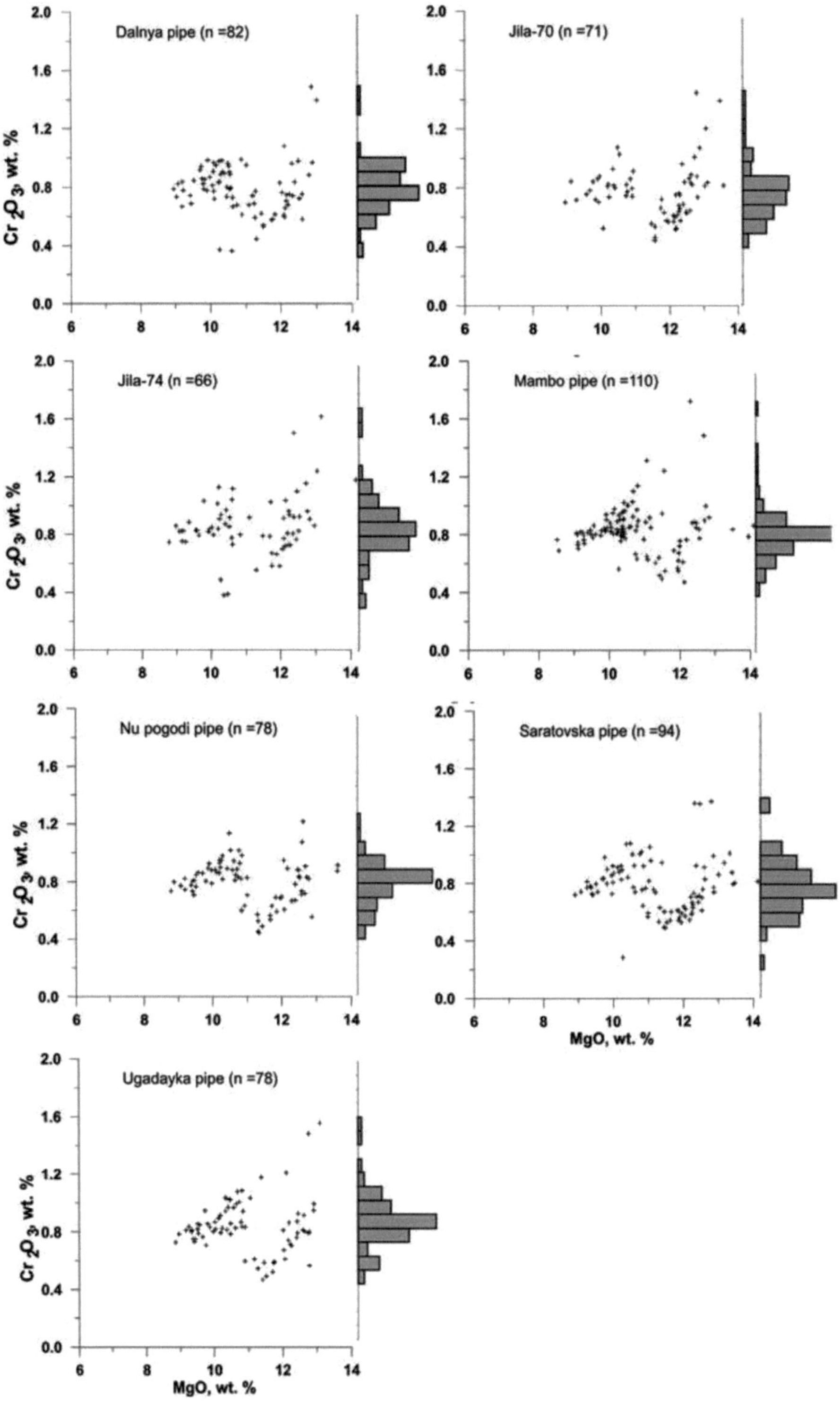

Figure 6.
MgO–Cr_2O_3 correlation diagrams combined with histograms of Cr_2O_3 distribution in Ilm from kimberlites in the Dalnya cluster.

3. The heterogeneous composition of Mg-ilmenite macrocrysts

Most Ilm grains are heterogeneous, with rims enriched in MgO. The Ilm zonal megacrysts are illustrated in **Figures** 7 and **8**, in which the rims are enriched with MgO. The gradual change in the primary composition of Ilm is especially clearly observed in polygranular megacrysts from the Mir pipe (**Figure 9**). Individual granules ranging in size from 100 μm to >1 mm, separated by microcracks, demonstrate a compositional change towards an increase in MgO content from the center to the edge (**Figure 9a, b, e, and f**). At the same time, there is a parallel decrease in the content of

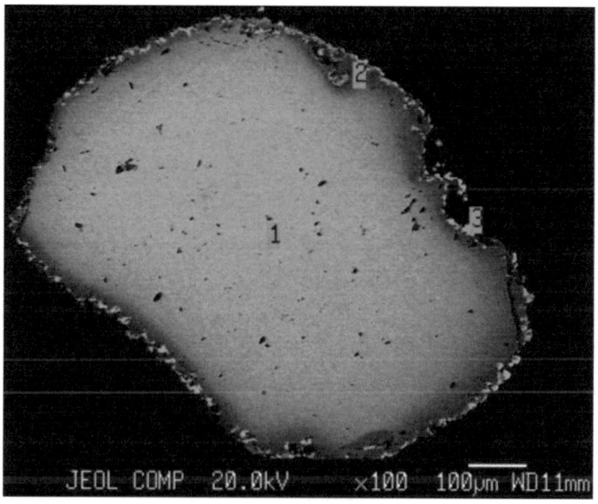

Figure 7.
Typical zoning of Ilm macrocryst: The marginal zone of the grain becomes more magnesian than the core. In the photo, the microcrystal Ilm from the kimberlite of the Malokuonamskaya pipe (Kuranakh field). Zoning: in the center (point 1) 5.16% MgO, 0.72% Cr_2O_3; In the rim (point 2) -11.8% MgO, 2.8% Cr_2O_3.

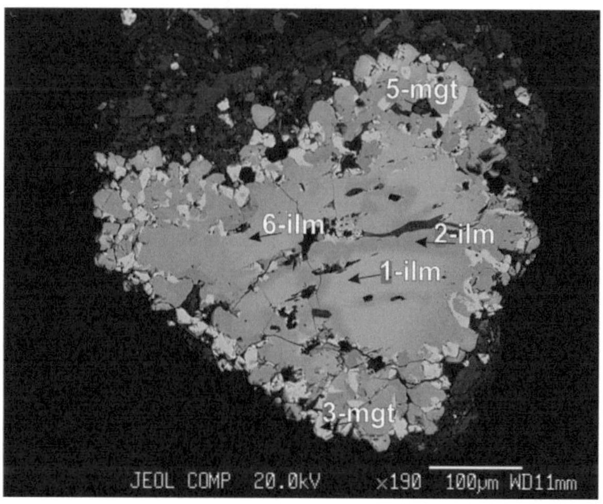

Figure 8.
Macrocryst Ilm from the kimberlite of the Universitetskaya pipe (Kuranakh field). In the center, relicts of high-Fe, low-Cr Ilm (point 1) are preserved, most of the grain is recrystallized in relatively high-Mg, high-Cr Ilm. Composition: Point 1 5.7% MgO, 0.4% Cr_2O_3; point 2–10.6% MgO, 1.5% Cr_2O_3; point 6–9.6% MgO, 0.9% Cr_2O_3.

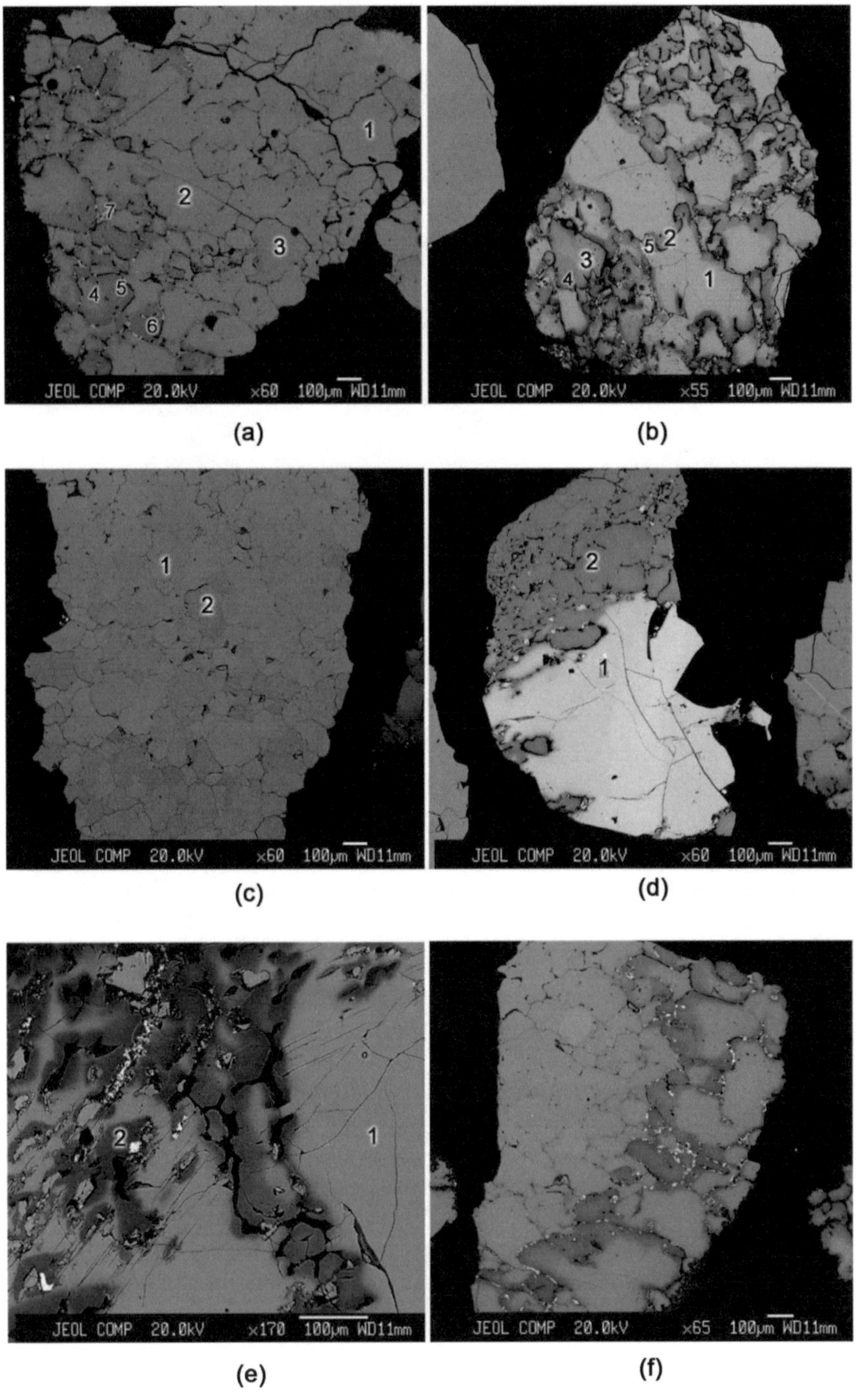

Figure 9.
(a) Uneven granulation of Ilm macrocryst. The formation of microcracks on the boundaries of granules. (b) the changing of Ilm composition is starting in the border zones of granules. (c) relatively uniform micro granulation of Ilm grains with the size of individual granules of 100–200 mμ. (d) Development of microcracks (granulation process) throughout the macrocrystal Ilm, accompanied by a change in composition on only one-half of the grain. (e, f) - The more intensive development of the processes of granulation and substitution takes place in linear microzones of fracturing.

Fe_2O_3 and an increase in the content of MnO, the content of Cr_2O_3 remains unchanged (**Table 3**). MnO-containing Ilm (up to 4.6 wt% MnO) and vein-like Ilm (~ 30 μm–**Figure 9a**–point 7; **Figure 9b**–point 5; see **Table 3**) are found in the intergranular space. it is possible that the granulation of Ilm macrocrysts is caused by deformation processes, during which deformed lherzolites were formed. It is important to bear in

Figures	**Figure 7**			**Figure 8**				**Figure 9A**		
Points	**1c**	**1**	**1r**	**1**	**2**	**3**	**4**	**1**	**2**	**3**
TiO_2	41.30	50.60	20.54	50.88	52.16	54.41	22.64	48.48	48.48	48.76
Al_2O_3	0.09	0.23	3.24	0.60	0.81	0.54	3.81	0.67	0.68	0.66
Cr_2O_3	0.04	1.18	0.41	0.41	0.77	2.42	1.14	0.17	0.15	0.14
Fe_2O_3	19.68	9.39	—	11.33	9.63	6.28	—	14.47	14.69	15.17
FeO	33.93	22.71	58.32	24.37	22.51	22.61	57.64	28.61	28.85	28.86
MnO	0.16	0.48	0.67	0.24	0.32	0.51	0.84	0.21	0.22	0.20
MgO	1.74	12.47	12.36	11.84	13.76	14.43	13.93	8.29	8.15	8.30
Total	96.94	97.05	95.55	99.67	99.95	101.20	99.99	100.9	101.22	102.09

Figures	**Figure 9a**				**Figure 9b**				
Points	**4**	**5**	**6**	**7**	**1**	**2**	**3**	**4**	**5**
TiO_2	48.94	52.20	54.74	51.04	38.07	48.28	43.76	49.83	47.76
Al_2O_3	0.73	1.36	0.10	0.05	0.60	0.26	0.69	0.33	0.00
Cr_2O_3	0.15	0.24	0.32	0.42	0.43	0.61	0.46	0.61	0.28
Fe_2O_3	14.52	7.11	5.00	1.01	31.79	14.91	22.51	12.69	4.00
FeO	27.50	27.84	28.77	40.70	25.29	27.14	26.67	26.26	38.19
MnO	0.23	0.36	0.46	4.51	0.16	0.27	0.18	0.33	4.55
MgO	9.13	10.51	11.21	0.35	4.92	8.98	7.01	10.22	0.08
Total	101.2	99.62	100.6	98.08	101.26	100.45	101.28	100.27	94.86

Figures	**Figure 9c**		**Figure 9d**		**Figure 9e**	
Points	**1**	**2**	**1**	**2**	**_1**	**2**
TiO_2	44.05	43.56	32.80	48.01	31.49	48.58
Al_2O_3	0.61	0.69	0.66	0.29	0.55	0.16
Cr_2O_3	0.14	0.19	3.76	5.17	4.15	5.60
Fe_2O_3	22.89	23.16	38.44	11.62	39.81	10.48
FeO	26.95	26.74	22.29	25.40	21.83	25.78
MnO	0.19	0.18	0.13	0.21	0.11	0.36
MgO	6.99	6.87	3.97	9.85	3.57	9.84
Total	101.82	101.39	102.05	100.55	101.51	100.8

Table 3.
*Chemical composition of the zoned macrocryst of Ilm from Zarnitsa (see **Figure 7**) and Komsomol'skaya pipes (**Figure 8**) and polygranular macrocrysts of Ilm from Mir pipe (**Figure 9**).*

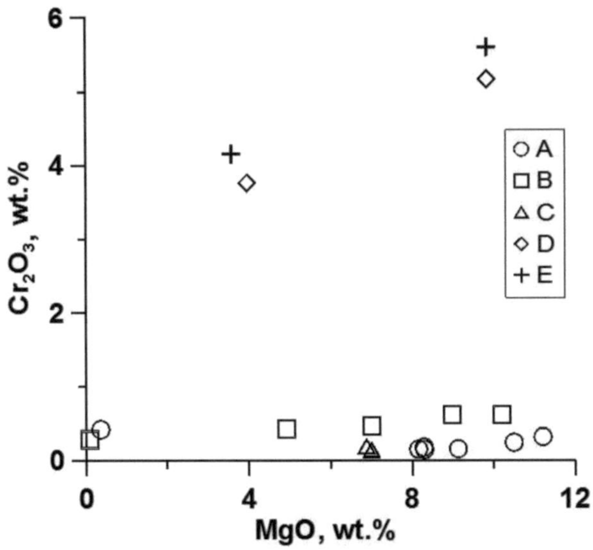

Figure 10.
A plot of the composition of Ilm in the coordinates of MgO-Cr_2O_3 at different points (see ***Figure 9****) of polygrainular grains from the Mir pipe.*

mind that despite the change in the MgO content in the recrystallized macrocrysts Ilm, the Cr_2O_3 content remains constant within the entire grain (**Figure 10**).

4. Ilm-bearing sheared peridotites

Compositional features of minerals from Ilm-bearing deformed peridotites provide important arguments for developing a model of Ilm genesis in kimberlites. The similarity of the compositions of olivine, garnet, and Mg-ilmenite from deformed peridotites and the corresponding megacrysts from kimberlite rocks [28–30] indicates a genetic relationship between them. We present a brief description of the petrographic and geochemical features of Ilm-bearing deformed lherzolite (sample 00–83) from the Udachnaya-Eastern kimberlite, described earlier in the article by Solov'eva et al. [31].

Sample 00–83 is an Ol-Phl-Ilm sheared peridotite xenolith with a fluidal porphyroclastic texture. This sample contains ~30 vol% olivine (Ol), ~40 vol% phlogopite (Phl), and ~ 30 vol% Ilm. Ol predominantly occurs as small (0.05–0.2 mm) euhedral neoblasts, but rare anhedral porphyroclasts (< 1 mm) are present. Laminar Phl porphyroclasts (< 2 mm) are deformed. Ilm occurs as thin lenses (≤0.5 cm wide and ≤ 4 cm long; **Figure 11**), which have polygranular textures (**Figure 9a**). In a similar manner to the polygranular Ilm macrocrysts, titaniferous-magnetite has precipitated in the interstitial space between Ilm granules and at the grain margins of Ilm lenses. Ilm from this xenolith is characterized by wide variations in MgO contents (8.6–12.5 wt%) with relatively constant Cr_2O_3 contents (i.e., 2.6–2.9 wt% –**Table 3**). Ar^{40}/Ar^{39} dating of Phl from this sample yields an age of 367.1 ± 1.4 Ma [32], which overlaps the age of host kimberlite (i.e., 367 Ma, [33]).

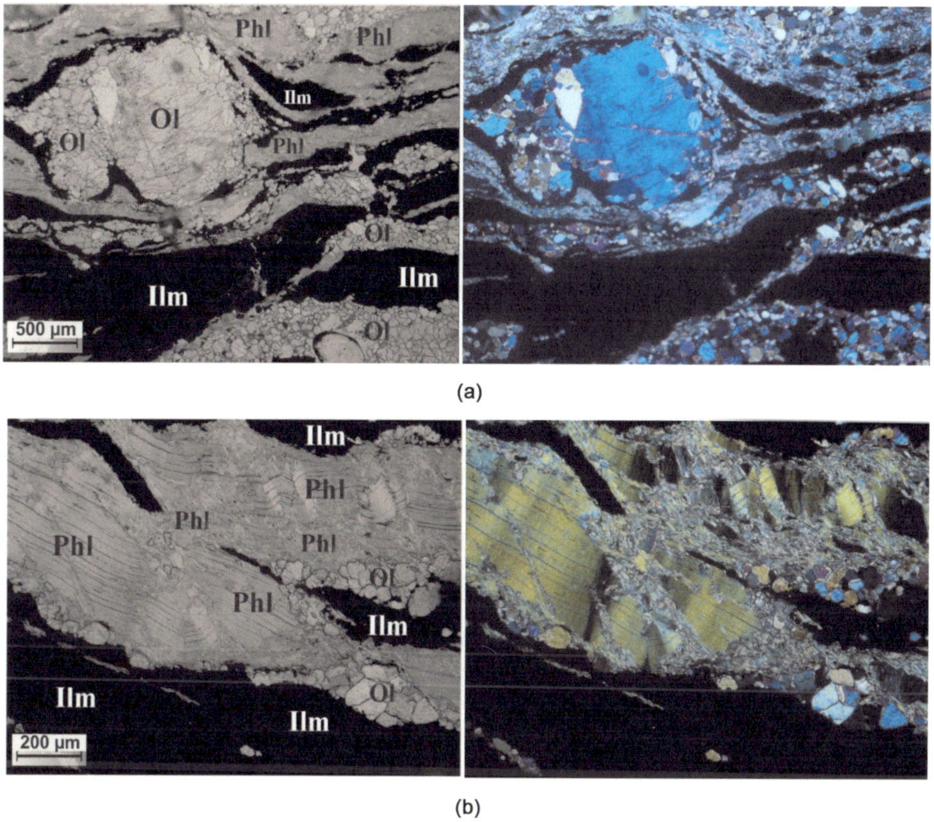

Figure 11.
Sample 00–83. Xenolith of deformed peridotite Phl-Ilm-Ol composition from the pipe Udachnaya-Eastern. Ilm forms micro-veins and micro-lenses. Photo of the section with crossed Nicol. The size of the photo is 6 x 4 mm.

5. Arguments for model of Mg-ilmenite crystallization

5.1 An asthenospheric and lithospheric source for kimberlites, and their megacryst suite

The similarity of Rb-Sr, Sm-Nd, and Lu-Hf isotope systematics, the same age of formation [11, 15, 16, 18, 28, 34] for kimberlites and low-Cr megacryst association of minerals (to which Ilm belongs) testify to a single primary asthenospheric source for them. The similar or almost identical compositions of Ilm in different pipes of one cluster can be accounted for by the existence of a common magmatic supply channel. Various clusters of pipes were fed via different channels of ascending kimberlitic melt, which therefore disintegrated and assimilated different mantle rocks. In the Orapa A/K-1 pipe, the Cr_2O_3 content of Ilm has been shown to be independent of the variation of other oxide components [35]. Two groups of Ilm are recognized in this pipe, with average Cr_2O_3 contents of 1.91 and 3.62 wt%, whereas the content of MgO remains virtually constant. The Ilm nodules from the same pipe, although showing discrete zoning in MgO and Fe_2O_3, are found to have homogeneous Cr_2O_3 contents. Ilm from the Monastery pipe (South Africa) can be divided into three groups [27] based on Cr_2O_3 and Nb contents, while they demonstrate the same trend in terms of major components. Thus, this feature of the behavior of Cr_2O_3 in Ilm is common for different kimberlite pipes. Moore et al. [27] suggested that there was a mixing of magmas or assimilation of host rocks in the magma chamber during Ilm crystallization. We suggest that the assimilation of lithospheric mantle rocks by the kimberlite melt might have occurred in the supply channel of kimberlite pipes. It appears that this peculiarity did not originate in the asthenosphere, but rather in the different channels and modes of ascent of the kimberlite melt, which led to the formation of the various clusters of pipes.

5.2 The presence and formation of Ilm/oxide melts

The presence of large Ilm megacrysts (up to 4 cm), their abundance (up to 3% of the total rock volume), sometimes found in pipes, and, finally, the existence of veinlets, Ilm lenses in deformed lherzolites (**Figure 11**)–all this indicates the existence of a melt Ilm composition. A number of researchers refer to the presence of such melts [18, 36–39]. The appearance of Ilm melts, judging by the veinlets in deformed lherzolites, is recorded at depths corresponding to the boundary between the asthenosphere and lithosphere. The liquidation of a high-Ti melt corresponding in composition to Ilm in the initial asthenospheric melt was caused by deformation processes and a change in the PT parameters during its ascent. We assume that the latter initiated the formation of deformed lherzolites and the ascent of the asthenospheric melt. Ilm crystallization from the kimberlite melt continued to the later stages of ascent and possibly during and after kimberlite emplacement into the upper crust, as indicated by the presence of small groundmass Ilm [40, 41].

5.3 The model of Ilm crystallization

It is commonly argued that fractional crystallization is the primary mechanism responsible for the formation of composition trends in minerals of the Cr-poor megacryst suite [3, 4, 27, 41, 42]. Geochemical data, as well as petrographic constraints (e.g., the abundance of Cpx inclusions in Ilm macrocrysts and Ilm-Cpx intergrowths), indicates that Ilm and Cpx were the final phases of Cr-poor megacryst suite to crystallize [1, 38]. However, the Ilm composition distributions considered above using the example of MgO-Cr_2O_3 plots showed that they cannot be readily explained by a process of fractional crystallization.

The features of the composition distribution of Ilm macrocrysts considered above, the heterogeneity of the composition of both individual macrocrysts (**Figures 2–6**) and polygranular megacrysts (**Figure 9**) were the basis for distinguishing three stages of Ilm crystallization, which occurred at the level of (1) asthenosphere (in the primary asthenospheric melt); (2) the lithosphere (in the melt, which changed its composition as a result of the capture and partial assimilation of rocks by the mantle lithosphere) and (3) the lithosphere and crust (as a result of changes in P–T–O crystallization parameters during ascent through the lithosphere and crust.

In the first stage, crystallization of minerals of the mega crystal low-Cr association of minerals took place, including Ilm. It is assumed that the leading crystallization mechanism was fractional crystallization. At the same time, Ilm and Cpx crystallized last, after Grt, Ol, and Opx.

The second stage of crystallization of Ilm occurred in a melt enriched in MgO and Cr_2O_3 (as a result of the assimilation of rocks of the lithospheric mantle), which was reflected in the corresponding graphs by the formation of the left branch of the Haggerty parabola (**Figure 3a**).

During the third stage, recrystallization of macrocrysts occurred as a result of an increase in fO_2 of the kimberlite melt as it ascended through the upper horizons of the lithosphere. This stage is reflected in the formation of heterogeneity in the composition of individual grains. Recrystallization of Ilm led to a decrease in the content of FeO and MnO with a corresponding increase in the content of MgO. Since the content of Cr_2O_3 remains unchanged, these changes in the composition are reflected in the plot of MgO and Cr_2O_3 by the formation of the right branch of the "Haggerty parabola". All three stages of Ilm crystallization occurred in different pipes (pipe clusters) in different ways, which is primarily due to a different section of the lithospheric mantle, with a different set of trapped and assimilated rocks of the lithospheric mantle. The formation of other Ilm compositional distribution patterns (e.g., "Step-like", and "Hockey stick") is attributed to different compositions of the entrained and partially assimilated lithospheric mantle material, and different ascent dynamics in each of the different kimberlite conduits (which were different for each different kimberlite cluster). Similar Ilm compositional distributions are also typical of other kimberlite provinces worldwide, and we infer that Ilm's three-stage crystallization model is responsible for these compositional distributions in all cases [4, 7–9, 27, 37]. These compositional features are attributed to the existence of a single magmatic conduit feeding all pipes of a given cluster, and different conduits feeding different clusters. Proto-kimberlite melt compositions evolved separately in each cluster (conduit) by the incorporation and partial assimilation of trapped fragments and minerals of the lithospheric mantle rocks.

Summing up, we come to the conclusion that the differences in Ilm compositions in individual pipes, pipe clusters are due to a different set of trapped and partially assimilated mantle xenoliths, or local heterogeneity of the lithospheric mantle. And thus, the similarity of Ilm compositions in the pipes of a particular kimberlite field can serve as a key to deciphering its structure (that is, identifying pipe clusters).

Acknowledgements

The author thanks the geological management of AK ALROSA for creating favorable conditions during the fieldwork, for financial assistance. The author is grateful to A.S. Ivanov and L.F. Suvorova for the microprobe analyses.

Author details

Sergey I. Kostrovitsky
Vinogradov Institute of Geochemistry, Siberian Branch of the Russian Academy of Science, Irkutsk, Russia

*Address all correspondence to: serkost@igc.irk.ru

References

[1] Garanin VK, Kudryavtceva GP, Soshkina LT. Ilmenite from Kimberlites. Moskow: MGU; 1984. p. 240

[2] Dawson JB. Kimberlites and their Xenoliths. Berlin, New York: Springer-Verlag; 1980. p. 252

[3] Mitchell RH. Magnesian ilmenite and its role in kimberlite petrogenesis. Journal of Geology. 1973;**81**:301-311

[4] Mitchell RH. Kimberlites: Mineralogy, Geochemistry, and Petrology. New York: Plenum Press; 1986. p. 442

[5] Eggler DH. Upper mantle oxidation state: Evidence from olivine-orthopyroxene-ilmenite. Assemblages. Geophysical Research Letters. 1983;**10**: 365-368

[6] Clarke DB, Mackay RM. Ilmenite-garnet-clinopyroxene nodule from Matsoku. Evidence for oxide-rich liquid immiscibility in kimberlites. The Canadian Mineralogist. 1990;**28**(pt 2): 229-239

[7] Haggerty SE. The chemistry and genesis of opaque minerals in kimberlite. Physics and Chemistry of the Earth. 1975;**9**:227-243

[8] Robles-Cruz SE, Watangua M, Isidoro L, Melgarejo JC, Galí S, Olimpio A. Contrasting compositions and textures of ilmenite in the Catoca kimberlite, Angola, and implications in exploration for diamond. Lithos. 2009; **112S**:966-975

[9] Harte B, Gurney JJ. The mode of formation of chromium-poor megacryst suites from kimberlites. Journal of Geology. 1981;**89**:749-753

[10] Bruin D. Multiple compositional megacryst groups from the Uintjiesberg and Witberg kimberlites, South Africa. South African Journal of Geology. 2005; **108**:233-246

[11] Tappe S, Steenfelt A, Nielsen TFN. Asthenospheric source of Neoproterozoic and Mesozoic kimberlites from the North Atlantic craton, West Greenland: New high precision U-Pb and Sr–Nd isotope data on perovskite. Chemical Geology. 2012;**320–321**:113-127

[12] Tappe S, Brand BN, Stracke A, van Acken D, Chuan-Zhou L, Strauss HM, et al. Plates or plumes in the origin of kimberlites: U/Pb perovskite and Sr-Nd-Hf-Os-C-O isotope constraints from superior craton. Chemical Geology. 2017;**455**:57-83. DOI: 10.1016/j.chemgeo.2016.08.19

[13] Gurney JJ, Jacob WRO, Dawson JB. Megacrysts from the monastery kimberlite pipe. In: Boyd FR, Meyer YOA, editors. The Mantle Sample: Inclusions in Kimberlites and Other Volcanics. Washington. (Proceedings of 2-nd IKC): Am. Geophys. Union; 1979. pp. 227-243

[14] Kostrovitsky SI. Geochemical Features of Minerals from Kimberlites. Novosibirsk: Nauka; 1986. p. 263

[15] Nowell GM, Pearson DG, Bell DR, Carlson RW, Smith CB, Noble SR. Hf isotope systematics of kimberlites and their megacrysts: New constraints on their source regions. Journal of Petrology. 2004;**45**(5):1583-1612. DOI: 10.1093/petrology/egh024

[16] Kopylova MG, Nowell GM, Pearson DG, Markovic G. Crystallization

of megacrysts from protokimberlitic fluids: Geochemical evidence from high-Cr megacrysts. n the Jericho kimberlite. Lithos. 2009;**112S**:284-295

[17] Giuliani A, Kamenetsky VS, Kendrick MA, Phillips D, Wyatt BA, Maas R. Oxide, sulphide and carbonate minerals in a mantle polymict breccia: Metasomatism by proto-kimberlite magmas, and relationship to the kimberlite megacrystic suite. Chemical Geology. 2013;**353**:4-18

[18] Kamenetsky VS, Belousova EA, Giuliani A, Kamenetsky MB, Goemann K, Griffin WL. Chemical abrasion of zircon and ilmenite megacrysts in the monastery kimberlite: Implications for the composition of kimberlite melts. Chemical Geology. 2014;**383**:76-85

[19] Moore A, Belousova E. Crystallization of Cr-poor and Cr-rich megacryst suites from the host kimberlite magma: Implications for mantle structure and the generation of kimberlite magmas. Contributions to Mineralogy and Petrology. 2005;**149**: 462-481

[20] Kargin AV, Sazonova LV, Nosova AA, Pervov VA, Minevrina EV, Khvostikov VA, et al. Sheared peridotite xenolith from the V. Grib kimberlite pipe, Arkhangelsk Diamond Province, Russia: Texture, composition, and origin. Geoscience Frontiers. 2017;**8**(4):653-669

[21] Tappe S, Pearson DG, Prelevic D. Kimberlite, carbonatite, and potassic magmatism as part of the geochemical cycle. Chemical Geology. 2013;**353**:1-3

[22] Tappe S, Katie A, Smart KA, Bogaard PV. 40Ar/39Ar geochronology and Sr-Nd-Hf-Pb isotope systematics of primitive alkaline basalts and lamprophyres from the SW Baltic shield. Geochimica et Cosmochimica Acta. 2015. DOI: 10.1016/j.gca.2015.10.006

[23] Kostrovitsky SI. Deciphering kimberlite field structure using Mg-ilmenite composition: Example of Daldyn field. (Yakutia). European Journal of Mineralogy. 2018;**30**(6): 1083-1094. DOI: 10.1127/ejm/2018/0030-2783

[24] Kostrovitsky SI, Yakovlev DA, Soltys A, Ivanov AS, Matsyuk SS, Robles-Cruz SE. A genetic relationship between magnesian ilmenite and kimberlites of the Yakutian diamond fields. Ore Geology Reviews. 2020;**120**: 1-16. DOI: 10.1016/j.oregeorev.2020.103419

[25] Ilupin IP, Kaminsky FV, Frantsesson EV. Geochemistry of Kimberlites. Moskow: Nedra; 1978. p. 352

[26] Genshaft YS, Ilupin IP, Kuligin BM, Vitozhents VI. Typomorphism of ilmenites of deep magmatic rocks. In: Composition and Properties of Deep Rocks of the earth's Crust and Upper Mantle of Platforms. Moskow: Science; 1983. pp. 95-190

[27] Moore RO, Griffin WL, Gurney JJ, Ryan CG, Cousens DR, Sie SH, et al. Trace element geochemistry of ilmenite megacrysts from the Monastery kimberlite, South Africa. Lithos. 1992;**29**: 1-18

[28] Kostrovitsky SI, Solov'eva LV, Gornova MA, Alymova NV, Yakovlev DA. About origin of megacrysts of garnet from kimberlites. Doklady Russian Academy of Sciences. 2008;**420**(2):225-230

[29] Nixon PH, Boyd FR. The discrete nodule (megacryst) association in

kimberlites from northern Lesotho. In: Nixon PH, editor. Lesotho Kimberlites. South Africa: Cape and Transvaal Printers; 1973. pp. 67-75

[30] Burgess SR, Harte B. Tracing lithosphere evolution through the analysis of heterogeneous G9/G10 garnet in peridotite xenoliths, II: REE chemistry. Journal of Petrology. 2004; **45**:609-634

[31] Solov'eva LV, Kostrovitsky SI, Kalashnikova TA, Ivanov AV. The Nature of Phlogopite–Ilmenite and Ilmenite Paragenesises in Deep-Seated Xenoliths from Udachnaya Kimberlite Pipe. Dokladi in Russian Academy of Sciences; 2019;**486**(2):223-227. DOI: 10.31857/S0869-56524862

[32] Solov'eva LV, Kalashnikova TV, Kostrovitsky SI, et al. Phlogopite and phlogopite–amphibole parageneses in the lithospheric mantle of the Birekte terrane (Siberian craton). Doklady Earth Sciences. 2017;**475**:822-827. DOI: 10.1134/S1028334X17070273

[33] Kinny PD, Griffin BJ, Heaman LM, Brakhfogel FF, Spetsius ZV. SHRIMP U-Pb ages of perovskite from Yakutian kimberlites. Russian Geology and Geophysics. 1997;**38**:97-105

[34] Woodhead J, Hergt J, Phillips D, Paton C. African kimberlites revisited: in situ Sr-isotope analysis of groundmass perovskite. Lithos. 2009;**112**:311-317

[35] Tollo RP. Petrology and Mineral Chemistry of Ultramafic and Related Inclusions from the Orapa a/K-1 Kimberlite Pipe, Botswana. Unpublished PhD thesis. Amherst, MA: University of Massachusetts; 1982

[36] Clarke DB, Mackay RM. Ilmenite-garnet-clinopyroxene nodule from Matsoku. Evidence for oxide-rich liquid immiscibility in kimberlites. The Canadian Mineralogist. 1990;**28**(pt 2): 229-239

[37] Wyatt BA, Baumgartner M, Anckar E, Grutter H. Compositional classification of kimberlitic and non-kimberlitic ilmenite. Lithos. 2004;**77**: 819-840

[38] Kostrovitsky SI, Malkovets VG, Verichev EM, Garanin VK, Suvorova LV. Megacrysts from the Grib kimberlite pipe (Arkhangelsk province, Russia). Lithos. 2004;**77**:511-523

[39] Kamenetsky VS, Yaxley GM. Carbonate–silicate liquid immiscibility in the mantle propels kimberlite magma ascent. Geochimica et Cosmochimica Acta. 2015;**158**:48-56

[40] Pasteris JD. The significance of groundmass ilmenite and megacryst ilmenite in kimberlites. Contributions to Mineralogy and Petrology. 1980;**75**(4): 315-325

[41] Agee JJ, Garrison Ir JR, Taylor LA. Petrogenesis of oxide minerals in kimberlite, Elliot County. Kentucky. American Mineralogist. 1982;**67**:28-42

[42] Griffin WL, Moore RO, Ryan CG, Gurney JJ, Win TT. Geochemistry of magnesian ilmenite megacrysts from southern African kimberlites. Russian Geology and Geophysics. 1997;**38**(2): 398-419

Chapter 9

Investigation of Accessory Minerals from the Blatná Granodiorite Suite, Bohemian Massif, Czech Republic

Miloš René

Abstract

The Central Bohemian magmatic complex belongs to the Central European Variscan belt. The granitic rocks of this plutonic complex are formed by several suites of granites, granodiorites, and tonalites, together with small bodies of gabbros, gabbro diorites, and diorites. The granodiorites of the Blatná suite are high-K, calc-alkaline to shoshonitic, and metaluminous to slightly peraluminous granitic rocks. Compared to the common I-type granites, granodiorites of the Blatná suite are enriched in Mg (1.0–3.4 wt.% MgO), Ba (838–2560 ppm), Sr. (257–506 ppm), and Zr (81–236 ppm). For granodiorites of the Blatná suite is assemblage of apatite, zircon, titanite, and allanite significant. Zircon contains low Hf concentrations (1.1–1.7 wt.% HfO_2). The composition of titanite ranges from 83 to 92 mol.% titanite end-member. Allanite is relatively Al-poor and displays $Fe_{ox.}$ ratio 0.2–0.5.

Keywords: granodiorite, I-type granite, accessory minerals, apatite, zircon, titanite, allanite, Bohemian Massif, Central European Variscides

1. Introduction

The main carriers of rare-earth elements (REE), uranium, thorium, and zirconium in granitic rocks are usually different accessory minerals, including apatite, zircon, monazite, xenotime, and allanite. However, more detailed information about the assemblage and composition of these accessories is often missing. The granitic rocks of the Blatná suite are part of relatively bigger the Central Bohemian magmatic complex. This magmatic complex represents one of the biggest Variscan magmatic bodies in the Central European Variscides. This magmatic complex is formed by several suites of granodiorites, tonalites, and granites. The Blatná suite is represented by the Blatná hornblende-bearing biotite granodiorites and the Červená hornblende-biotite granodiorites. The accessory minerals assemblage is in a both granodiorites represented by apatite, zircon, titanite and allanite. The presented study is concentrated on comprehensive petrological and geochemical description of the Blatná suite and detailed investigation of accessory minerals assemblage that occurred in this magmatic suite, which is represented by the occurrence of apatite, zircon, titanite, and allanite.

2. Geological setting

The Central Bohemian plutonic complex is a large composite magmatic body that occurred in the central part of the Bohemian Massif between Prague and Klatovy. This plutonic complex represents according to its mineralogical and geochemical compositions the most fractionated Variscan magmatic body in the Bohemian Massif (**Figure 1**). The most widespread rock types that occurred in this complex are amphibole-biotite granodiorites accompanied by biotite granites, tonalites, and melagranites (durbachites) [1–3]. The Blatná suite occurred in the south-west part of magmatic complex is formed by the Blatná hornblende-bearing biotite granodiorites and the Červená hornblende-biotite granodiorites. Other petrographic varieties of the Blatná suite are the Klatovy and Kozárovice granodiorites and tonalites occurring in the southwestern part of the Central Bohemian magmatic complex (**Figure 1**). The Blatná granodiorites intruded during the Variscan magmatic event (346.7 ± 1.6 Ma, U/Pb TIMS analyses on zircon) [2].

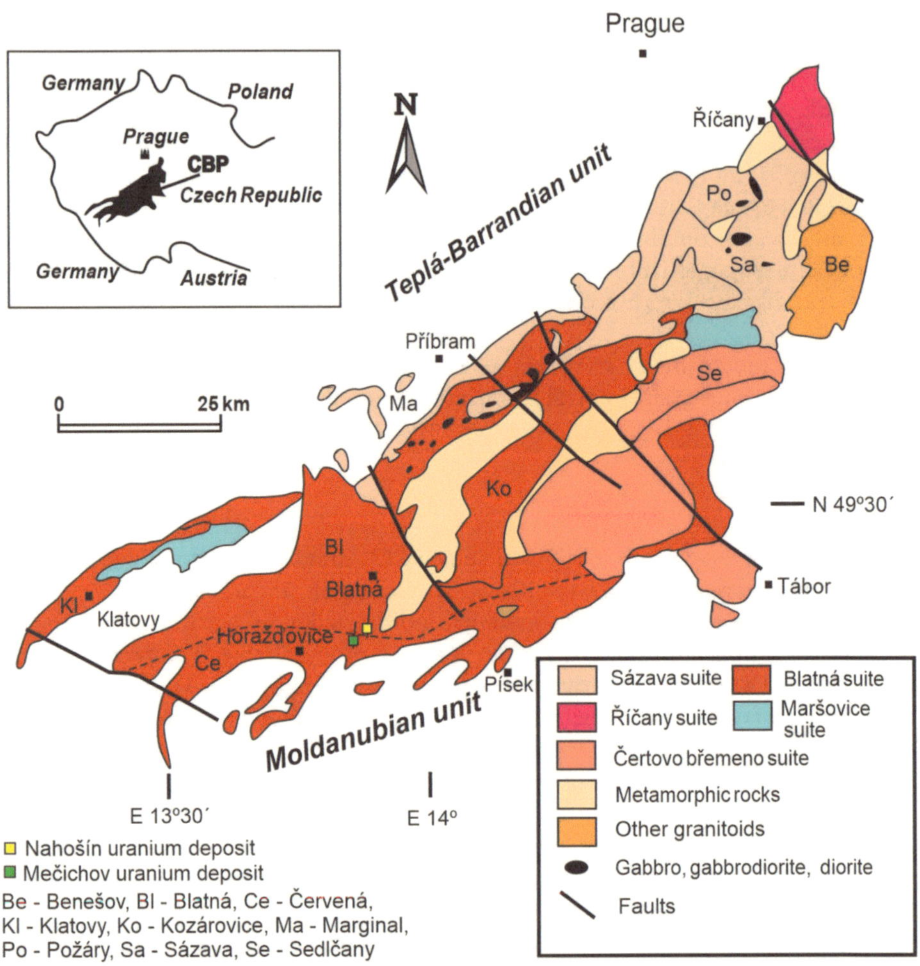

Figure 1.
Geologic map of the Central Bohemian Plutonic complex, modified from [1].

3. Materials and methods

Detailed mineralogical and geochemical investigations of the Blatná and Červená granodiorites were carried out on a representative suite of the 37 rock samples which were taken predominantly from boreholes performed by the Czechoslovak Uranium industry (ČSUP, recently DIAMO) during their exploration activities (1978–1989) in this area [4, 5]. The contents of major elements were determined by a standard XRF method, using the Philips PW 1410 spectrometer at the Geochemical laboratories of the Czechoslovak Uranium Industry (Stráž under Ralsko, Northern Bohemia). The FeO content was measured via titration, whereas the H_2O content was determined gravimetrically. The contents of selected trace elements were determined also by a standard XRF method, using the Philips PW 1410 spectrometer at the chemical laboratory of the Unigeo Brno Ltd. in Brno, Moravia. The content of U and Th was determined by gamma spectrometry using a multichannel spectrometer at Geophysics Brno Ltd., also in Brno, Moravia. The content of REE was quantified by inductively coupled plasma mass spectrometry (ICP MS) at Activation Laboratories Ltd., Ancaster, Canada, using a Perkin Elmer Sciex ELAN 6100 ICP mass spectrometer, following standard sample preparation procedures involving lithium metaborate/tetraborate fusion and acid decomposition. All chemical analyses were calibrated against international reference materials.

Approximately 140 quantitative electron microprobe analyses of apatite, zircon, allanite, titanite, and selected rock-forming minerals (plagioclase, K-feldspar, and biotite) were collected from representative samples of the Blatná suite. All these minerals were analyzed in polished thin sections. The back-scattered electron (BSE) images were acquired to study the internal structure of mineral aggregates and individual mineral grains. The abundances of all chemical elements were determined using a CAMECA SX 100 electron probe micro-analyzer (EPMA) operated in wavelength-dispersive mode at the Department of Geological Sciences, Masaryk University in Brno. The accelerating voltage and beam currents were 15 kV and 20 or 40 nA, respectively, and the beam diameter was 1–5 μm. The peak count time was 20 s, and the background time was 10 s for major elements. For the trace elements, the times were 40–60 s on the peaks, and 20–40 s on the background positions. The following standards, X-ray lines and crystals (in parentheses), were used: AlK_α, sanidine (TAP); CaK_α, fluorapatite (PET); CeL_α, $CePO_4$ (PET); ClK_α, vanadinite (LPET); DyL_α, $DyPO_4$ (LLIF); ErL_α, $ErPO_4$ (PET); EuL_β, (LLIF); FK_α, topaz (PC1); FeK_α, almandine (LLIF); GdL_β, $GdPO_4$ (LLIF); HfM_α, Hf (TAP); KK_α, sanidine (TAP); LaL_α, $LaPO_4$ (PET); MgK_α, Mg_2SiO_4 (TAP); MnK_α, spessartine (LLIF); NaK_α, albite (PET); NbL_α, columbite, Ivigtut (LPET); NdL_β, $NdPO_4$ (LLIF); PK_α, fluorapatite (PET); PbM_α, vanadinite (PET); PrL_β, $PrPO_4$ (LLIF); RbL_α, RbCl (LTAP); SK_α, $SrSO_4$ (LPET); ScKα, ScP_5O_{14} (PET); SiK_α, sanidine (TAP); SmL_β, $SmPO_4$ (LLIF); SrL_α, $SrSO_4$ (TAP); TaM_α, $CrTa_2O_6$ (TAP); TbL_α, $TbPO_4$ (LLIF); ThM_α, $CaTh(PO_4)_2$ (PET); TiK_α, anatase (PET); UM_β, metallic U (PET); VK_β, vanadinite (LPET); YL_α, YPO_4 (PET); YbL_α, $YbPO_4$ (LLIF); and ZrL_α, zircon (TAP). The raw data were corrected using the PAP matrix corrections [6]. The detections limits were approximately 400–500 ppm for Y, 600 ppm for Zr, 500–800 ppm for REE, and 600–700 ppm for U and Th.

Apatite structural formula was calculated on the basis of 13 oxygen. The calculation of mineral formulas for end-member F-, Cl, and OH-apatites was performed according to Piccoli and Candela [7]. The formula of titanite was calculated on the basis of 1 Si as suggested by Harlov et al. [8]. Allanite formula was calculated on the basis of 12.5 oxygen and eight cations per formula using WinEpclas software developed by Yavuz and Yildirim [9].

4. Results

4.1 Petrology of the Blatná suite

Granitic rocks of the Blatná suite are formed by the Blatná hornblende-bearing biotite granodiorites and the Červená hornblende-biotite granodiorites. The Blatná granodiorites are medium-grained, usually equigranular rocks. Major components of these granodiorites are biotite (12–18 vol.%) formed by phlogopite-eastonite (Fe/Fe + Mg 0.48–0.51, Al^{4+} 2.4–2.6 apfu (atoms per formula unit), Ti 0.20–0.43 apfu), plagioclase (An_{23-31}) (40–42 vol.%), quartz (25–28 vol.%), K-feldspar (10–17 vol.%), and magnesiohornblende (0.2–0.7 vol.%) (**Figure 2**). The relatively rarely occurring porphyric variety contains K-feldspar phenocrysts, up to 1–2 cm big. Accessory minerals are represented by apatite, zircon, magnetite, titanite, and rare allanite.

The Červená granodiorites are medium-grained, equigranular to slightly porphyritic rocks, containing biotite (15–17 vol.%) formed by eastonite (Fe/Fe + Mg 0.44–0.47, Al^{4+} 2.5–2.6 apfu, and Ti 0.28–0.38 apfu), plagioclase (An_{22-40}) (28–41 vol.%), quartz (22–23 vol.%), K-feldspar (9–19 vol.%), and 1–2 vol.% hornblende (magnesiohornblende, actinolite) (**Figure 3**). Accessory minerals are represented by apatite, zircon, titanite, magnetite, and allanite. The Červená granodiorites display, in some cases, a strong planar fabric.

4.2 Chemical composition of the Blatná suite

The representative chemical analyses of the Blatná suite are presented in **Table 1**. These granodiorites are high-K, calc-alkaline to shoshonitic, and metaluminous to slightly peraluminous rocks (A/CNK = mol. $Al_2O_3/(CaO + Na_2O + K_2O)$) = 0.8–1.2. Compared to the common I-type granites [10, 11], these granodiorites are enriched in

Figure 2.
Microphotograph of the Blatná granodiorite (Bt, biotite; Kfs, K-feldspar; Pl, plagioclase; Qz, quartz). Crossed polarizers.

Figure 3.
Microphotograph of the Červená granodiorite (Bt, biotite; Kfs, K-feldspar; Pl, plagioclase; Qz, quartz). Crossed polarizers.

Mg (1.0–3.4 wt.% MgO), Ba (838–2560 ppm), Sr. (257–506 ppm), Zr (81–236 ppm), Th (2–32 ppm), and U (2–15 ppm).

4.3 Accessory minerals association and textures

The REE-, Zr-, and Y-bearing accessories in granodiorites of the Blatná suite are represented by apatite, zircon, and relatively rare occurred titanite and allanite. Apatite occurs usually in form of euhedral and subhedral grains (20–50 μm, rarely up to 100–120 μm) (**Figure 4**). Zircon usually occurs as small euhedral and subhedral grains (10–15 μm, rarely 50–70 μm). Both minerals are usually enclosed in biotite flakes. Apatite and zircon are sometimes zoned (**Figures 5** and **6**). Titanite occurs in relatively bigger, 100–200 μm, subhedral to euhedral grains (**Figure 7**). Allanite grains are 200–600 μm large, usually euhedral, interstitially grown between major mineral phases in the granodiorite groundmass (**Figure 8**). Its grains sometimes exhibit complex growth/alteration textures (**Figure 9**).

4.4 Apatite composition

All analyzed apatites contain more F (3.0–4.5 wt.%) and less Cl (0.0–0.2 wt.%) (**Table 2**). Their content of Fe (0.02–0.27 wt.% FeO) and Mn (0.02–0.12 wt.% MnO) are low. Their contents of sulfur and natrium are also low (0.01–0.28 wt.% SO_3, 0.01–0.09 wt.% Na_2O). The concentrations of U and Th in analyzed apatite are low (0.01–0.09 wt.% UO_2, 0.01–0.05 wt.% ThO_2). The concentrations of REE and Y are usually under microprobe detection limits. In analyzed apatite, grains were found La/Y ratios from 0.0 to 0.4. Zonation of analyzed apatites, that could be observed in the backscattered images (BSE) is very rare and is coupled with different concentrations of Y and REE. Light parties of analyzed apatite grain are enriched in Y (0.31 wt. % Y_2O_3) and REE (0.13 wt.% Ce_2O_3) (**Figure 5**).

Sample	R-704	R-708	R-782	R-794
Locality	Nahošín	Nahošín	Mečichov	Mečichov
Rock wt %	Biotite granodiorite	Biotite granodiorite	Amphibole-biotite granodiorite	Amphibole-biotite granodiorite
SiO_2	69.52	68.50	63.38	63.26
TiO_2	0.60	0.64	0.71	0.67
Al_2O_3	15.66	15.59	16.02	16.28
Fe_2O_3	0.01	0.26	0.80	0.83
FeO	2.08	2.45	3.55	3.46
MnO	0.04	0.05	0.07	0.07
MgO	1.27	1.47	2.65	2.65
CaO	1.98	2.40	3.25	3.45
Na_2O	3.26	3.12	3.06	3.25
K_2O	4.20	3.86	4.11	3.89
P_2O_5	0.30	0.30	0.26	0.22
H_2O^+	0.83	0.72	0.86	0.64
H_2O^-	0.00	0.02	0.20	0.00
CO_2	0.14	0.19	0.14	0.00
Total	99.89	99.57	99.06	98.67
A/CNK	1.26	1.14	1.04	1.03
ppm				
Ba	1165	1167	1562	1284
Rb	161	160	101	105
Sr	410	421	466	261
Nb	11	9	11	14
Zr	150	170	223	195
Y	23	21	29	29
U	9	9	8	8
Th	18	21	15	13

Table 1.
Representative chemical analyses of granodiorites.

4.5 Zircon composition

The analyzed zircons contain low Hf concentrations (1.1–1.7 wt.% HfO_2) (**Table 3**). The proportion of the hafnium end member indicated by atomic ratio Hf/(Zr + Hf) varies from 0.010 to 0.015 (**Figure 10**). The concentration of Y in analyzed zircon is partly variable and varies from 0.01 to 0.43 wt.% Y_2O_3. All analyzed zircons display lower concentrations of U (0.03–0.49 wt.% UO_2) and Th (0.01–0.19 wt.% ThO_2).

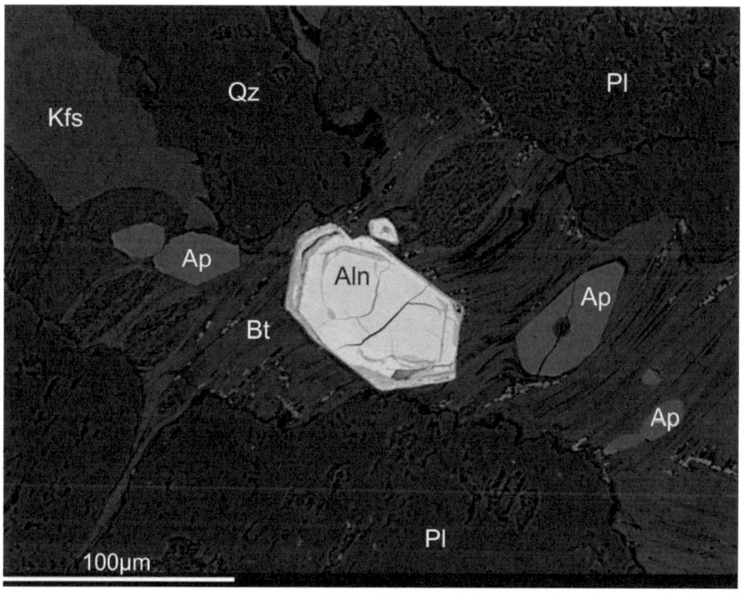

Figure 4.
Back-scattered electron (BSE) image of allanite from the Blatná granodiorite (Aln, allanite; Ap, apatite; Bt, biotite; Kfs, K-feldspar; Pl, plagioclase; Qz, quartz).

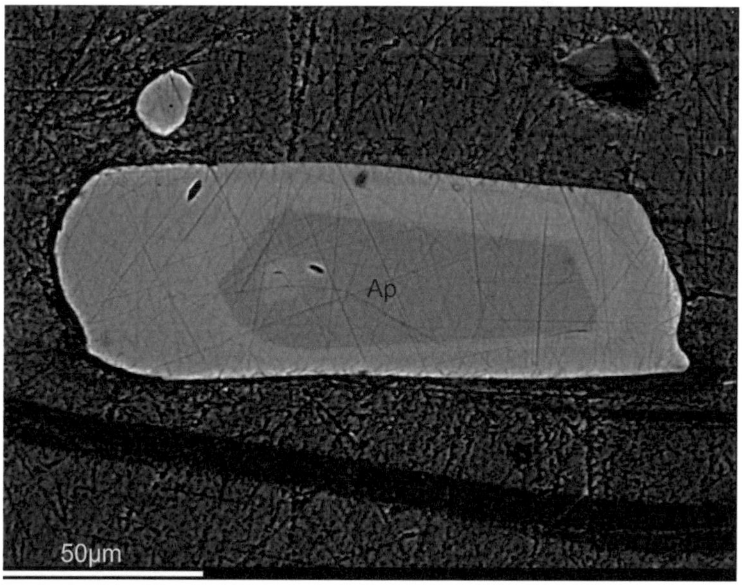

Figure 5.
BSE image of zoned apatite from the Červená granodiorite.

4.6 Titanite composition

The composition of titanite ranges from 83 to 92 mol.% titanite end-member (**Table 4**). The Al and Fe^{3+} contents range from 0.06 to 0.16 apfu and from 0.01 to

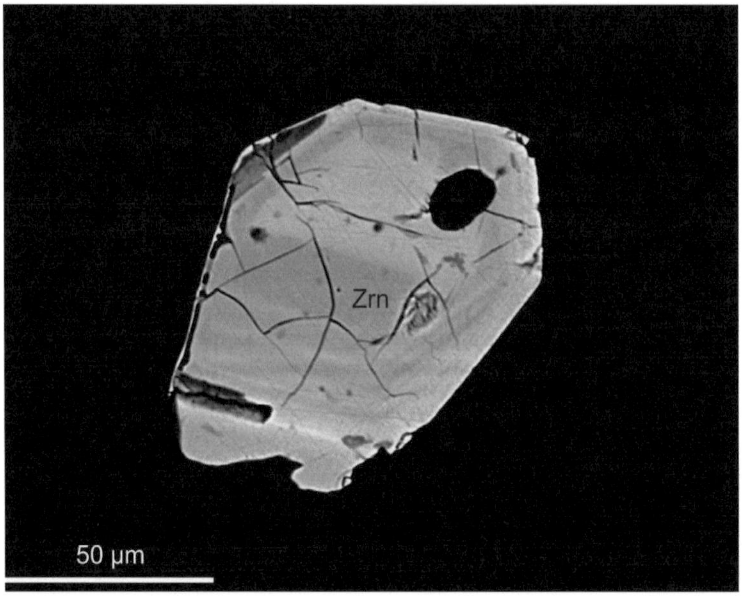

Figure 6.
BSE image of zoned zircon from the Červená granodiorite.

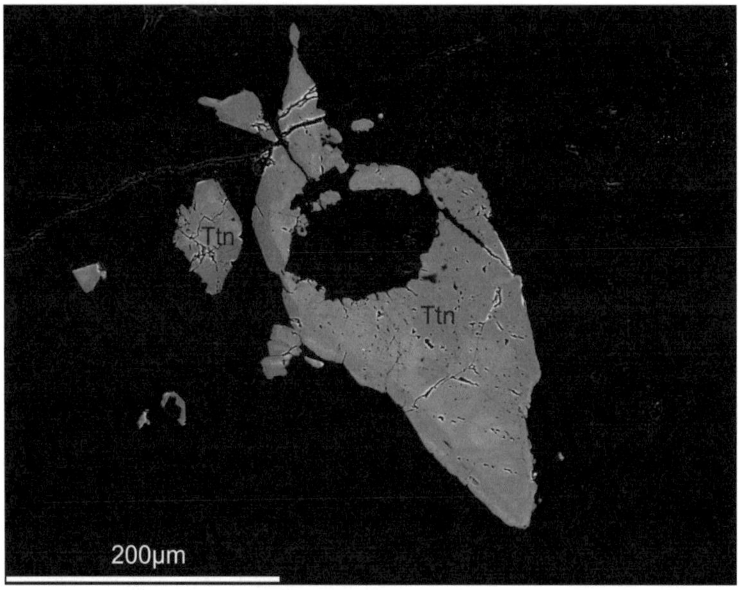

Figure 7.
BSE image of titanite (Ttn) from the Červená granodiorite.

0.04 apfu, respectively. Analyzed titanites show Al + Fe^{3+} excess over F. This excess indicates the occurrence of the coupled substitution of Al + Fe^3 (**Figure 11**). The content of (Al + Fe^{3+})-F component ranges from 6 to 13 mol.%. The content of (Al + Fe^{3+})-OH component is lower and ranges from 0 to 5 mol.%. The content of REE in analyzed titanites is usually low, however the content of Ce_2O_3 ranges from 0.11 to 0.33 wt.% and content of Nd_2 O_3 ranges from 0.02 to 0.30 wt.%.

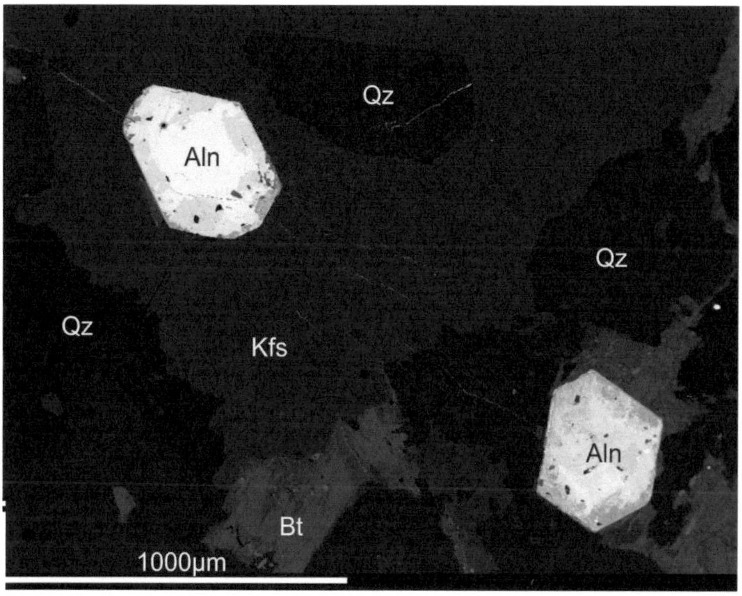

Figure 8.
BSE image of allanite from the Blatná granodiorite (Aln, allanite; Bt, biotite; Kfs, K-feldspar; Qz, quartz).

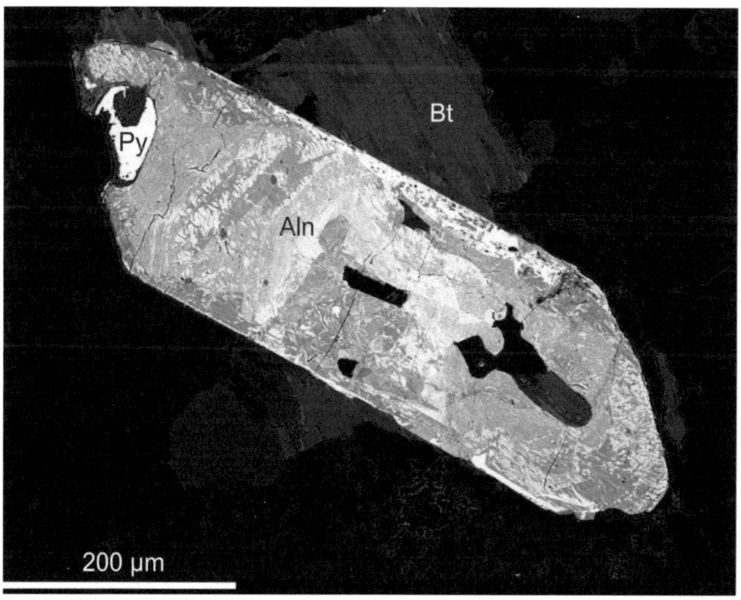

Figure 9.
BSE image of altered allanite from the Blatná granodiorite (Aln, allanite; Bt, biotite; Py, pyrite).

4.7 Allanite composition

The originally magmatic allanite contains 31.4–32.4 wt.% SiO_2, 10.3–13.3 wt.% CaO, 9.8–15.1 wt.% FeO, 0.1–2.7 wt.% ThO_2, and 17.5–23.2 wt.% REE_2O_3 (**Table 5**). Analyzed allanites are relatively enriched by Mn, containing 0.23–0.65 wt.% MnO. All analyzed allanites display variable distribution of REE, with the preference of Ce over

Sample	R-704-5	R-712-44-	R-782-6	R-782-7	R-981-10
Locality	**Nahošín**	**Nahošín**	**Mečichov**	**Mečichov**	**Mečichov**
Rock wt.%	**Blatná**	**Blatná**	**Červená**	**Červená**	**Červená**
SO_3	0.03	0.01	0.02	b.d.l.	b.d.l.
P_2O_5	40.95	41.40	40.54	40.89	40.67
SiO_2	0.81	0.48	0.07	0.45	0.12
La_2O_3	b.d.l.	0.08	b.d.l.	0.02	0.04
Ce_2O_3	0.18	0.26	0.02	0.13	0.04
Pr_2O_3	0.14	0.13	b.d.l.	b.d.l.	b.d.l.
Nd_2O_3	0.16	0.27	0.03	0.12	0.08
Sm_2O_3	0.17	0.11	0.03	0.06	b.d.l.
Gd_2O_3	0.15	0.07	0.04	0.13	b.d.l.
Dy_2O_3	0.07	0.08	0.02	0.01	b.d.l.
Er_2O_3	0.05	0.05	0.05	0.06	b.d.l.
Y_2O_3	0.63	0.38	0.05	0.31	0.08
ThO_2	0.01	0.03	0.04	b.d.l.	b.d.l.
UO_2	0.04	0.07	b.d.l.	b.d.l.	b.d.l.
CaO	53.74	53.82	55.55	54.50	56.06
FeO	0.13	0.13	0.12	0.19	0.01
MnO	0.02	0.08	0.04	0.07	0.03
SrO	b.d.l.	b.d.l.	0.02	0.04	b.d.l.
Na_2O	0.05	0.02	b.d.l.	b.d.l.	0.08
SO_3	0.03	0.01	0.02	b.d.l.	b.d.l.
F	3.71	3.57	3.14	3.09	3.46
Cl	0.02	0.02	0.08	0.05	0.14
O=F,Cl	1.57	1.51	1.34	1.31	1.49
Total					
X_{Fap}	0.985	0.948	0.834	0.820	0.919
X_{Clap}	0.003	0.003	0.012	0.007	0.021
X_{OHap}	0.012	0.049	0.154	0.173	0.060

b.d.l., below detection limit.

Table 2.
Representative microprobe analyses of apatite.

La. Cerium is thus the predominant lanthanide, thus these allanites could be classified as allanite-(Ce). On the plot proposed by Petrík et al. [12], the analyzed allanites are located between allanite and ferriallanite (**Figure 12**). The Al values range between 1.44 and 2.12 apfu. It can be also observed, that the individual points are located between isolines 0.2 and 0.5 $Fe_{ox.} = Fe^{3+}/(Fe^{3+} + Fe^{2+})$. The values of these points calculated by the method Armbruster et al. [13] are partly lower (0.18–0.43).

Sample	R-704-1	R-704-2	R-704-15	R-782-3	R-977-15
Locality	**Nahošín**	**Nahošín**	**Nahošín**	**Mečichov**	**Mečichov**
Rock wt.%	**Blatná**	**Blatná**	**Červená**	**Červená**	**Červená**
SiO_2	32.51	32.34	32.63	32.47	32.23
Al_2O_3	0.01	b.d.l.	b.d.l.	b.d.l.	0.01
ZrO_2	64.92	64.15	64.61	66.36	65.82
HfO_2	1.22	1.20	1.47	1.23	1.39
CaO	0.04	0.06	0.02	0.03	0.01
FeO	b.d.l.	0.11	0.24	0.05	0.32
P_2O_5	0.11	0.10	0.06	0.01	b.d.l.
Sc_2O_3	0.01	0.03	0.04	0.04	b.d.l.
Y_2O_3	0.23	0.37	0.21	b.d.l.	b.d.l.
La_2O_3	0.04	0.01	b.d.l.	b.d.l.	b.d.l.
Ce_2O_3	0.05	0.05	b.d.l.	0.06	b.d.l.
Dy_2O_3	0.05	0.10	0.05	0.01	0.01
Er_2O_3	0.04	0.10	0.02	0.06	0.06
Yb_2O_3	0.09	0.05	0.01	0.05	0.01
UO_2	0.20	0.25	0.23	0.19	0.18
ThO_2	0.02	0.11	0.12	0.07	0.04
Total	99.54	99.03	99.71	100.63	100.08
apfu, O = 4					
Si	1.002	1.003	1.005	0.994	0.993
Al	0.000	0.000	0.000	0.000	0.000
Zr	0.970	0.970	0.970	0.990	0.988
Hf	0.011	0.011	0.013	0.011	0.012
Ca	0.001	0.002	0.001	0.001	0.000
Fe	0.000	0.003	0.006	0.001	0.008
P	0.003	0.003	0.002	0.000	0.000
Sc	0.000	0.001	0.001	0.001	0.000
Y	0.004	0.006	0.003	0.000	0.000
La	0,000	0.000	0.000	0.000	0.000
Ce	0.001	0.001	0.000	0.001	0.000
Dy	0.000	0.001	0.000	0.000	0.000
Er	0.000	0.001	0.000	0.001	0.001
Yb	0.001	0.000	0.000	0.000	0.000
U	0.001	0.002	0.002	0.001	0.001
Th	0.000	0.001	0.001	0.000	0.000

b.d.l., below detection limit.

Table 3.
Representative microprobe analyses of zircon.

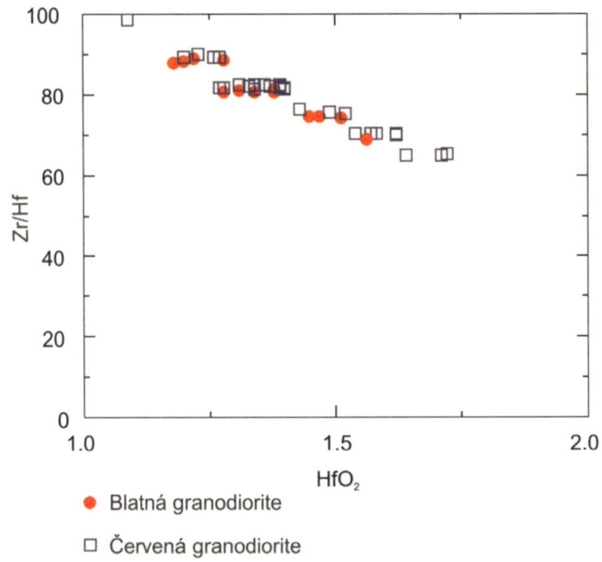

Figure 10.
Chemical composition of zircon from granodiorites of the Blatná suite.

The altered allanites are enriched in Si (40.6–42.0 wt.% SiO_2 and Th (3.1–6.4 wt.% ThO_2), depleted in Fe (2.2–4.2 wt.% FeO) and Ca (5.8–8.6 wt.% CaO). The altered allanites display also a lower total analytical sum, which indicate postmagmatic alterations.

5. Discussion

5.1 Fractionation of the Blatná suite

The high-K calc-alkaline to shoshonitic granitic rocks of the Blatná suite could be classified as hybrid H-granites in the sense of granite classification proposed by Castro et al. [14]. According to the primary Rb/Sr. ratio and the Nd-isotope ratios, the origin of this suite could be coupled either by mixing of different magmas with distinct isotopic features and/or by crustal contamination of more basic magmas. According to some other interpretation, granitic rocks of the Blatná suite are products of fractionation mantle-derived magmas and their mixing with relatively heated metamorphic rocks of the Moldanubian Zone [15]. The recently preferred explanation of the Blatná suite origin is coupled with remelting of a heterogeneous earth crust composed of immature greywackes rich in the Cambrian volcanogenic detritus [2]. An additional important process was variable mixing with slightly enriched mantle-derived monzonitic magmas, which also may have supplied the extra heat needed for the crustal anatexis [1].

5.2 Substitution in apatite

There are systematic and distinctive differences in Fe, Mn, REE, F, and Cl contents in apatites from I- and S-type granitic rocks [7, 16]. For I-type, granitic rocks are

DOI: http://dx.doi.org/10.5772/intechopen.102628

Sample	R-704-10	R-704-12	R-712-52	R-782-1	R-782-10
Locality	Nahošín	Nahošín	Nahošín	Mečichov	Mečichov
Rock wt.%	Blatná	Blatná	Blatná	Červená	Červená
SiO_2	30.38	30.63	30.36	31.09	30.93
TiO_2	37.42	36.23	36.70	35.60	34.68
Nb_2O_5	0.42	0.24	0.10	0.31	0.05
Ta_2O_5	0.03	b.d.l.	0.11	0.08	0.03
Al_2O_3	1.50	2.07	2.11	2.36	4.30
FeO	0.64	0.87	0.62	0.42	0.52
MnO	0.09	0.07	0.07	0.07	0.02
CaO	28.68	28.77	29.26	28.15	29.05
MgO	b.d.l.	0.01	0.01	b.d.l.	0.03
Na_2O	0.02	0.03	0.01	0.02	0.02
K_2O	0.02	b.d.l.	b.d.l.	b.d.l.	0.02
La_2O_3	b.d.l.	b.d.l.	b.d.l.	b.d.l.	b.d.l.
Ce_2O_3	0.18	0.12	0.16	0.20	0.14
Pr_2O_3	b.d.l.	0.02	0.01	b.d.l.	0.08
Nd_2O_3	0.11	0.03	b.d.l.	0.13	b.d.l.
Y_2O_3	0.09	0.09	0.11	0.28	0.02
ZrO_2	0.03	0.02	b.d.l.	b.d.l.	b.d.l.
ThO_2	0.05	b.d.l.	b.d.l.	0.06	b.d.l.
V_2O_3	0.39	0.17	0.24	0.25	0.19
F	0.71	0.84	0.63	0.80	1.33
O=F	0.30	0.35	0.27	0.34	0.56
Total	100.83	100.37	100.57	99.87	101.47
Si	1.000	1.000	1.000	1.000	1.000
Ti	0.926	0.890	0.909	0.861	0.843
Nb	0.006	0.004	0.001	0.005	0.001
Ta	0.000	0.000	0.001	0.001	0.000
Al	0.058	0.080	0.082	0.089	0.164
Fe^{3+}	0.018	0.024	0.017	0.011	0.014
Mn	0.003	0.002	0.002	0.002	0.001
Ca	1.011	1.006	1.033	0.970	1.006
Mg	0.000	0.000	0.000	0.000	0.001
Na	0.001	0.002	0.001	0.001	0.001
K	0.001	0.000	0.000	0.000	0.001
La	0.000	0.000	0.000	0.000	0.000
Ce	0.002	0.002	0.002	0.002	0.002
Pr	0.000	0.000	0.000	0.000	0.001

Sample	R-704-10	R-704-12	R-712-52	R-782-1	R-782-10
Locality	**Nahošín**	**Nahošín**	**Nahošín**	**Mečichov**	**Mečichov**
Rock wt.%	**Blatná**	**Blatná**	**Blatná**	**Červená**	**Červená**
Nd	0.001	0.000	0.000	0.001	0.000
Y	0.002	0.002	0.002	0.005	0.000
Zr	0.000	0.000	0.000	0.000	0.000
Th	0.000	0.000	0.000	0.000	0.000
V	0.010	0.004	0.006	0.006	0.005
F	0.074	0.087	0.066	0.081	0.136
OH	0.002	0.017	0.033	0.019	0.042
X(Ttn)	0.92415	0.89537	0.90179	0.89594	0.82566
X(Al,Fe^{3+}-F)	0.07385	0.08753	0.06548	0.08429	0.13320
X(Al,Fe^{3+}-OH)	0.00200	0.01710	0.03274	0.01977	0.04114

b.d.l., below detection limit.

Table 4.
Representative microprobe analyses of titanite.

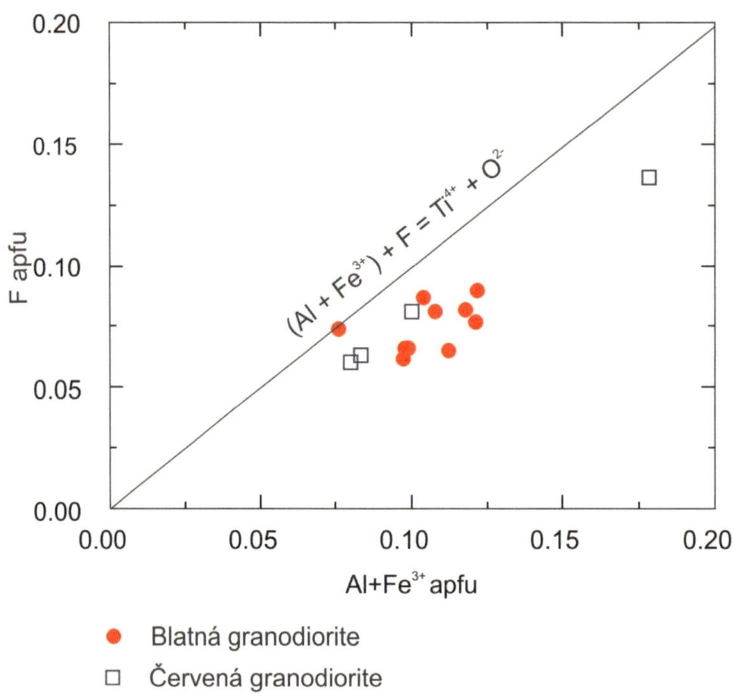

Figure 11.
Chemical composition of titanite from granodiorites of the Blatná suite.

significant a lower content of Fe and Mn, higher contents of LREE, and lower content of Y [17]. The content of Fe in analyzed apatites is 0.02–0.27 wt.% FeO and content of Mn in these apatites is 0.03–0.12 wt.% MnO. The increase of Mn content in apatites

Sample	**R-704-27**	**R-704-28**	**R-704-35**	**R-977-26**
Locality	**Nahošín**	**Nahošín**	**Nahošín**	**Mečichov**
Rock wt.%	**Blatná**	**Blatná**	**Blatná**	**Červená**
SiO_2	31.44	31.43	31.85	32.53
TiO_2	1.10	1.18	1.64	0.39
Al_2O_3	13.80	13.64	15.19	19.59
FeO	15.14	15.03	12.85	9.77
MnO	0.53	0.41	0.47	0.35
MgO	1.45	1.41	1.15	0.68
CaO	10.74	10.93	11.05	13.83
Na_2O	0.06	0.01	0.06	b.d.l.
La_2O_3	6.18	6.39	3.04	3.32
Ce_2O_3	11.49	11.58	9.83	8.65
Pr_2O_3	1.14	1.14	1.46	1.02
Nd_2O_3	3.56	3.66	5.95	4.42
Sm_2O_3	0.31	0.29	0.91	0.62
Gd_2O_3	0.21	0.08	0.53	0.42
Tb_2O_3	b.d.l.	0.05	0.08	0.04
Y_2O_3	0.16	0.13	0.52	0.31
ThO_2	0.80	0.82	1.93	0.06
UO_2	0.04	0.04	0.09	0.03
F	0.25	0.21	0.17	0.08
O=F	0.11	0.09	0.07	0.03
Total	98.30	98.34	98.70	96.22
apfu, O = 12.5				
Si	2.992	2.998	3018	2.986
Al(IV)	0.008	0.002	0.,000	0.014
Total-T-site	3.000	3.000	3.018	3.000
Ti	0.079	0.085	0.117	0.027
Al(VI)	1.540	1.531	1.696	2.106
Fe^{3+}	0.482	0.452	0.184	0.216
Fe^{2+}	0.693	0.731	0.834	0.534
Mn^{2+}	0.000	0.000	0.006	0.024
Mg	0.206	0.201	0.162	0.093
Total-M-site	3.000	3.000	3.000	3.000
Mn^{2+}	0.043	0.033	0.032	0.003
Fe^{2+}	0.030	0.016	0.000	0.000
Ca	1.095	1.117	1.122	1.360
Na	0.011	0.002	0.011	0.000

Sample	R-704-27	R-704-28	R-704-35	R-977-26
Locality	**Nahošín**	**Nahošín**	**Nahošín**	**Mečichov**
Rock wt.%	**Blatná**	**Blatná**	**Blatná**	**Červená**
REE	0.803	0.814	0.774	0.635
Th	0.017	0.018	0.042	0.001
U	0.001	0.001	0.002	0.001
Total-A-site	2.000	2.000	1.982	2.000
F	0.075	0.063	0.051	0.023
$F_{ox.}$	0.40	0.38	0.18	0.29

b.d.l., below detection limit.

Table 5.
Representative microprobe analyses of allanite.

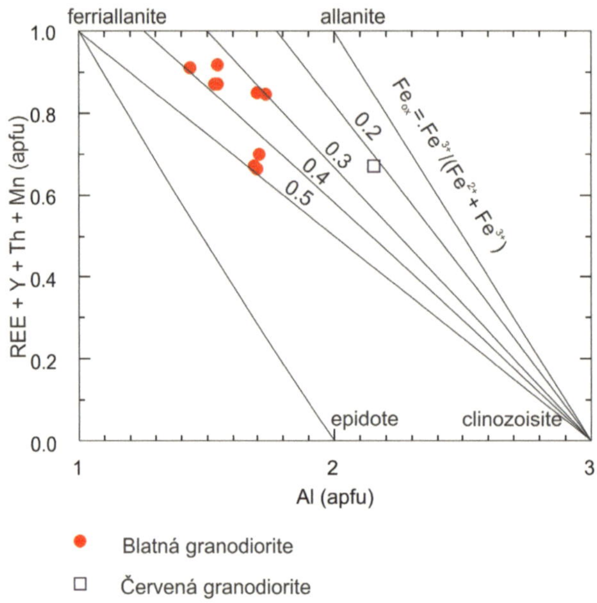

Figure 12.
The plot of total REE + Y + Th + Mn vs. Al contoured with isolines of the ratio $F_{ox.} = Fe^{3+}/(Fe^{2+} + Fe^{3+})$ illustrating the chemical relationships in the system allanite-ferriallanite-epidote-clinozoisite according Petrík et al. [12].

from S-type granitic rocks is a function of an increase of Mn/Fe and Mn/Ca ratios with fractionation [16]. The La/Y ratio in analyzed apatites is 0.01–0.40. This ratio is partly comparable with the La/Y ratio for I-type granitic rocks (0.2–3.25) according to Sha and Chappell [17].

5.3 Substitution in zircon

The most common trace element in zircon is hafnium. The HfO_2 contents in granitic rocks usually range from 0.5 to 9.9 wt.%, with a median of 1.5 wt.% [18]. In similar granitic rocks from the Bohemian Massif, their contents usually range between 0.8 and 2.2 wt.% HfO_2 [19, 20]. The content of Y in zircon from granitic rocks is usually 0.2–0.7 wt.% Y_2O_3 [21]. In similar, biotite granites that form the Moldanubian batholith zircon contains 0.1–0.9 wt.% Y_2O_3 [19]. The content of ThO_2 in F-low biotite granites from the Krušné Hory/Erzgebirge batholith is partly higher (up to 1.3 wt.% ThO_2) [20], whereas its content in zircon from two-mica granites of the Moldanubian batholith is similar (0.01–0.2 wt.% ThO_2) [19].

5.4 Substitution in titanite

Titanite is, according to the variability of its chemical composition, suggested as a highly sensitive indicator of oxygen and water fugacity [22–24]. The chemical composition of analyzed titanite shows that the substitution (Al, Fe^{3+}) + F = Ti^{4+} O^{2-} is the most significant in analyzed titanites. According to their F, Al, and Fe^{3+} concentrations, the analyzed titanites could be considered as low-Al titanites, according to Oberti et al. [25]. Their low F and Al content could be well compared with the contents of both elements in similar magmatic titanites [26].

5.5 Substitution in allanite

For allanite, two main substitutions occur, namely the epidote-allanite and the allanite-ferriallanite substitutions [27, 28]. For analyzed allanites from granodiorites of the Blatná suite, the allanite-ferriallanite substitution is significant. In the other Variscan granitic rocks from the Bohemian Massif allanite was found in some granites and granodiorites of the Moldanubian batholith [29] and from lamprophyres of the Krkonoše-Jizera composite pluton [30]. Allanites from both magmatic bodies display similar chemical compositions and also similar values of $Fe_{ox.}$ = 0.3–0.5.

6. Conclusions

The granodiorites of the Blatná suite contain 12–18 vol.% of biotite, 28–42 vol.% plagioclase, 22–28 vol.% quartz, 9–19 vol.% K-feldspar, and 0.2–1.2 vol.% hornblende. These granodiorites are high-K, calc-alkaline to shoshonitic rocks.

The REE-, Zr-, and Y-bearing accessories in granodiorites of the Blatná suite are represented by apatite, zircon, and relatively rare occurred titanite and allanite. All analyzed apatites contain more F (3.0–4.5 wt.%) and less Cl (0.0–0.2 wt.%). Apatite zonation is very rare and coupled with different concentrations of Y and REE. The analyzed zircons contain low Hf concentrations (1.1–1.7 wt.% HfO_2). The composition of analyzed titanite ranges from 83 to 92 mol.% titanite end-member. The analyzed allanites display variable distribution of REE, with the preference of Ce over La. Allanite is relatively Al-poor and displays $Fe_{ox.}$ ratio 0.2–0.5.

Acknowledgements

This study was carried out thanks to the support of the long-term conceptual development research organization RVO 67985891. I am grateful to R. Škoda, R. Čopjaková, and J. Haifler from the Department of geological sciences of Masaryk

University for technical assistance by electron microprobe analyses of selected minerals (allanite, apatite, titanite, and zircon).

Conflict of interest

The author declares no conflict of interest.

Author details

Miloš René
Institute of Rock Structure and Mechanics, v.v.i., Czech Academy of Sciences, Prague, Czech Republic

*Address all correspondence to: rene@irsm.cas.cz

References

[1] Janoušek V, Bowes DR, Rogers G, Farrow CM, Jelínek E. Modelling diverse processes in the petrogenesis of a composite pluton. The Central Bohemian Pluton, Central European Variscides. Journal of Petrology. 2000;**41**:511-543. DOI: 10.1093/petrology/41.4.511

[2] Janoušek V, Wiegand BA, Žák J. Dating the offset of Variscan crustal exhumation in the core of the Bohemian Massif: New U/Pb single zircon ages from the high-K calc-alkaline granodiorites of the Blatná suite, Central Bohemian Plutonic Complex. Journal of Geological Society. 2010;**167**: 347-360. DOI: 10.1144/0016-7649 2009-008

[3] Žák J, Verner K, Holub FV, Kabele P, Chlupáčová M, Halodová P. Magmatic to solid state fabrics in syntectonic granitoids recording early Carboniferous orogenic collapse in the Bohemian Massif. Journal of Structural Geology. 2012;**36**:27-42. DOI: 10.1016/j. jsg.2011.12.011

[4] Habásko J, Litochleb J, Pletánek Z. The new findings about uranium mineralization in granitoids of the SW part of the Central Bohemian Pluton. In: Sborník Symposia Hornická Příbram ve vědě a technice. Příbram: Geologie; 1980. pp. 67-87 (in Czech)

[5] Litochleb J, Kotlovský P. Geological building and mineralization of the Nahošín uranium deposit. In: Sborník symposia Hornická Příbram ve vědě a technice. Příbram: ČSUP; 1988. pp. 91-101 (in Czech)

[6] Pouchou JL, Pichoir F. PAP (φ-ρ-Z) procedure for improved quantitative microanalysis. In: Armstrong JT, editor. Microbeam Analysis. San Francisco: San Francisco Press; 1985. pp. 104-106

[7] Piccoli PM, Candela PA. Apatite in igneous systems. Reviews in Mineralogy and Geochemistry. 2002;**48**:285-292. DOI: 10.2138/rmg.2002.48.6

[8] Harlov D, Tropper P, Seifert W, Nijland T, Förster HJ. Formation of Al-rich titanite ($CaTiSiO_4O$-$CaAlSiO_4OH$) reaction rims on ilmenite in metamorphic rocks as a function of fH_2O and fO_2. Lithos. 2006;**88**:72-84. DOI: 10.1006/j.Lithos.2005.08.00

[9] Yavuz F, Yildirim DK. A windows program for calculation and classification of epidote-supergroup mineral. Periodico di Mineralogia. 2018; **87**:269-285. DOI: 10.2451/2018/PM7808

[10] Collins WJ, Beams SD, White AJR, Chappell BW. Nature and origin of A-type granites with particular reference to Southeastern Australia. Contributions to Mineralogy and Petrology. 1982;**80**: 189-200. DOI: 10.1007/BF00374895

[11] Whalen JB, Currie KI, Chappell BW. A-type granites: Geochemical characteristics discrimination and petrogenesis. Contributions to Mineralogy and Petrology. 1987;**95**: 407-419. DOI: 10.1007s/BF00402202

[12] Petrík I, Broska I, Lipka J, Siman P. Granitoid allanite-(Ce): Substitution relations, redox conditions and REE distribution (on example of I-type granitoids, Western Carpathians, Slovakia). Geologica Carpathica. 1995; **46**:79-94

[13] Armbruster T, Bonazzi P, Akasaka M, Bermanec V, Chopin C, Gierė R, et al. Recommended nomenclature of epidote-group minerals. European Journal of Mineralogy. 2006; **18**:551-567. DOI: 10.11027/0935-1221/ 2006/0018-0551

[14] Castro A, Moreno-Ventas I, de la Rosa JD. H-type (hybrid granitoids): A proposed revision of the granite-type classification and nomenclature. Earth Science Reviews. 1991;**31**:237-253. DOI: 10.1016/0012-8252(91)90020-G

[15] René M. Granitoids as a result of mixing of mantle-derived rocks with metasediments from upper crust. In: Terra Nova, Abstracts Supplement 1. Oxford: Oxford University Press; 1997. p. 466

[16] Broska I, Williams T, Uher P, Konečný P, Leichmann J. The geochemistry of phosphorus in different granite suites of the Western Carpathians, Slovakia: The role of apatite and P-bearing feldspar. Chemical Geology. 2004;**205**:1-15. DOI: 10.1019/j.chemgeo.2003.09.004

[17] Sha LK, Chappell BW. Apatite chemical composition, determined by electron microprobe and laser-ablation inductive plasma mass spectrometry, as a probe into granite petrogenesis. Geochimica et Cosmochimica Acta. 1999;**63**:3861-3881. DOI: 10.1016/s0016-7037(99)00210-0

[18] Wang X, Griffin WI, Chen J. Hf contents and Zr/Hf ratio in granitic zircons. Geochemical Journal. 2010;**44**: 65-72

[19] René M. Occurrence of Th, U, Y, Zr and REE-bearing accessory minerals in granites and their petrogenetic significance. In: Basik M, Bogdaska H, editors. Granite, Occurrence, Mineralogy and Origin. 1st ed. New York: Nova Science Publishers, Inc.; 2012. pp. 27-56

[20] Breiter K. Monazite and zircon as major carriers of Th, U, and Y in peraluminous granites: Examples from the Bohemian Massif. Mineralogy and Petrology. 2016;**110**:767-785. DOI: 10.1007/s710-016-0448-0

[21] Bělousova EA, Griffin WL, O'Reilly SY, Fisher NI. Igneous zircon: Trace element composition as an indicator of source rock type. Contributions to Mineralogy and Petrology. 2002;**143**:602-622. DOI: 10.1007/s00410-002-0364-7

[22] Wones DR. Significance of the assemblage titanite + magnetite + quartz in granitic rocks. American Mineralogist. 1989;**74**:744-749

[23] Piccoli P, Candela P, Rivers M. Interpreting magmatic processes from accessory phases: Titanite—A small scale recorder of large-scale processes. Transactions of the Royal Society of Edinburgh: Earth Sciences. 2000;**91**: 257-267. DOI: 10.1017/s026359330007422

[24] Troitzsch U, Ellis DJ. Thermodynamic properties and stability of AlF-bearing titanite $CaTiSiO_4$–$CaAlFSiO_4$. Contributions to Mineralogy and Petrology. 2002;**142**:543-563. DOI: 10.1007/s004100100309

[25] Oberti R, Smith DC, Rossi G, Caucia F. The crystal chemistry of high-aluminium titanite. European Journal of Mineralogy. 1991;**3**:777-792. DOI: 10.1127/ejm/3/5/0777

[26] Enami M, Suzuki K, Liou JG, Bird DK. Al-Fe^{3+} and F-OH substitutions in titanite and constraints on their P–T dependence. European Journal of Mineralogy. 1993;**5**:219-231. DOI: 10.1127/ejm/5/2/2019

[27] Deer WA, Howie RA, Zussman J. Rock-Forming Minerals: Disilicates and Ringsilicates, Volume 1B. 2nd ed. London: The Geological Society Publishing House; 1986. pp. 1-629

[28] Gierė R, Sorensen SS. Allanite and other REE-rich epidote-group minerals. Reviews in Mineralogy and

Geochemistry. 2004;**56**:431-493. DOI: 10.2138/gsrmg.56.1.431

[29] René M. Allanite from granitic rocks of the Moldanubian batholith (Central European Variscan belt). In: Aide M, editor. Rare Earth Elements and their Minerals. 1st ed. London: IntechOpen Ltd.; 2020. pp. 21-30. DOI: 10.5772/intechopen.86356

[30] Gros K, Slaby E, Jokubauskas P, Sláma J, Kozub-Budzyń G. Allanite geochemical response to hydrothermal alteration by alkaline, low-temperature fluids. Minerals. 2020;**10**:1-30. DOI: 10.3390/min10050392

Chapter 10

Mineralogy of Peralkaline Silicic Volcanics: Information from Kone Volcano, Ethiopian Rift Valley

Dereje Ayalew and Bekele Abebe

Abstract

The presented paper described in detail the mineralogy of silicic peralkaline eruptives from Kone volcano within the Ethiopian rift system, which is formed predominantly by rhyolite with some small occurrence of trachyte. The majority of eruptive rocks in the Kone volcanic area are phenocryst-poor. The studied rocks contain alkali feldspars (anorthoclase and sanidine), quartz, clinopyroxene (hedenbergite), aenigmatite and olivine (fayalite), accompanied by rare Fe-To oxides (ilmenite) and apatite. All these minerals are described in detail. These data are very interesting for all researchers, who study similar eruptive rocks.

Keywords: comendite, Ethiopia, Kone, rift

1. Introduction

The Ethiopian rift valley, forming a part of the East African rift system, is the largest active continental rift on Earth (**Figure 1**). This rift shows transitional character between continental rift and seafloor spreading [1] as it is underlain by thinned crust intruded by mafic dykes. The Ethiopian rift is characterized by numerous volcanic centres most of which are still active as evidenced by strong fumarolic activity and episodes of deformation, such as rapid uplift [2]. Kone is one of these silicic volcanic edifices.

Kone volcano (also previously known as Gariboldi volcano [3, 4]) is composed of a series of silicic cones with a summit caldera. It is located (8.8 N, 39.69 E) at the end of the Ethiopian rift valley, near the junction with Afar depression (**Figure 1**). Kone covers an area of 250 km^2. The caldera is elliptical in shape (5 × 7.5 km wide structure) trending E-W [5, 6]. The rim of the caldera rises about 100 m above the caldera floor. The summit of the volcano has an elevation of 1619 m above mean sea level. Roughly N-S-trending regional normal faults and fissures, forming a part of the Wonji fault belt, cut across the caldera and its flanks, especially the eastern side of the volcano.

Despite their petrological and volcanological interest, there is very little published data on the mineralogy of peralkaline silicic volcanics. Here, we present electron microprobe analyses for comendites of Kone volcano within the Ethiopian rift valley (**Figure 2**). The aim of this paper is to document in detail the phenocryst compositions of cemendites.

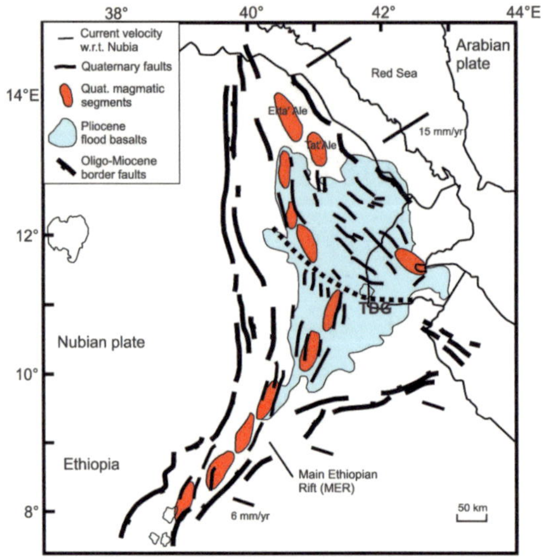

Figure 1.
Location of the active quaternary magmatic segments - zones of dense faulting and aligned eruptive centres which are the current locus of strain within the Ethiopian rift (southern Red Sea and main Ethiopian rift). Arrows show plate motions relative to stable Nubia. Danakil block is a microplate between the Nubian and Arabian plates. TGD is Tendaho-Goba'ad discontinuity marking the active and ancient boundary between the east African and the Red Sea rifts.

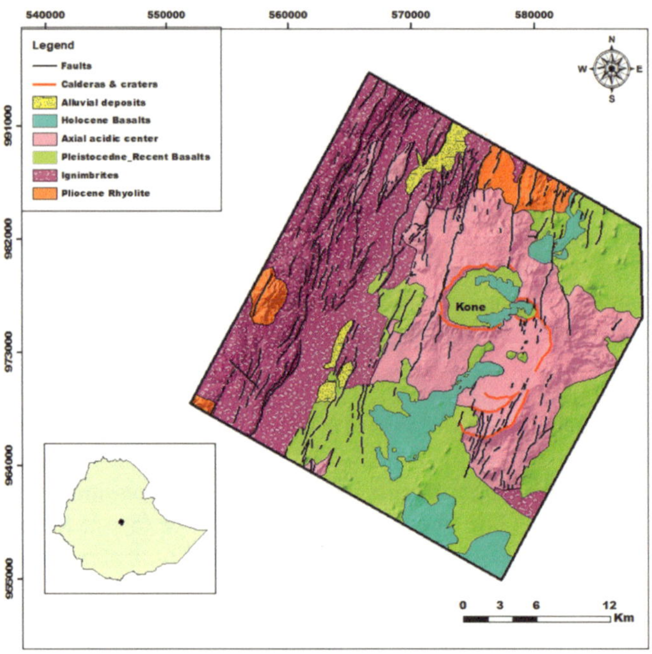

Figure 2.
Simplified geological map of Kone volcano and its surroundings.

2. Geologic background

Volcanic activity commenced around 45 Ma in southern Ethiopia [7], resulting in volumetrically significant basaltic flows and associated rhyolites. Nevertheless, the peak of magmatism had occurred c. ~30 Ma ago, resulting from the impinging Afar mantle plume at the base of the Ethiopian lithosphere and leading to flooding basalt eruptions in Ethiopia and Yemen [8]. The Yemen plateau basalts were united to their Ethiopian counterparts prior to the opening of the Red Sea basin. At ~25 Ma continental rifting commenced in the southern Red Sea [9]. In Southern Ethiopia, extension began ~18 Ma ago [10] and was accompanied by basaltic magmatism, active for about seven to eight million years [7]. The southern Ethiopian rift propagated northward, reaching the present central MER ~14 Ma ago and ultimately joining the southern Red Sea rift ~11 Ma ago [11]. Contemporaneously to the connection between the main Ethiopian and Red Sea rifts, a flood basalt event occurred in this area. Beginning in the late Miocene and continuing throughout the Pliocene, silicic volcanic centres emerged from the rift floor [6]. Progressive weakening of the lithosphere in Afar, associated with heating and the thermomechanical erosion of the lower crust generated by the Afar mantle plume, resulted in the onset of oceanic rifting at 5.3 Ma [9]. Oceanic rifting is still active in Afar, whereas it has not commenced in the MER yet [11].

The Pliocene-Pleistocene boundary was marked by a change in the stress field, giving rise to oblique rifting [12]. Since then, an extension has been localized in narrow (50 km long, maximum 20 km wide) en-echelon arranged segments on the rift floor [6], with a system of bounding faults that are referred to as the Wonji Fault Belt [12]. Moreover, these segments have been the locus of volcanic activity throughout the Quaternary and are thus referred to as magmatic segments by [6]. Volcanic activity associated with the magmatic segment was initially characterized by large volumes of felsic lavas. When these faults reached the upper mantle in recent times, basaltic volcanism commenced [12]. At present, volcanic activity within the magmatic segment is dominated by fissural basalt eruptions [1, 6].

3. Eruptions at Kone volcano

Volcanological investigation reveals that three silicic eruptive episodes have occurred at Kone volcano [3, 4, 13, 14], followed by a basaltic phase (**Figure 2**). The first eruptive phase produced lavas, forming the main topographic expression of the volcano. The second phase of activity formed ignimbrite outflow sheets. The third silicic eruptive phase built widespread pumice deposits, varying in thickness from 50 cm to over 20 m. A separate, recent basalt eruption, erupted during the first half of the 19th century, formed cinder cones and associated lava flows inside and outside the caldera [15].

4. Analytical methods

Phenocryst and glass matrix compositions were determined by electron microprobe at Université de Loraine (Nancy, France). Prior to analyses, samples were coated with a thin film of carbon. Analytical conditions were an accelerating voltage of 15 kV, a probe current of 20 nA, and a beam spot diameter of 5 μm.

5. Mineralogy of Kone silicic eruptives

Kone volcano is predominantly rhyolite in composition with minor trachyte [13]. The majority of kone eruptive rocks are phenocryst-poor with total modal contents of <6%. Given the textural diversity of Kone silicic volcanics (lava, ignimbrite and pumice) we have only selected samples from the first phase (lava) and second phase (ignimbrite) eruptives to illustrate the mineralogy of the silicic melt. The dominant mineral assemblage, with decreasing order of abundance, is alkali feldspar, quartz, clinopyroxene, aenigmatite and olivine, accompanied by rare Fe-Ti oxides and apatite.

Alkali feldspar is overwhelmingly the most abundant phenocryst and forms tabular crystals. Most of the crystals are fragmented. Quartz occurs as rounded microphenocrysts. Clinopyroxene forms euhedral to subhedral pale green crystals. The Na-Fe-Ti silicate aenigmatite is a distinctive phenocryst constituent, recognized by its blood-red to black pleochroism. Olivine occurs as colorless, partly resorbed grains. Fe-Ti oxides form discrete equant crystals, but more commonly occur as inclusions within clinopyroxene. Small euhedral apatite prisms are a common accessory mineral. Quartz and alkali feldspar are common groundmass phases.

Sample	$7C_c$	$7C_r$	11B	5D	4A	4A	$4A_m$	$3A_m$	3A	$3A_m$
SiO_2	67.69	66.64	67.00	67.28	67.78	67.57	67.93	67.32	67.65	67.90
Al_2O_3	18.16	17.85	17.86	19.07	18.69	18.30	18.43	18.30	18.65	17.96
FeO	0.63	1.30	0.97	0.41	0.61	0.42	0.64	0.86	0.47	0.69
CaO	0.00	0.00	0.01	0.04	0.01	0.02	0.04	0.00	0.06	0.00
Na_2O	7.99	6.93	7.10	8.21	7.59	7.85	7.65	7.29	8.29	6.93
K_2O	5.39	7.31	7.03	4.99	5.80	6.19	6.34	6.55	4.80	6.90
Total	99.87	100.03	99.97	99.99	100.49	100.34	101.03	100.31	99.93	100.38
Structural formula (8 oxygen)										
Si	3.02	3.01	3.01	2.99	3.01	3.01	3.01	3.01	3.01	3.03
Al	0.96	0.95	0.95	1.00	0.98	0.96	0.96	0.96	0.98	0.94
Fe	0.02	0.05	0.04	0.02	0.02	0.02	0.02	0.03	0.02	0.03
Ca	0.00	0.00	0.00	0.00	0.00	0.00	0.00	0.00	0.00	0.00
Na	0.69	0.61	0.62	0.71	0.65	0.68	0.66	0.63	0.71	0.60
K	0.31	0.42	0.40	0.28	0.33	0.35	0.36	0.37	0.27	0.39
Total	5.00	5.03	5.02	5.00	4.99	5.02	5.01	5.01	4.99	4.99
%An	0.00	0.00	0.05	0.17	0.07	0.09	0.19	0.02	0.30	0.00
%Ab	69.27	59.02	60.51	71.32	66.50	65.76	64.61	62.84	72.18	60.40
%Or	30.72	40.98	39.44	28.50	33.43	34.15	35.20	37.14	27.52	39.60

The symbol C, R and M represent core, rim and matrix, respectively.

Table 1.
Composition of alkali feldspar from Kone silicic eruptives.

Phenocryst compositions of Kone silicic volcanics are reported in **Tables 1–4**. Compositions of alkali feldspar span the anorthoclase-sanidine boundary (**Figure 3a**). There is no notable compositional variation between individual samples (**Table 1**). Zoning is uncommon. Where present is usually normal, whereby the rim is richer in K_2O than the core (**Table 1**). Microlites of alkali feldspar have slightly more potassic composition than the alkali feldspar phenocrysts. It is noteworthy that analyses of alkali feldspar matrix are broadly comparable to those of alkali feldspar phenocryst rim. Clinopyroxene is generally hedenbergite (**Figure 3b**). Hedenbergite is essentially unzoned (**Table 2**). Aenigmatite shows very restricted variation in TiO_2, FeO and Na_2O (**Table 3**). Olivine appears almost pure fayalite and

Sample	**7C**	**11B**	**5D**	**4A**	**3A**
SiO_2	50.08	49.09	47.64	48.73	49.26
Al_2O_3	0.18	0.11	0.13	0.10	0.13
TiO_2	0.41	0.34	0.44	0.21	0.23
FeO	28.46	28.95	29.13	28.26	28.87
MnO	1.29	1.13	1.69	1.48	1.33
MgO	0.40	0.69	1.06	1.77	0.70
CaO	15.44	15.87	18.39	17.79	16.19
Na_2O	3.65	3.30	1.51	1.44	2.86
K_2O	0.00	0.00	0.00	0.00	0.00
TOTAL	100.18	99.47	100.02	99.86	99.59
Structural formula (6 oxygen)					
Si	2.01	1.98	1.94	1.98	2.00
Al^{IV}	0.00	0.01	0.01	0.00	0.00
T	2.01	1.99	1.95	1.99	2.00
Al^{VI}	0.01	0.00	0.00	0.00	0.00
Fe^{3+}	0.24	0.27	0.20	0.13	0.21
Ti	0.01	0.01	0.01	0.01	0.01
Mg	0.02	0.04	0.06	0.11	0.04
Fe^{2+}	0.71	0.71	0.79	0.82	0.76
Mn	0.04	0.04	0.06	0.05	0.05
M1	1.05	1.07	1.13	1.13	1.07
Ca	0.66	0.69	0.80	0.77	0.70
Na	0.28	0.26	0.12	0.11	0.22
K	0.00	0.00	0.00	0.00	0.00
T + M1 + M2	4.00	4.00	4.00	4.00	4.00
%En	1.43	2.37	3.36	5.67	2.39
%Fs	59.15	58.21	54.74	53.38	57.84
%Wo	39.42	39.42	41.90	40.95	39.77

Table 2.
Composition of clinopyroxene in Kone silicic volcanics.

Sample	7C	11B	$3A_C$	$3A_R$	$3A_M$
SiO_2	41.20	41.29	40.56	51.406	50.777
Al_2O_3	0.36	0.19	0.4673	0.1813	0.0887
TiO_2	8.56	8.73	8.8818	2.2242	1.3667
FeO	41.18	41.02	41.664	31.074	29.278
MnO	1.42	1.07	1.1592	1.231	1.9151
MgO	0.23	0.30	0.238	0.5307	1.4578
CaO	0.29	0.20	0.3953	0.3743	0.6536
Cr_2O_3	0.00	0.01	0.0002	0.0002	0.0078
NiO	0.00	0.09	0.0745	0.0062	0.0002
Na_2O	7.19	7.01	7.0118	5.7041	6.6961
K_2O	0.00	0.00	0.0001	1.526	1.4608
Total	100.44	99.894	100.45	94.258	93.702
Structural formula (6 oxygen)					
Si	1.6794	1.695	1.6576	2.1756	2.123
Al^{IV}	0.0175	0.009	0.0225	0	0
T	1.6969	1.7039	1.6801	2.1756	2.123
Al^{VI}	0	0	0	0.009	0.0044
Fe^{3+}	0.6677	0.6195	0.672	0.0486	0.2842
Ti	0.2623	0.2695	0.273	0.0708	0.043
Cr	6E-06	0.0002	6E-06	7E-06	0.0003
Ni	3E-06	0.0028	0.0024	0.0002	7E-06
Mg	0.0139	0.0183	0.0145	0.0335	0.0908
Fe^{2+}	0.7292	0.7821	0.745	1.0507	0.7366
Mn	0.0492	0.0373	0.0401	0.0441	0.0678
M1	1.7222	1.7298	1.747	1.257	1.227
Ca	0.0125	0.0086	0.0173	0.017	0.0293
Na	0.5685	0.5576	0.5556	0.4681	0.5428
K	5E-06	5E-06	5E-06	0.0824	0.0779
T + M1 + M2	4.00	4.00	4.00	4.00	4.00

The symbol C, R and M represent core, rim and matrix, respectively.

Table 3.
Compositions of aenigmatite from Kone silicic eruptives.

has a high content of MnO (4.33 wt.%, **Table 4**), which is a characteristic feature of peralkaline rhyolites [16]. Fe-Ti oxides are essentially ilmenite (**Table 4**). It is noteworthy that similar phenocryst and groundmass mineralogies occur.

Matrix glass has a rhyolite composition with SiO_2 content of 69.3 wt.% (**Table 4**). It has high K_2O content (5.7 wt.%), but extremely low Na_2O content (2.6 wt.%). The low Na_2O content may be due to deutric mobilization.

Sample	5D	Sample	5D	4A/5–2	4A/3–2	5D
	Fayalite		**Ilmenite**	**Ilmenite**	**Apatite**	**Glass**
No	34	SiO2	0.00	0.03	3.26	69.28
SiO2	29.07	Al2O3	0.00	0.00	0.00	9.62
FeO	64.98	TiO2	56.20	51.85	0.00	0.57
MnO	4.33	FeO	39.70	44.04	0.91	6.84
MgO	0.74	MnO	2.70	2.08	0.06	0.27
CaO	0.35	MgO	0.11	0.00	0.00	0.02
Total	99.47	CaO	0.07	0.05	46.84	0.03
		Na2O	0.02	0.00	0.40	2.59
		K2O	0.01	0.00	0.00	5.70
		P2O5	0.00	0.01	29.71	0.00
		Total	98.79	98.06	81.18	94.90

Table 4.
Compositions of olivine, ilmenite, apatite and glass matric in Kone silicic eruptives.

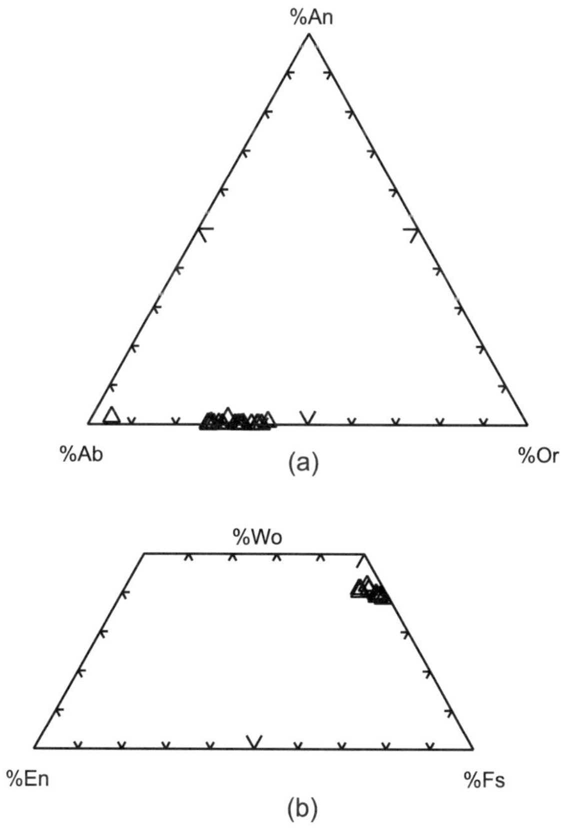

Figure 3.
Compositional ranges for alkali feldspar (a) and Ca-rich clinopyroxene (b) in silicic volcanics of Kone volcano.

6. Peralkaline affinity of Kone silicic volcanics

Kone silicic eruptives are characterized by the phenocryst assemblage of alkali feldspar, quartz, hedenbergite, aenigmatite and fayalite. This assemblage is accompanied by rare ilmenite and apatite. The appearance of aenigmatite in silicic volcanics has been attributed to reflect a peralkaline affinity [17]. Peralkaline magmas likely form when fractional crystallization removes a high proportion of plagioclase relative to mafic minerals [18]. The coexistence of aenigmatite and hednbergite is thought to be a typical feature of the mildly peralkaline silicic rocks such as comendite and trachycomendite [19].

The presence of quartz as a major phenocryst phase in Kone silicic volcanics indicates an advanced stage of fractionation (i.e., the more SiO_2-rich magma). Furthermore, the SiO_2 content of the glass matrix (69.3 wt.%, **Table 4**) shows a rhyolite composition. Thus, the phenocryst mineralogies reflect the SiO_2 content of the magma from which they crystallized. On this basis, the alkali feldspar + hedenbergite + aenigmatite + fayalite assemblage only crystallizes in the most evolved magmas (rhyolites) with the appearance of quartz as a phenocryst phase.

On the basis of the mineral assemblage, the Kone silicic volcanics can be regarded as more mildly peralkaline rhyolite (comendite) as evidenced by the appearance of hedenbergite, which appears to cease crystallizing in strongly peralkaline rhyolite (pantellerite), generally coinciding with the crystallization of aegirine or aegirine-augite [19]. This is probably related to the low fO_2, at or close to FMQ, at which the magmas evolve [20]. We conclude that the alkali feldspar + quartz + hedenbergite + aenigmatite + fayalite assemblage is essentially restricted to the mildly peralkaline rhyolite (comendite), but close to the comendite-pantellerite boundary as crystallization of aenigmatite is restricted to pantellerite [16].

7. Conclusions

Kone silicic eruptives are characterized by the phenocryst assemblage of alkali feldspar (anorthoclase and sanidine), quartz, hedenbergite, aenigmatite and fayalite, accompanied by rare ilmenite and apatite. The existence of aenigmatite is a typical feature of the peralkaline silicic rocks. We conclude that the alkali feldspar + quartz + hedenbergite + aenigmatite + fayalite assemblage is essentially restricted to the mildly peralkaline rhyolite (comendite) as hedenbergite appears to cease crystallizing in strongly peralkaline rhyolite (pantellerite), generally coinciding with the crystallization of aegirine or aegirine-augite.

Acknowledgements

Funding has been provided by CNRS (Centre National des Recherches Scientifiques, France). We are grateful to the School of Earth Sciences of Addis Ababa University for logistic support.

Author details

Dereje Ayalew* and Bekele Abebe
School of Earth Sciences, Addis Ababa University, Addis Ababa, Ethiopia

*Address all correspondence to: dereayal@yahoo.com; dereje.ayalew@aau.edu.et

References

[1] Ayalew D, Pik P, Bellahsen N, France L, Yirgu G. Differential fractionation of rhyolites during the course of crustal extension, western Afar (Ethiopian rift). Geochemistry, Geophysics, Geosystems. 2019;**20**(2):571-593

[2] Hunt JA, Zafu A, Mather TA, Pyle DM, Barry PH. Spatially variable CO_2 degassing in the main Ethiopian rift: Implications for magma storage, volatile transport, and rift-related emissions. Geochemistry, Geophysics, Geosystems. 2017;**18**:3714-3737

[3] Mohr P. The Ethiopian rift system. Bulletin of the Geophysical Observatory. 1962;**5**:33-62

[4] Cole JW. Gariboldi volcanic complex, Ethiopia. Bulletin Volcanologique. 1969;**33**:566-578

[5] Acocella V, Korme T, Salvini F, Funiciello R. Elliptic calderas in the Ethiopian rift: Control of pre-existing structures. Journal of Volcanology and Geothermal Research. 2003;**119**:189-203

[6] Kurz T, Gloguen R, Ebinger C, Casey M, Abebe B. Deformation distribution and type in the main Ethiopian rift (MER): A remote sensing study. Journal of African Earth Sciences. 2007;**48**(2-3):100-114

[7] George R, Rogers N, Kelley S. Earliest magmatism in Ethiopia: Evidence for two mantle plumes in one flood basalt province. Geology. 1998;**26**:923-926

[8] Hofmann C, Courtillot V, Féraud G, Rochette P, Yirgu G, Ketefo E, et al. Timing of the Ethiopian flood basalt event and implications for plume birth and global change. Nature. 1997;**389**:838-841

[9] Hart W, Wolde Gabriel G, Walter R, Mertzman S. Basaltic volcanism in Ethiopia: Constraints on continental rifting and mantle interactions. Journal of Geophysical Research. 1989;**94**:7731-7748

[10] Keranen K, Klemperer SL. Discontinuous and diachronous evolution of the main Ethiopian rift: Implications for development of continental rifts. Earth and Planetary Science Letters. 2008;**265**:96-111

[11] Wolfenden E, Ebinger C, Yirgu G, Renne PR, Kelley SP. Evolution of a volcanic rifted margin: Southern Red Sea, Ethiopia. Geological Society of America Bulletin. 2005;**117**:846-864

[12] Boccaletti M, Mazzuoli R, Bonini M, Trua T, Abebe B. Plio-quaternary volcanotectonic activity in the northern sector of the Main Ethiopian rift: Relationships with oblique rifting. Journal of African Earth Sciences. 1999;**29**:679-698

[13] Rampey ML, Oppeheimer C, Pyle DM, Yirgu G. Caldera-forming eruptions of the quaternary Kone volcanic complex, Ethiopia. Journal of African Earth Sciences. 2010;**58**:51-66

[14] Rampey ML, Oppenheimer C, Pyle DM, Yirgu G. Physical volcanology of the Gubisa formation, Kone volcanic complex, Ethiopia. Journal of African Earth Sciences. 2014;**96**:212-219

[15] Ayalew D, Jung S, Romer RL, Kersten F, Pfänder JA, Garbe-Schönberg D. Petrogenesis and origin of modern Ethiopian rift basalts: Constraints from isotope and trace element geochemistry. Lithos. 2016;**258-259**:1-14

[16] Macdonald R, White J, Baginski B, Leat P. Mineral stability in peralkaline silicic rocks: Information from trachytes of the Menengai volcano, Kenya. Lithos. 2011;**125**:553-568

[17] Ayalew D, Pyle D, Ferguson D. Effusive Badi silicic volcano (Central Afar, Ethiopian rift); sparse evidence for pyroclastic rocks. Volcanology. 2021. DOI: 10.5772/Intechopen.98558

[18] Ayalew D, Barbey P, Marty B, Reisberg L, Yirgu G, Pik R. Source, genesis, and timing of giant ignimbrite deposits associated with Ethiopian continental flood basalts. Geochimica et Cosmochimica Acta. 2002;**66**(8):1429-1448

[19] Peccerillo A, Barberio MR, Yirgu G, Ayalew D, Berbieri M, Wu TW. Relationships between mafic and peralkaline silicic magmatism in continental rift settings: A petrological, geochemical and isotopic study of the Gedemsa volcano, central Ethiopian rift. Journal of Petrology. 2003;**44**(11):2003-2032

[20] Di Carlo I, Rotolo SG, Scaillet B, Buccheri V, Pichavant M. Phase equilibrium constraints on pre-eruptive conditions of recent felsic explosive volcanism at Pantelleria Island, Italy. Journal of Petrology. 2010;**5**:2245-2276

Section 3

Sedimentary Mineralogy

Chapter 11

Use of Natural Safiot Clay for the Removal of Chemical Substances from Aqueous Solutions by Adsorption: A Combined Experimental and Theoretical Study

Aziz El Kassimi, Mohammadine El Haddad, Rachid Laamari, Mamoune El Himri, Youness Achour and Hicham Yazid

Abstract

The main objective of this work was to investigate the potential of Natural Safiot Clay (NSC), as an adsorbent for the removal of two cationic dyes such as Basic Blue 9 (BB9) and Basic Yellow 28 (BY28) from single and binary systems in aqueous solutions. For this, the effects of three factors controlling the adsorption process, such as initial dye concentration, adsorbent dose, and initial pH on the adsorption extent, were investigated and examined. The natural safiot clay was characterized using the following technique: energy-dispersive X-ray spectroscopy (EDX), scanning electron microscopy (SEM), DRX, and Fourier transform infrared (FT-IR) and pH of the point of zero charge (pH_{ZPC}). Energy-dispersive X-ray spectroscopy results indicate high percentages of Silica and Alumina. FT-IR spectrum identified kaolinite as the major mineral phase in the presence of quartz, calcite, and dolomite. The quantum theoretical study confirms the experimental results, through the study of the global and local reactivity and the electrophilicity power of the dyes. The electrophilicity power of dyes affects the removal efficiency. The theoretical study proves that BB9 (ω = 6.178) is more electrophilic than BY28 (ω = 2.480) and more interactions with surface sites. The results of the molecular dynamics simulation indicate that the dyes are adsorbed parallel to the surface of natural Safi clay (kaolinite), implying the strong interaction with the kaolinite atoms. All the results of quantum chemistry calculations and simulations of molecular dynamics are in perfect agreement with the results of the experimental study.

Keywords: adsorption, dyes; natural safiot clay, characterization, density functional theory

1. Introduction

The growth of humanity and the development of science and technology are causing an environmental disorder due to the pollution of water by a number of pollutants including synthetic dyes, which have a complex molecular structure, which makes them more stable and difficult. These dyes are discharged with the liquid effluents, which are most of the time directly discharged into watercourses without prior treatment [1]. These colored discharges pose a great threat to human health and the environment because many of the dyes are toxic. Therefore, it is very important to develop efficient methods for the removal of dyes from aqueous media.

Traditional processes such as biological processes give unsatisfactory results due to the composition of these releases of toxic and dyestuffs, which are difficult to biodegrade; on the other hand, physicochemical processes, which include coagulation-flocculation, oxidation, and membrane filtration, have been widely studied and have revealed a high efficiency in water discoloration [2, 3]. These techniques have proven to be very effective, but their high cost has prompted many researchers to try other cheaper and abundant materials. Adsorption, as a flexible, simple, and inexpensive approach, can be used for the removal of pollutant chemical species. Every day researchers describe new ways and elaborate a new adsorbent on the basis of good selectivity for a chemical species or for the low cost of the process. The elimination of dyes in aqueous solutions by adsorption on different solid materials, in particular on activated carbon, has been the subject of much work [4–6]. The adsorption of dyes on activated carbon has been found to be very effective, but its use remains limited due to the difficulties of its regeneration and its high cost [7].

For this reason, the use of clay as an adsorbent is of great interest because of its effectiveness, its accessible cost, and its abundance. From this perspective, lot of research has been done on clays and all have shown their depollution efficiency toward dyes [8, 9]. The importance given to these materials is granted to their abundant availability in nature and their great capacity of retention of various pollutants, which is offered by their structure in sheets, which gives a large specific surface toward the adsorption [10, 11].

In this context, our choice focused on clay, which is an adsorbent material found in abundance in Morocco, is effective and more economical. The use of clays for the depollution of waters contaminated by dyes requires a good knowledge of their mineralogical characteristics and of the mechanism of adsorption of these pollutants. On the other hand, Natural Safiot Clay (NSC) materials could be an attractive alternative for the adsorption of various pollutants from wastewater due to their low cost, their lamellar structure, which provides high specific surface areas, thermal stability, high cation exchange capacity, abundance, and high adsorption capacity [12–14].

Generally, most of the studies carried out to eliminate cationic dyes are carried out on single dyes [15], which lead us to study the possibility of eliminating a mixture of two dyes at the same time, close to the real conditions of effluents in the environment, because industrial discharges are a complex mixture of several pollutants.

Recently and with computer development, quantum chemistry calculations are widely used in studies of dye adsorption [16, 17]. A lot of research has been carried out on this theoretical approach; in particular since the appearance of their efficiencies in the study of adsorption mechanisms on the one hand, and the study of behavior of dyes with respect to the adsorbent surface on the other hand [18, 19]. The theoretical reactivity indices based on density functional theory (DFT) have become a powerful and informative tool for studying organic reactivity and for describing intermolecular interactions [20, 21].

The purpose of this study was to demonstrate the ability of Natural Safiot Clay to remove some mixture industrial dyes such as Basic blue 9 and Basic yellow 28 from single and binary aqueous solutions. This allows optimizing the cost of the process and the quantity of the adsorbent used in the adsorption process to have a multiple removal efficiencies of different pollutants instead of one [22, 23]. In this study, the possibility of adsorption of Basic blue 9 (BB9) and Basic yellow 28 (BY28) in the single and binary system has been studied experimentally and theoretically using density functional theory (DFT) and molecular dynamics simulations (MDS), In order to explain the competitiveness between the two dyes on active sites and their reactivity, check if the experimental results are in good correlation with the theoretical results.

2. Materials and methods

2.1 Preparation of natural safiot clay

The Natural Safiot Clay (NSC) used in this work is collected from a natural basin in the region of Safi in Morocco for removal of Basic blue 9 and Basic yellow 28 from aqueous solutions and used without any prior activation. Samples were ground and sieved to obtain very fine particle sizes and washed with distilled water to ensure the removal of dust and any soluble impurities may exist.

2.2 Preparation of dyes solutions mixture

Basic blue 9 (BB9) and Basic yellow 28 (BY28) as representative cationic dyes were purchased from Sigma-Aldrich with a purity of 99% and used without further purification. The chemical structures of the studied dyes are given in **Figure 1**.

2.3 Experimentation conditions of adsorption studies

For single and binary system, several stock solutions 100 mg/L of BB9 and BY28 dyes have been prepared by mixing calculated volumes of the stock solutions of each dye and accurately diluting it with distilled water. The NSC and mixture of dyes are shaken in batch experiments at various parameters such as the amount of natural safiot clay (5 mg–35 mg), initial dye concentration (10 mg–40 mg/L), and initial solution pH (2–12). The working solutions' pH was adjusted to the desired values with dilute HCl (0.1 M) or NaOH (0.1 M) using a pH-Meter HANNA 5222. After stirring of a prescribed contact time, the solution is filtrated using filter syringe, and the maximum absorbance value of BB9 and BY28 is measured using spectrophotometer UV–Visible (JENWAY 6300) at 663 and 438 nm, respectively, as shown in **Figure 2**.

The percentage removal and the quantity adsorbed q_e (mg/g) of dye on NSC were calculated using the following equation:

$$\%Removaldye = \frac{C_0 - C_e}{C_0} * 100 \tag{1}$$

$$q_e(mg/g) = \frac{C_0 - C_e}{m} * V \tag{2}$$

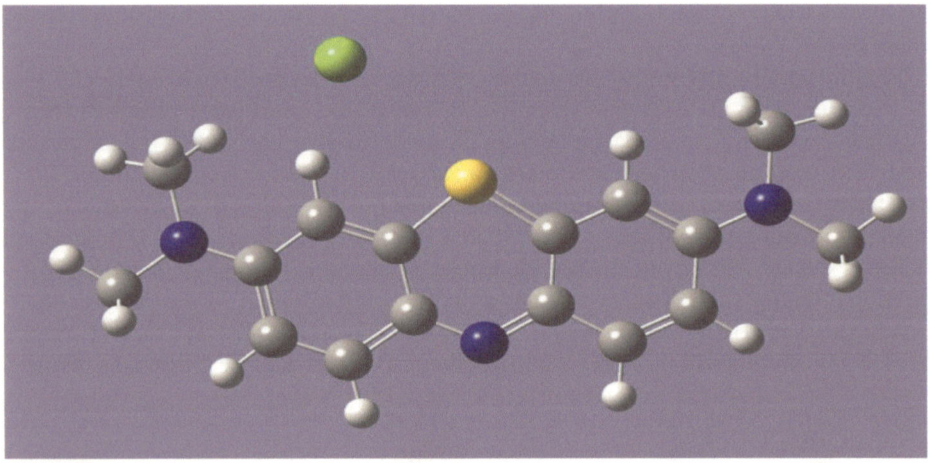

(c) BY28

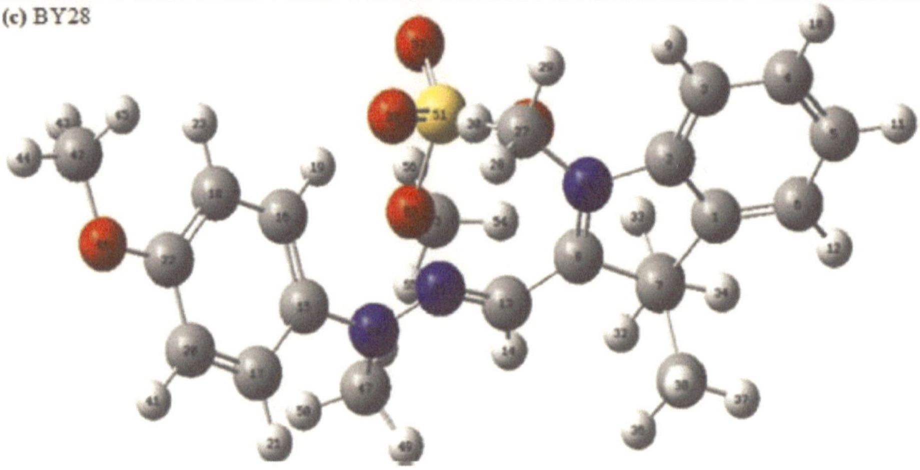

Figure 1.
Chemical structures of BB9 and BY28.

Where C_0 (mg/L) and C_e (mg/L) represent the concentration of BB9 and BY28 at initial and equilibrium, respectively, V (L) is the volume of solution, and W (g) is the weight of adsorbent used.

2.4 Surface characterization

The surface morphology, chemical composition, and the nature of functional groups responsible for adsorption of dyes onto the NSC adsorbent were studied by Scanning Electron Microscopy (SEM), Energy Dispersive X-ray spectroscopy (EDX), X-Ray Diffraction (XRD), and Fourier Transform Infrared (FT-IR).

2.5 Computational details

Density functional theory (DFT) is one of the most important tools of quantum chemistry of understanding popular qualitative chemical concepts such as energy of highest occupied molecular orbital (E_{HOMO}) and the lowest unoccupied molecular

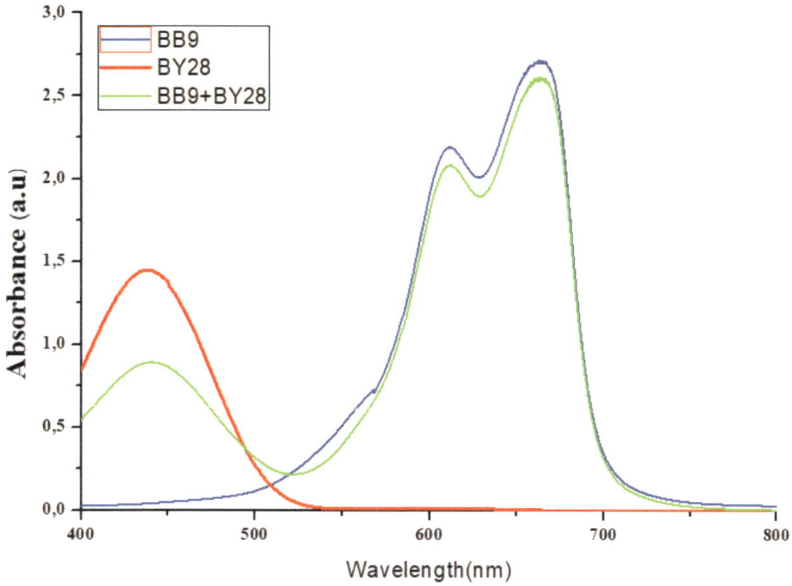

Figure 2.
Absorption spectra of BB9, BY28, and their mixture.

orbital (E_{LUMO}), dipole potential (μ), hardness (η), softness (S), electrophilicity index (ω), and local reactivity descriptors such as Parr function P(r) [24, 25]. All computations are carried out with the Gaussian 09 program. The geometries of dyes, BB9 and BY28, are optimized using density functional theory (DFT) at the B3LYP/6 G-31G (d) level. Optimizations are carried out using the Berny analytical gradient optimization method. The geometries optimized are characterized by positive vibrational frequency definite Hessian matrices [26].

When the values of E_{HOMO} and E_{LUMO} are known, one can determine through the following expressions [27] the values of the electronic chemical potential μ, the absolute hardness η, and the softness S as:

$$\mu = \frac{E_H + E_L}{2} \tag{3}$$

$$\eta = (E_L - E_H) \tag{4}$$

The global softness (S) introduced is the inverse of the global hardness [28]:

$$S = \frac{1}{\eta} \tag{5}$$

Using Parr's definition [29], the electrophilicity ω index is given by:

$$\omega = \frac{\mu^2}{2\eta} \tag{6}$$

Based on this idea, Domingo et al. [30] have introduced an empirical (relative) nucleophilicity index N, based on the HOMO energies obtained within the Kohn–Sham scheme and defined as:

$$N = E_{HOMO}(Nu) - E_{HOMO(TCE.)} \tag{7}$$

The HOMO energy of Tetracyanoethylene is -0.3351 a.u. at the same level of theory.

The electrophilic P_K^+ and nucleophilic P_K^- Parr functions, which allow for the characterization of the electrophilic and nucleophilic centers of a molecule, were obtained through the analysis of the Mulliken ASD (Atomic Spin Density) of the radical anion and the radical cation, respectively. These indices were obtained by single-point energy calculations over the optimized neutral geometries using the restricted B3LYP formalism for radical species. The results obtained will be compared with the experimental data.

The adsorption progress of the studied dyes on kaolinite surface is performed using Materials Studio (MS) 8.0 software developed by Accelrys Inc. The kaolinite crystal was optimized (a = 5.196 Å, b = 9.007 Å, c = 7.372 Å, and α = 93.029°, β = 105.983°, γ = 89.866) and cleaved along the (001) plane, a vacuum slab with 10 Å thickness was built. The final structure was enlarged to $(4 \times 2 \times 1)$ to provide a large surface for the interaction of the dyes [31].

3. Results and discussion

3.1 Characterization of natural safiot clay

3.1.1 Energy-dispersive X-ray spectroscopy

The spectrum of chemical constitution of natural safiot clay adsorbent is given in **Figure 3**. The EDX spectrum of **Figure 2** presents well-defined peaks, confirms the presence of the following chemical elements: Si, Al, Mg, Fe, K, P, S, O, Ca, C. These results confirm those found by the analysis XRF (**Table 1**), which also reveals the presence of these elements in the form of oxides: SiO_2, Al_2O_3, Fe_2O_3, MgO, Na_2O, CaO, K_2O, TiO_2. The atomic and mass percentages of the elements are summarized in **Table 2**. The predominance of silicon and oxygen peaks is clearly observed, which confirms the majority presence of kaolinite and quartz in the sample studied.

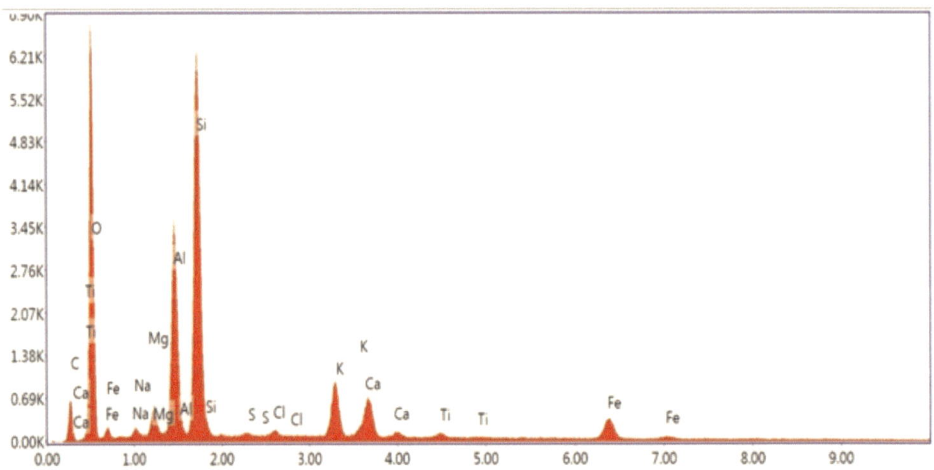

Figure 3.
EDX spectrum of natural safiot clay.

Dyes	Langmuir			Freundlich			Dubinin–Radushkevich			
	q_{max}	R_L	R^2	K_F	1/n	R^2	q_{max}	$K_{D\text{-}R}$	E	R^2
BB9 (S)	68.49	0.038	0.992	37.89	0.314	0.852	59.03	$1\ 10^{-7}$	2236	0.925
BB9 (B)	41.15	0.015	0.996	25.71	0.222	0.566	44.78	$1\ 10^{-7}$	2236	0.897
BY28 (S)	166.67	0.363	0.761	10.50	0.976	0.955	75.33	$2\ 10^{-6}$	500	0.994
BY28 (B)	21.32	0.064	0.986	10.07	0.260	0.718	18.18	$6\ 10^{-7}$	912.8	0.512

Table 1.
Isotherm constants for BB 9 and BY28 in single and binary systems.

Eléments	% atomique	% massique
O	57.87	46.70
C	12.96	7.85
Na	0.59	0.68
Mg	1.25	1.53
Al	7.50	10.20
Si	13.78	19.52
K	2.18	4.30
Ca	1.70	3.43
Ti	0.19	0.47
Fe	1.72	4.85

Table 2.
Atomic and mass percentage of the natural safiot clay constituents.

3.1.2 Fourier transform infrared spectroscopy

Fourier transform infrared (FTIR) analysis was applied to determine the functional groups present on the surface of natural safiot clay and understand its adsorption mechanism. FT-IR spectra of NS clay in the range of 400 cm–4000 cm^{-1} are taken to obtain information on the nature of functional groups at the surface of the adsorbent. The spectrum of natural safiot clay is shown in **Figure 4**. The band that stretches between 3200 and 3700 cm^{-1} shows a peak with two shoulders at 3407 cm^{-1} and 3610 cm^{-1} corresponding to the vibrations of elongation of the hydroxyl group –OH linked to the water of constitution. In addition to the vibrations of deformation of the O-H bond due to the water molecules adsorbed between the sheets located at 1639 cm^{-1}. The bands that appear approximately around 3430 cm^{-1} and 1630 cm^{-1} correspond respectively to the vibrations of elongation and deformation of the OH group of the adsorbed water [32]. While the characteristic bands of carbonates are detected at 1436 cm^{-1} and 2521 cm^{-1} [33].

An intense absorption band at 900 cm–1200 cm^{-1} is centered on 1030 cm^{-1}, it characterizes the valence vibrations of the Si-O bond [34]. The bands between 795 and 748 cm^{-1}, coming from the Si-O-Al bond, also give way to a band around 778.4 cm^{-1} [35]. The absorption band located at 1030 cm^{-1} is in agreement with the X-ray

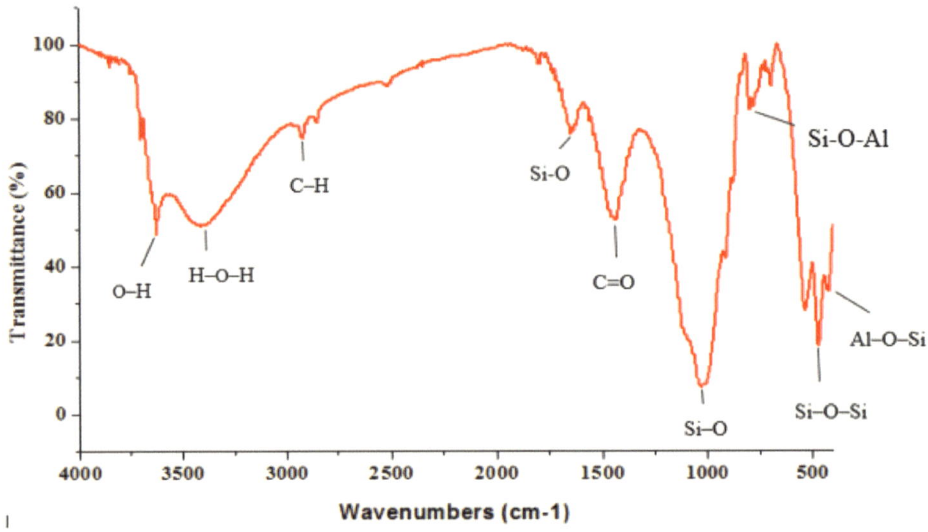

Figure 4.
FT-IR spectrum of natural safiot clay.

fluorescence indicating the presence of kaolinite in natural clay. However, the absorption bands at 423, 480, 534, 694, and 797 cm^{-1} correspond to quartz [36]. These results are in agreement with those found from XRF. They confirm the presence of quartz, carbonate, kaolinite, and dolomite in the clay studied.

3.1.3 Scanning electronic microscopy analysis

Scanning electronic microscopy (SEM) technique was carried out in order to observe the morphology, structure, and distribution of the grains of our adsorbent material studied. **Figure 5a** and **b** show the SEM micrographs of natural safiot clay before and after adsorption. The scanning electron microscope image (**Figure 5a**) shows aggregates of kaolin grains in spherical form and of heterogeneous size, the interstices between the grains form pores. We also observe large irregularly shaped cavities; this confirms the heterogeneous composition of our clay revealed by the XRD (kaolinite + calcite + vermiculite). In contrast, **Figure 5b** shows that the NSC surface is more homogeneous and saturated after adsorption.

3.1.4 XRD analysis

X-ray diffraction analysis allows us to identify the different mineralogical phases contained in our material. The X-ray diffraction patterns of NSC adsorbent are illustrated in **Figure 6**. This result demonstrated the principal presence of kaolinite characterized by an intense peak at 2θ = 26.63° (2θ = 26.63°; d = 3.343 A°) and a series of peaks with varying intensities at 12.33°, 19.75°, 40.85° and those of quartz at 20.87°, 37.92°, and 42. 440°. In addition, the reflections spectrum at 2θ = 8.62°, 29.47°, and 30.81° confirm the presence of illite, calcite, and dolomite, respectively. We see that the diagram also shows the presence of the peak corresponding to the following

DOI: http://dx.doi.org/10.5772/intechopen.101605

(a)

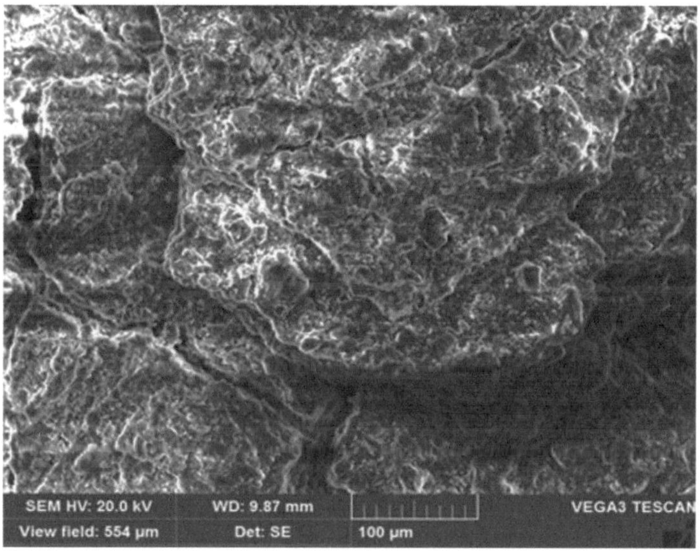

(a)

Figure 5.
SEM micrograph of the natural Safi clay before (a) and after (b) adsorption.

minerals: kaolinite, calcite, and vermiculite, which implies that our clay is heterogeneous.

3.1.5 Determination of pH zero-point charge

The pH zero-point charge (pHzpc) plays an important role in the adsorption process. The point of zero charge (PZC) of our clay was determined using the pH drift method [37]. Six vials containing solutions of pH in the range of 2–12 (pH_i) and 50 mg of NSC are shaken for 24 h at room temperature, and the final pH was measured. The difference between the initial and final pH ($\Delta pH = pH_i - pH_f$) was plotted against the initial pH (pH_i) and the point where $\Delta pH = 0$ was taken as the point of zero charge. As shown in **Figure 7**, the pHpzc of NSC was determined to be 7.2.

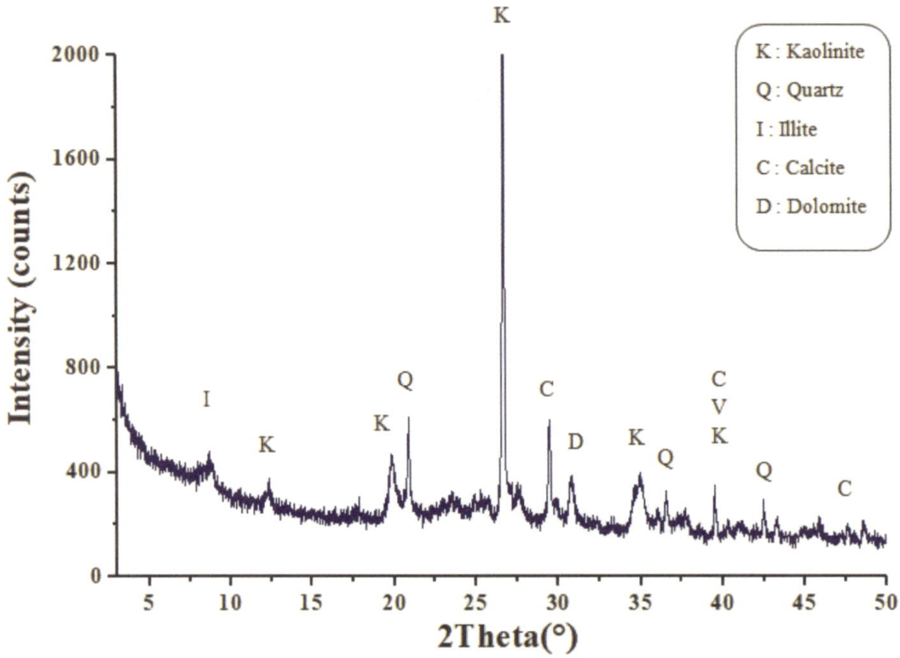

Figure 6.
X-ray diffraction of natural safiot clay.

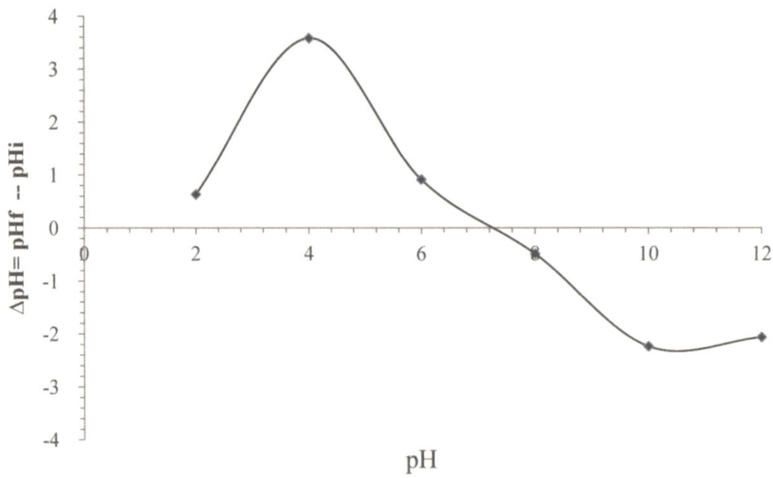

Figure 7.
pHzpc values for natural safiot clay.

3.2 Experimental setup: single and binary adsorption studies

3.2.1 Effect of adsorbent dose

It is recognized that the effect of the adsorbent dose on the adsorption process is also considered to be one of the most important parameters that must be optimized,

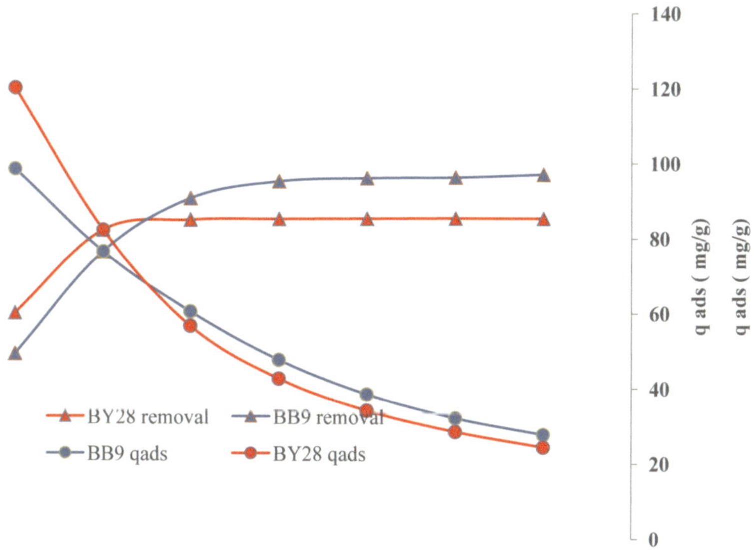

Figure 8.
Effect of adsorbent amount on the removal efficiency and adsorption capacity of basic blue 9 and basic yellow 28.

since the mass of adsorbent has an effect on the adsorption capacity for a given initial concentration of the adsorbate under the operating conditions. The adsorption of BB9 and BY28 in single and mixture on natural safiot clay is studied by varying the mass of adsorbent from 5 to 35 mg in 50 mL solution of 20 mg/L dye concentration at a constant stirring rate of 60 minutes. From **Figure 8**, it can be observed that removal efficiency of the dye increases from 49.17% to 97.03% for BB9 and from 59.97% to 85.32% for BY28 as adsorbent dose is an increase from 5 to 35 mg. This is because of the extra number of adsorption sites accessible with an increase in the adsorbent dose. On the other hand, the dye uptake capacity reduces from 98.34 to 27.72 mg/g and from 119.9 to 24.38 mg/g for BB9 and BY28, respectively. This can be attributed to the unsaturation of adsorption sites through the adsorption reaction with increasing adsorbent dosage [38, 39]. Another important reason is that at high adsorbent dosage, the available dye molecules are deficient to completely cover the available binding sites on the natural safiot clay, which results in low solute uptake [40, 41]. Similar results have been reported previously by other researchers for the adsorption of dyes by different material [42–44]. The optimum adsorbent dose is fixed conveniently at 30 mg per 50 mL of solution dye for the following studies.

3.2.2 Effect of initial dye concentration

The effect of initial concentrations of BB9 and BY28 dyes is examined at different initial concentrations ranging from 10 to 40 mg/L on the adsorption capacity and removal efficiency onto natural safiot clay. As seen from **Figure 9**, the adsorption capacity increases from 22.58 to 58.45 mg/g for BB9 and of 12.68 to 58.89 mg/g for BY28. In this case, the % removal decreases from 98.97% to 87.67% and from 87.67% to 70.08% for BB9 and BY28, respectively. These results indicate that the adsorption sites of NSC adsorbent for dyes adsorption are still unsaturated

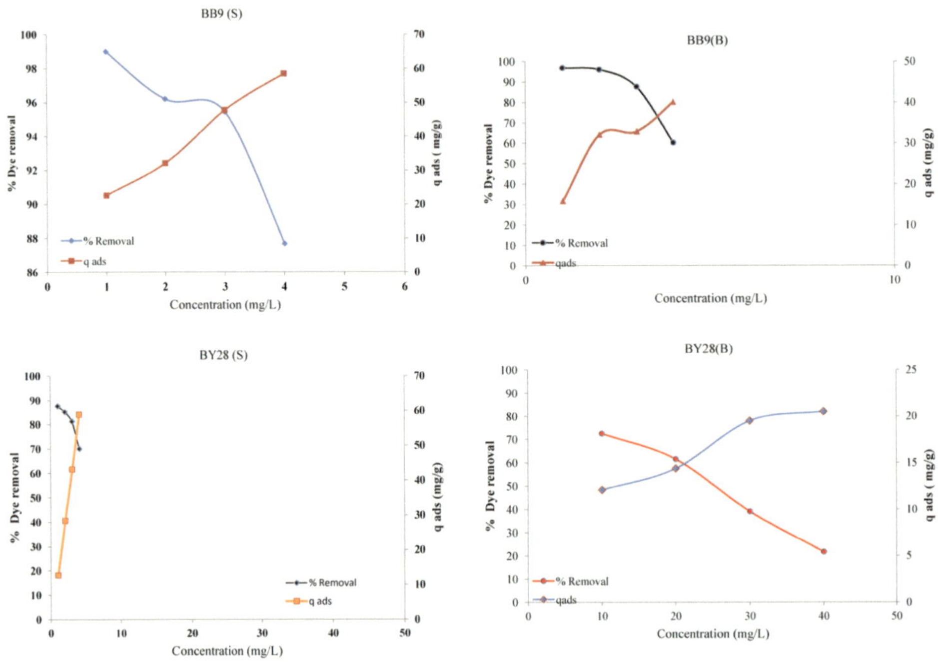

Figure 9.
Effect of the initial dye concentration on the adsorption capacity and removal efficiency of BB9 and BY28, in single (S) and binary (B) system onto NSC.

within the dye concentration range. In addition, increasing initial dyes' concentrations increases the number of collisions between dyes ions and the surface area of NSC adsorbent, which enhances the adsorption process [45]. Similar results had been reported by Auta and Hameed [46] for MB dye removal onto modified ball clay chitosan composite.

3.2.3 Effect of initial dye pH

The initial pH of the aqueous solution is important parameter controlling the adsorption process, where it affects both the degree of ionization of the dye and the surface properties of the adsorbent. The effect of initial pH of dye solution on the percentage removal of dye is studied by varying the initial pH from 2 to 12 under constant process parameters onto NSC, and results are presented in **Figure 10**. It is revealed that pH has practically a small effect on the percentage removal of the two basic dyes in simple system. In binary system, BY28 percentage removal is increased from 34.74% to 71.14% when pH is varied from 2 to 12 and also for BB9 percentage removal increase from 90.86% to 96.75%. These results are explained by pH zero-point charge pH_{ZPC}, the pHzpc of any adsorbent is a very important characteristic that determines the pH at which the surface has net electrical neutrality. In explaining this behavior by the fact that the negative charge dominates the adsorbent's surface in the basic medium. Thus, an electrostatic attraction exists between the negative charges of OH deposited on the clay surface and the positive charges of the dyes. Moreover, an electrostatic attraction between BB9 and BY28 dyes and the positive charge on the surface of NSC at low pH are evident [47, 48].

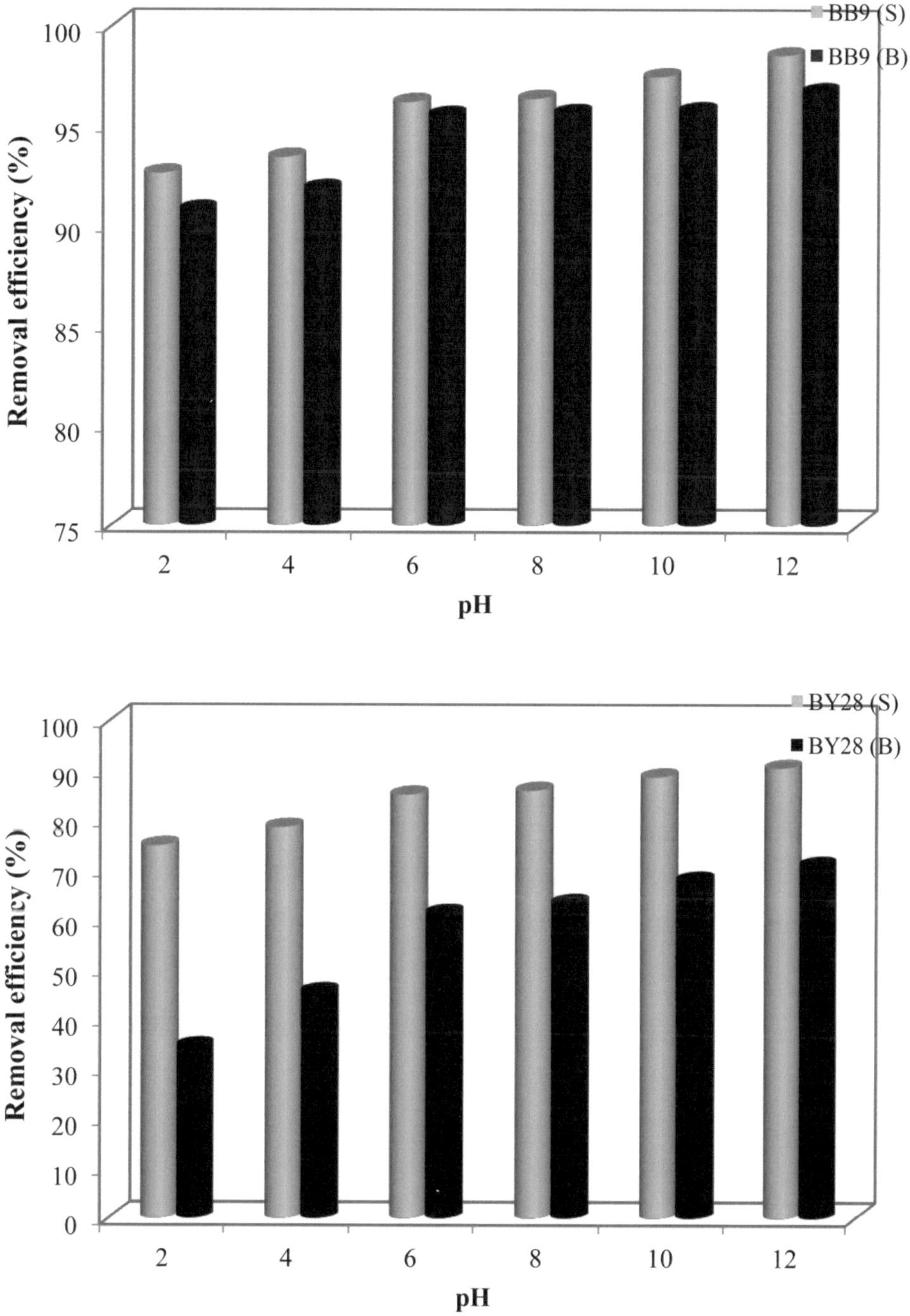

Figure 10.
Effect of initial dye pH on the removal efficiency (%) of BB9 and BY28 dyes in single and binary solutions.

Consequently, the mechanism proposed can be described by the following equation.

$$\text{Si-OH} + OH^- \leftrightarrow \text{Si-O}^- + H_2O \tag{8}$$

$$\text{Si-OH}^- + \text{BB9}^+ \text{ or BY28+} \leftrightarrow \text{Si-O}^- + \text{BB9 or Si-OH}^- + \text{BY28} \quad (9)$$

Similarly, the montmorillonite, bentonite clay, and montmorillonite/$CoFe_2O_4$ composite adsorption capacities were studied as a function of pH, and it was observed that maximum basic blue 9 dye adsorption was in acidic pH range [49–51].

3.2.4 Competitive adsorption between BB9 and BY28

BB9 and BY28 adsorption in single and binary adsorption systems onto NSC is studied and is illustrated in **Figure 11**. The removal efficiency decreased in binary systems (as compared with single dye systems), the reduction was from 96.20% to 95.57% and from 85.32% to 61.48% for BB9 and BY28, respectively. For these results it is clear in the binary system, BB9 dye is most dominant and BY28 is most recessive dye. The values of R% also show that adsorption of BB9 and BY28 is reduced by the presence of other dyes in solutions within reduction percentage of 0.66% and 27.94% for BB9 and BY28, respectively. This behavior can be explained by the competitive adsorption between BB9 and BY28 for active sites with that BB9 dye is the first to be adsorbed in the active sites and by that BB9 is more electrophilic than BY28. This result will be demonstrated by following quantum chemicals study.

3.3 Adsorption isotherms

Adsorption isotherms play an important role in the determination of the maximum adsorption capacity and the identification of the type of adsorption. The results of the adsorption experiments were analyzed per the well-known models of Langmuir, Freundlich, and Dubinin–Radushkevich (D-R):

The Langmuir isotherm is valid for monolayer adsorption on surface containing a finite number of identical sites [52]. The linear form of the Langmuir isotherm can be represented by the following equation:

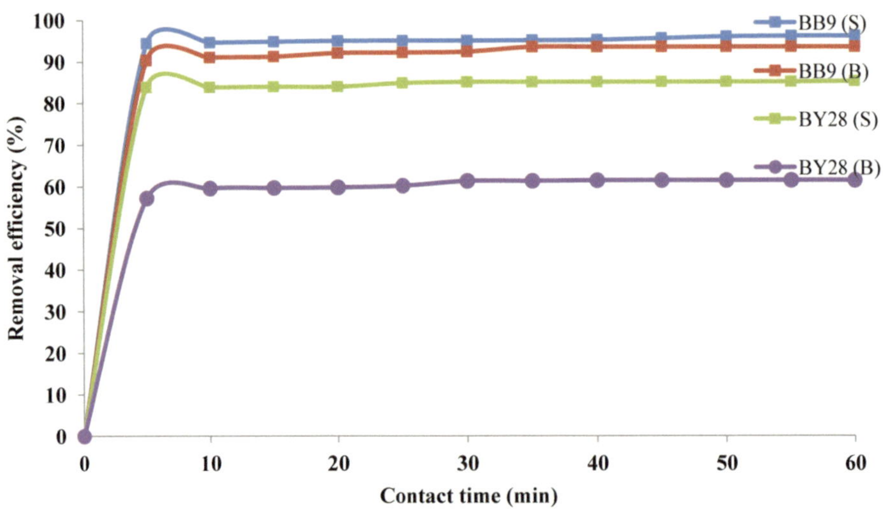

Figure 11.
Adsorption competition of BB9 and BY28 onto NSC adsorbent sites.

$$\frac{C_e}{q_e} = \frac{1}{q_m K_L} + \frac{1}{q_m} C_e \quad (10)$$

Where Ce (mg/L) represents the equilibrium concentration of the adsorbate, qe the amount adsorbed at equilibrium (mg/g), K_L (L/mg) and qm (mg/g) are the Langmuir constant and the maximum amount of adsorbate, respectively.

To confirm the favorability of the adsorption process, the separation factor R_L was calculated by the following Equation [53]:

$$R_L = \frac{1}{1 + K_L C_0} \quad (11)$$

where the adsorption process is unfavorable ($R_L > 1$), linear ($R_L = 1$), favorable ($0 < R_L < 1$), or irreversible ($R_L = 0$). Here, R_L values for the adsorption of dyes are less than 1 and greater than 0, indicating favorable adsorption.

Freundlich isotherm model is an empirical equation based on sorption on a heterogeneous surface or surface supporting sites of varied affinities [54]. The linearized Freundlich model is represented by the following equation:

$$\log(q_e) = \log(K_f) + \frac{1}{n}\log(C_e) \quad (12)$$

where K_f (mg/g) is the measurement of adsorption capacity, and 1/n is the adsorption intensity of the adsorbent.

The Dubinin–Radushkevich model is a more generalized model as compared with the Langmuir isotherm and often used to estimate the characteristic porosity and the apparent free energy of adsorption [55]. The linearized Dubinin–Radushkevich (D-R) isotherm model is represented by the following equation:

$$ln(q_e) - ln(q_m) - B\varepsilon^2 \quad (13)$$

where q_m is the theoretical saturation capacity (mg/g), B is the D-R constant related to the sorption energy (mol^2/kJ^2), and ε represents the Polanyi potential (J/mol), which is determined by:

$$\varepsilon = RTLn\left(1 + \frac{1}{C_e}\right) \quad (14)$$

R is the universal gas constant (8.314 J mol^{-1} K^{-1}), and T is the absolute temperature (K). The mean free energy of adsorption E calculated from B using the following relation:

$$E = \frac{1}{\sqrt{2B}} \quad (15)$$

The main parameters, characterizing each model as well as the coefficients of determination (R^2), are grouped in **Table 1**. Comparison with Freundlich and Dubinin–Radushkevich model shows the high correlation coefficient of Langmuir isotherm for both dyes in single and mixture systems. This result suggests that the dye was homogeneously adsorbed on a monolayer surface of the adsorbent.

The value of parameter 1/n of the Freundlich equation gives an indication of the validity of the adsorption of the adsorbent adsorbate system. The values of 1/n presented in **Table 1** are between 0 and 1 indicating that the adsorption of the two dyes on our prepared adsorbent material (NSC) is favorable.

The magnitude of E is useful for estimating the type of adsorption process. The found values of E for BB9 and BY28 in the single and binary system are less than 8 kJ mol^{-1}, knowing that energy values less than 8 kJ mol^{-1} indicate physisorption and energy values varying from 8 to 16 kJ mol^{-1} indicate chimisorption. Therefore, the adsorption type of BB9 and BY28 onto NSC has been defined as physical adsorption (physisorption). This confirms the results following the study of the influence of pH.

3.4 Kinetics of adsorption

The kinetic of adsorption is an important characteristic in evaluating the efficiency of adsorption process. Three kinetics models (pseudo-first order, pseudo-second order, and intraparticle diffusion) were utilized to test the experimental data and predict the controlling mechanism of dye adsorption process.

3.4.1 Pseudo-first-order model

The linearized form of pseudo-first-order rate expression is given as:

$$\log\left(q_e - q_t\right) = \log\left(q_e\right) - \frac{k_1}{2.303}t \tag{16}$$

Where qe and qt are the amount of dye adsorbed on sorbent (mg/g) at equilibrium and time t, respectively, $k_1$corresponds to the reaction rate constant of pseudo-first-order (min^{-1}), and t is time (min) [56]. The values of qe and k_1 were calculated from the slope and intercept of the plots of the log (qe-qt) vs. t.

3.4.2 Pseudo-second-order model

Pseudo-second-order rate expression reaction model is expressed as (linearized form) [57]:

$$\frac{t}{q_t} = \frac{1}{k_2 q_e^2} + \frac{1}{q_e}t \tag{17}$$

Where k_2 is the pseudo-second-order rate constant (g/mg.min). A plot of t/qt and t should give a linear relationship if the biosorption follows pseudo-second-order model. The q_e and k_2 can be calculated from the slope and intercept of the plot.

3.4.3 Intraparticle diffusion model

The intraparticle diffusion model is based on the theory proposed by Weber and Morris [58]. The Weber and Morris equation is (18):

$$q_t = k_{id}t^{1/2} + C \tag{18}$$

Dyes concentration (mg/l)			Pseudo-first order			Pseudo-second order			Intraparticle diffusion		
		qe,exp	qe,cal	K_1	R^2	qe,cal	K_2	R^2	Ki	C	R^2
BB9 (S)	10	15.48	7.998	0.028	0.752	15.78	0.198	0.999	0.095	16.20	0.965
	20	32.07	0.857	0.037	0.668	31.34	0.156	1.000	0.114	30.24	0.976
	30	47.72	1.795	0.074	0.403	47.85	0.109	1.000	0.306	45.66	0.794
	40	58.45	8.707	0.061	0.783	58.82	0.022	0.999	0.927	51.51	0.971
BY28(S)	10	12.69	0.129	0,112	0,485	12.73	0,470	1.000	0,100	12.04	0,542
	20	28.39	0.823	0,070	0,826	28.49	0,246	1.000	0,098	27.72	0,798
	30	43.14	2.960	0,023	0,913	42.37	0,071	0,999	0,316	39.86	0,905
	40	58.89	96.11	0,285	0,858	59.88	0,020	0,999	1.499	49,17	0,825
BB9(B)	10	15.80	23.22	0.041	0.484	15.82	1.051	1.000	0.036	15.54	0.756
	20	32.01	1.376	0.054	0.892	32.05	0.191	1.000	0.217	30.45	0.843
	30	43.82	8.586	0.064	0.937	44.05	0.031	0.999	0.881	37.27	0.982
	40	48.00	51.54	0.059	0.812	50.76	0.003	0.960	4.923	9.712	0.954
BY28(B)	10	12.08	3.088	0.177	0.813	12.16	0.209	0.999	0.131	11.19	0.873
	20	14.37	12.58	0.217	0.851	14.45	0.200	0.999	0.112	13.58	0.879
	30	19.49	12.36	0.165	0.898	19.68	0.087	0.999	0.423	16.69	0.649
	40	20.49	1.173	0.170	0.756	20.57	0.332	1.000	0.167	19.42	0.502

Table 3.
Kinetics parameters for the adsorptive removal onto NSC adsorbent of BB9 and BY28 dyes and their mixture.

Where q_t is the adsorption capacity (mg/g) at time t (min), k_{id} is the intraparticle diffusion rate constant (mg/g.min), and C (mg/g) is a constant, which provides the information regarding the thickness of the boundary layer. The values of K_{id} and C were calculated from the slope and intercept of the plots of the q_t against $t^{1/2}$.

The conformity between the experimental data and the predicted model is based on the values of the correlation coefficients (R^2), hence the value R^2 closest to unity will indicate the adequate model to correctly describe the kinetics of adsorption of the dye.

Table 3 summarizes the rate constants and correlation coefficients (R^2) of the three kinetic models. The fitting of the kinetic data in the pseudo-second-order equation showed excellent linearity with high correlation coefficient ($R^2 > 0.999$), and the good agreement between the experimental and calculated equilibrium adsorption for the pseudo-second-order model confirms that this one describes correctly the adsorption kinetics. Similar results have been observed in the adsorption of basic dyes onto Moroccan Clay [59] and in the adsorption of Methylene Blue (MB) by montmorillonite clay [60].

3.5 Quantum chemical studies

3.5.1 Global reactivity descriptors

The global chemical reactivity descriptors, energy gap (ΔE), dipole moment (μ), hardness (η), softness (S), nucleophilicity (N), and electrophilicity index (ω) witch

Dyes	BB9	BY28
E_T (u.a)	−1643.18	−1754.96
μ (Debye)	11.351	7.573
E_{HOMO} (eV)	−4.373	−5.585
E_{LUMO} (eV)	−3.210	−2.387
μ (eV)	−3.791	−3.986
η (eV)	1.163	3.198
S (eV)	0.859	0.312
ω (eV)	6.178	2.480
N (eV)	4.747	3.535

Table 4.
Quantum chemical parameters of the studied dyes calculated at B3LYP/6 G-31G (d).

calculated from HOMO and LUMO energies and are obtained at the level of theory B3LYP/6 G-31G(d) and summarized in **Table 4**.

Energy gap ($\Delta Egap = E_{HOMO} - E_{LUMO}$): The energy gap between the HOMO and LUMO is very important in determining the chemical reactivity of the molecule dyes toward the adsorption on the adsorbent surface. On the other hand, the decrease in the value of ΔEgap increases the reactivity of the molecule, which facilitates adsorption and increases the adsorption efficiency. It can be seen from **Figure 12** that the BB9 dye shows a lower ΔEgap (ΔEgap = 1.163 eV) compared with the BY28 dye, which has a difference of 3.198 eV, which clearly means that the molecule of BB9 is more reactive than BY28, Therefore, the BB9 dye will be adsorbed firstly. This conclusion is in agreement with the experimental results.

Dipole moment (μ): The dipole moment (μ) is another important electronic parameter, provides information on the polarity of the whole molecule. The high

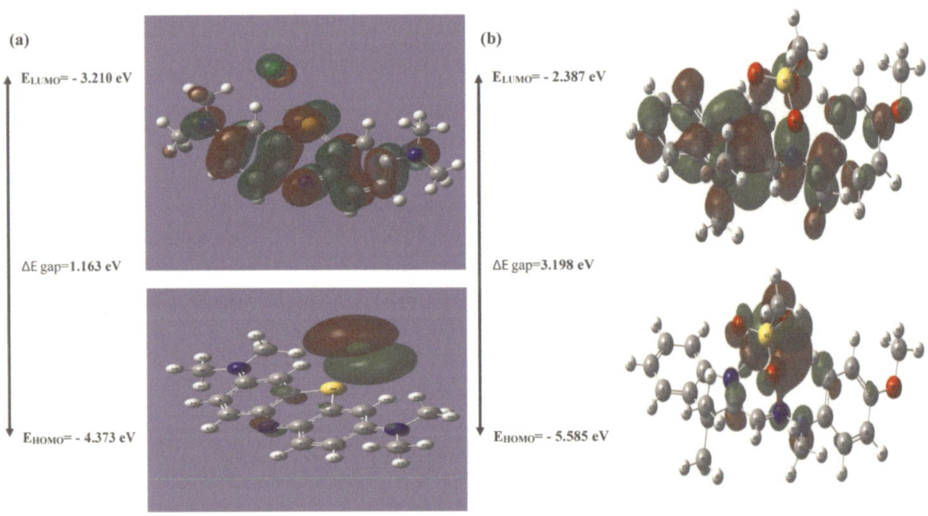

Figure 12.
Highest occupied molecular orbital (HOMO) and the lowest unoccupied molecular orbital (LUMO) density of BB9 (a) and BY28 (b) BY DFT at the B3LYP/6 G-31G(d).

molecular polarity probably gives rise to great chemical reactivity. It is clearly established in the literature that molecules with high dipole moments are more reactive, and their action results in a significant elimination efficiency. In our case the high dipole moment value of BB9 (11.351 D) probably increases the adsorption between the BB9 dye and the surface of natural clay compared with BY28 (7.573 D), which explains the adsorption efficiency higher than BB9 when compared with BY28 and confirms the experimental results.

Hardness (η) and softness (S): The stability and reactivity of a molecule are determined by the calculation of two important parameters: the global hardness (η) and the softness (S). The resistance of a molecule to deformation is determined by the chemical hardness, a hard molecule has a high energy gap. In addition, a soft molecule has a low energy gap. It is important to note that electronic systems with hard molecules have the least tendency to react while systems with soft molecules have a higher tendency to react.

The high percentage of elimination of a molecule is linked to a low value for chemical hardness and a high value for softness. In the present work, the values of the global hardness (η) and the softness (σ) presented in **Table 5** clearly show that the BB9 dye has the lowest value of the hardness (η = 1.163 eV) and the higher value of the softness (S = 0.859 eV), which explains their significant elimination percentage compared with BY28, these results are in good agreement with the experimental results.

Global electrophilicity index (ω): The global electrophilicity index (ω) represents the capacity of the dyes to accept electrons. More reactive nucleophilic is characterized by lower value of ω, and conversely more reactive electrophilic is characterized by a higher value of ω. From **Table 5**, we notice that the electrophilicity value of BB9 (ω = 6.178) is greater than that of BY28 (ω = 2.48); this indicates that the molecule of BB9 is more electrophilic than that of BY28. Consequently, BB9 will be adsorbed first followed by BY28.

Dye	N° of atoms	Atoms	P_k^+	P_k^-	ω_k	N_k
BB9	1	C	0.117052	−0.091655	0.72314726	—
	2	C	−0.083178	0.109989	—	0.52134786
	3	C	0.123453	−0.0193705	0.76269263	0.09195176
	4	C	−0.0825946	0.0874818	−0.51026944	0.4152761
	5	C	0.1346605	−0.0391588	0.83193257	0.18588682
	6	C	−0.044701	0.0617533	−0.27616278	0.29314292
	7	C	−0.0622537	0.0077143	−0.38459903	0.03661978
	8	C	0.1288878	−0.049104	0.79626883	0.23309669
	9	C	−0.0873976	0.0988698	−0.53994237	0.46933494
	11	C	0.1294818	−0.031639	0.79993856	0.15019033
	12	C	−0.045136	0.063713	—	0.30199962
	13	C	0.1402246	−0.0370003	0.86630758	0.17564042
	19	N	0.3640112	−0.0556138	2.24885996	—
	20	S	0.1232636	−0.0137067	0.76152252	—
	21	N	0.0351644	0.09916571	0.21724566	0.47073963
	22	N	0.0643843	0.15539811	0.39776621	0.73767483

Dye	N° of atoms	Atoms	P_k^+	P_k^-	ω_k	N_k
	23	C	−0.0038013	−0.006641	—	—
	27	C	−0.0035767	−0.00689	—	—
	31	C	−0.0061814	−0.005169	—	—
	35	C	−0.0061321	−0.010607	—	—
	39	Cl	0.0012179	0.5782508	0.00752419	2.74495655
BY28	1	C	0.040255	0.066561	0.105689697	0.23709028
	2	C	−0.000773	0.00887	—	0.03159494
	3	C	0.045305	0.037015	0.118214337	0.13184743
	4	C	−0.025946	−0.014818	—	—
	5	C	0.066052	0.091588	0.172349484	0.32623646
	6	C	−0.021701	−0.037533	—	—
	7	C	−0.036225	0.001143	—	0.00407137
	8	C	0.458878	−0.089904	1.197350365	—
	13	C	−0.173976	0.168698	—	0.60090228
	15	C	−0.014488	0.091639	—	0.32641812
	16	C	0.022465	0.020003	0.058617925	0.07125069
	17	C	0.03011	−0.006138	0.078566023	—
	18	C	−0.011644	0.016571	—	0.05902590
	20	C	−0.013843	0.039811	—	0.14180678
	22	C	0.026365	0.07067	0.068794195	0.25172654
	24	N	0.121797	0.125083	0.317804912	0.44554565
	25	N	0.362411	−0.068898	0.945639022	—
	26	N	0.059207	0.13669	0.154488825	0.48688978
	27	C	−0.006573	−0.003929	—	—
	31	C	0.019834	0.001078	0.051752856	0.00383984
	35	C	0.016503	0.000443	0.043061278	0.00157797
	39	O	0.000958	0.10219	0.002499709	0.36400078
	40	O	0.00195	0.081069	0.005088135	0.28876778
	42	C	−0.000314	−0.005534	—	—

Table 5.
Theoretical prediction of reactive sites using Parr function for BB9 and BY28 dyes.

3.5.2 Local molecular reactivity

The local reactivity site of the studied dyes has been analyzed by evaluating Parr functions (PF). The PF is used to obtain the detail information of local reactivity of each atom in the molecule. Domingo proposed the Parr functions P(r) [61], which are given by the following equations:

$$P^{+}(r) = \rho_s^{ra}(r) \quad \text{for nucleophilic attack} \tag{19}$$

$$P^{-}(r) = \rho_s^{rc}(r) \quad \text{for electrophilic attack} \tag{20}$$

With these electrophilic and nucleophilic Parr functions are at hand, the local electrophilicity ω_K and the local nucleophilicity N_K indices will be redefined as follows:

$$\omega_K = \omega P_K^{+} \tag{21}$$

$$N_K = \mathrm{N}P_K^{-} \tag{22}$$

Table 5 shows that at the DFT level, the most susceptible site to a nucleophilic attack for BB9 is located on sulfur, nitrogen, and benzene ring. In the case of an electrophilic attack, the most reactive site is on Cl39. For BY28 the more susceptible sites to nucleophilic attacks are nitrogen and C8 atoms, while N26 and C13 are the most susceptible sites for electrophilic. The results indicated that the BB9 dye has more and strong electrophilic sites than BY28; consequently, BB9 has a high affinity for NSC than BY28.

3.5.3 MC and MD simulations

In this study, Monte Carlo simulations were performed to study the adsorption and orientation of dyes on charged surfaces based on (001) kaolinite surface and all-atom models. The most stable low-energy adsorption configurations of the studied dyes are shown in **Figure 13**. It is clear that the three dyes examined BB9, BB41, and BY28 are adsorbed almost parallel to the plane to maximize surface and contact coverage. These adsorption configurations indicate that there are strong interactions between the studied dyes and the kaolinite atoms. This facilitates their adsorption to the surface of the kaolinite (001) by blocking a maximum of sites and ensuring a great influence on the removal efficiency.

The outputs and descriptors calculated by the Monte Carlo simulation are presented in **Table 6**. The parameters presented in **Table 6** include total energy, in kcal mol^{-1}, of the substrate–adsorbate configuration. As can be seen from **Table 6**, BB9 gives the maximum adsorption energy in negative value found during the simulation process. High values of adsorption energy indicate that BB9 molecule will give the highest removal efficiency and strong interaction between a kaolinite substrate and the studied dye. These results are in good agreement with experimental findings.

To further confirm our results, we have performed the energy fluctuation curves as obtained from MD simulations; the equilibration of the system is confirmed by the stable mean values of energy fluctuations, as shown in **Table 7**. The mean square displacement (MSD) and the diffusion coefficient were calculated after 100,000 steps. The obtained data included in **Table 7** show that the diffusion coefficient of the free water molecules was more pronounced (5.85×10^{-6} cm^2/s) than the water with BB9 and BY28 molecules. Much smaller diffusion coefficients obtained for water with BB9 were caused by the strong interaction between (water + BB9) and the kaolinite surface, which decreased the mobility of the water [62].

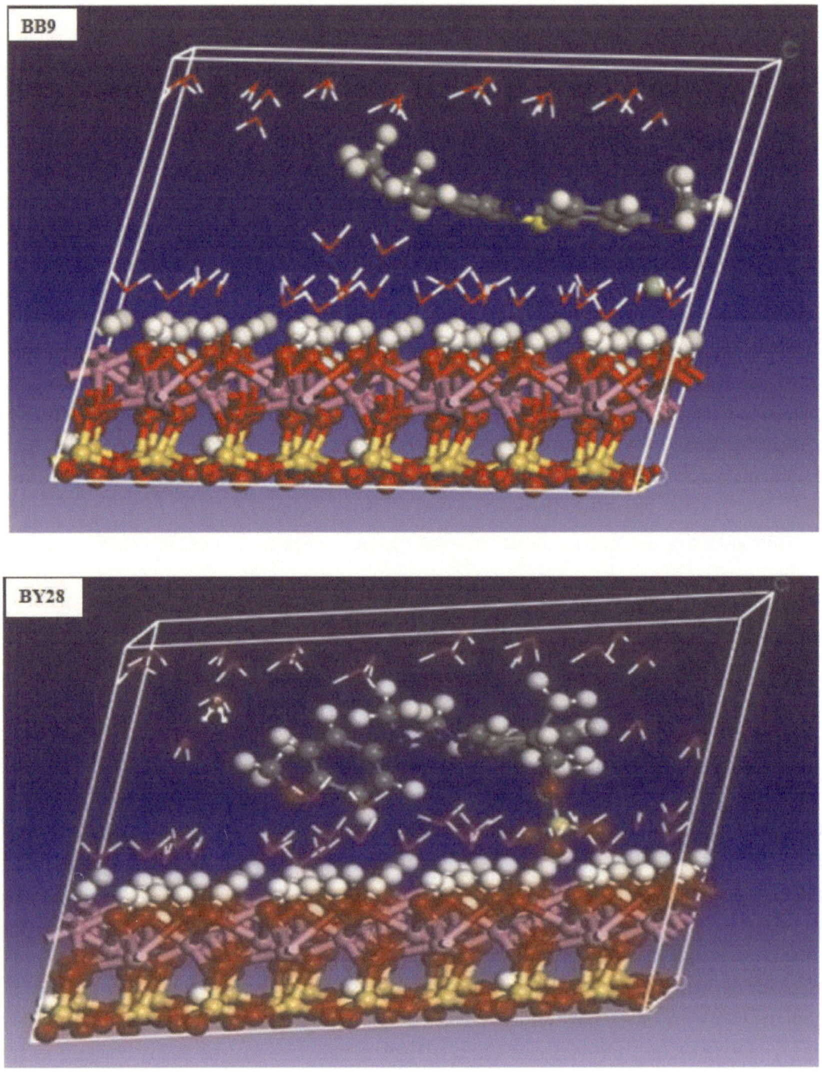

Figure 13.
The most stable low-energy configuration for the adsorption of the dyes on kaolinite surface obtained through the Monte Carlo simulations.

System	Total energy	Adsorption Energy	Rigid adsorption energy	Deformation energy	Ead/Ni Dye
BB9/(001)	−111.4	−317.53	−150.77	−166.5	−317.53
BY28/(001)	−213.8	−146.92	−117.07	−29.89	−147.18

Table 6.
Outputs and descriptors calculated by the Monte Carlo simulations for the lowest adsorption configurations of tested dyes on kaolinite (001) surface (in kcal/Mol).

Conditions	Diffusion coefficient of water (10^{-6} cm^2/s)
Free water molecules	5.850
Water with BY28	2.430
Water with BB9	0.450

Table 7.
Calculated diffusion coefficient of free water and water with dyes in kaolinite surface.

4. Conclusion

The results obtained in this research indicated that natural safiot clay was considered low cost, eco-friendly, and a promising alternative adsorbent for removing BB9 and BY28 dyes in single and binary systems from an aqueous solution. The percentage of dye removed increased with increasing adsorbent dosage, decreased with increasing initial dye concentration, and varied with dye solution pH. The rate of adsorption was found to obey pseudo-second-order kinetics model with a good correlation coefficient indicating toward chemisorption, and the Langmuir isotherm represented the equilibrium adsorption and shows monolayer homogeneous surface of adsorbent for single and binary mixture.

All the theoretical parameters of DFT calculations show that the two dyes can be adsorbed and that the BB9 molecule has the best percentage of elimination. From this calculation, we note that: The dye removal efficiency increases with the highest values of ω, E_{LUMO}, μ, and S and the lowest values of ΔEgap, E_{HOMO} and η. Another important conclusion that we can draw from these calculations is that the theoretical results from reactivity descriptors show that nitrogen, sulfur, carbon, and nitrogen atoms of BB9 and BY28 are the main adsorption sites, respectively. Finally, this study displays a good correlation between the theoretical and experimental data, which confirm the reliability of the quantum chemical methods to study the competitive adsorption of two cationic dyes onto NSC surfaces, in the single and binary system. DFT calculations, Monte Carlo method, and Molecular dynamics simulations support the experimental findings.

Author details

Aziz El Kassimi*, Mohammadine El Haddad, Rachid Laamari, Mamoune El Himri, Youness Achour and Hicham Yazid
Poly-disciplinary Faculty of Safi, Cadi Ayyad University, Laboratory of Analytical and Molecular Chemistry, Safi, Morocco

*Address all correspondence to: azizelkassimi2012@gmail.com

References

[1] Anuar FI, Hadibarata T, Syafrudin M, Fona Z. Removal of Procion Red MX-5B from aqueous solution by adsorption on Parkia speciosa (stink bean) peel powder. Biointerface Research in Applied Chemistry. 2020;**10**:4774-4779

[2] Katheresan V, Kansedo J, Lau SY. Efficiency of various recent wastewater dye removal methods: A review. Journal of Environmental Chemical Engineering. 2018;**6**:4676-4697

[3] Slimani R, El Ouahabi I, Abidi F, El Haddad M, Regti A, Laamari MR, et al. Calcined eggshells as a new biosorbent to remove basic dye from aqueous solutions: thermodynamics, kinetics, isotherms and error analysis. Journal of the Taiwan Institute of Chemical Engineers. 2014;**45**(4):1578-1587

[4] El Haddad M, Mamouni R, Saffaj N, Lazar S. Evaluation of performance of animal bone meal as a new low cost adsorbent for the removal of a cationic dye Rhodamine B from aqueous solutions. Journal of Saudi Chemical Society. 2016;**20**:S53-S59

[5] Regti A, El Ayouchia HB, Laamari MR, Stiriba SE, Anane H, El Haddad M. Experimental and theoretical study using DFT method for the competitive adsorption of two cationic dyes from wastewaters. Applied Surface Science. 2016;**390**:311-319

[6] Somsesta N, Sricharoenchaikul V, Aht-Ong D. Adsorption removal of methylene blue onto activated carbon/cellulose biocomposite films: Equilibrium and kinetic studies. Materials Chemistry and Physics. 2020;**240**:122221

[7] Kristanti RA, Hadibarata T, Al Qahtani HMS. Adsorption of bisphenol A on oil palm biomass activated carbon: characterization, isotherm, kinetic and thermodynamic studies. Biointerface Research in Applied Chemistry. 2019;**9**: 4217-4224

[8] Sözüdoğru O, Fil BA, Boncukcuoğlu R, Aladağ E, Kul S. Adsorptive removal of cationic (BY2) dye from aqueous solutions onto Turkish clay: Isotherm, kinetic, and thermodynamic analysis. Particulate Science and Technology. 2016;**34**(1):103-111

[9] Mukherjee K, Kedia A, Rao KJ, Dhir S, Paria S. Adsorption enhancement of methylene blue dye at kaolinite clay-water interface influenced by electrolyte solutions. RSC Advances. 2015;**5**: 30654-30659

[10] Turabik M. Adsorption of basic dyes from single and binary component systems onto bentonite: Simultaneous analysis of Basic Red 46 and Basic Yellow 28 by first order derivative spectrophotometric analysis method. Journal of Hazardous Materials. 2008; **158**(1):52-64

[11] Lee YC, Kim JY, Shin HJ. Removal of malachite green (MG) from aqueous solutions by adsorption, precipitation and alkaline fading using talc. Separation Science and Technology. 2013;**48**: 1093-1101

[12] Tsai WT, Chang CY, Ing CH, Chang CF. Adsorption of acid dyes from aqueous solution on activated bleaching earth. Journal of Colloid and Interface Science. 2004;**275**:72-78

[13] Gürses A, Karaca S, Doar C, Bayrak R, Açıkyıldız M, Yalçın M. Determination of adsorptive properties of clay/water system: methylene blue sorption. Journal of Colloid and Interface Science. 2004;**269**:310-314

[14] Kausar A, Iqbal M, Javed A, Aftab K, Nazli ZH, Bhatti HN, et al. Dyes adsorption using clay and modified clay: A review. Journal of Molecular Liquids. 2018;**256**:395-407

[15] Ferfera-Harrar H, Benhalima T, Sadi A. Development of functional chitosan-based superabsorbent hydrogel nanocomposites for adsorptive removal of Basic Red 46 textile dye. Polymer Bulletin. 2021:1-32

[16] El Kassimi A, Achour Y, El Himri M, Laamari R, El Haddad M. Removal of two cationic dyes from aqueous solutions by adsorption onto local clay: Experimental and theoretical study using DFT method. International Journal of Environmental Analytical Chemistry. 2021:1-22

[17] Nakarmi A, Bourdo SE, Ruhl L, Kanel S, Nadagouda M, Alla PK, et al. Benign zinc oxide betaine-modified biochar nanocomposites for phosphate removal from aqueous solutions. Journal of Environmental Management. 2020; **272**:111048

[18] Dehmani Y, Lgaz H, Alrashdi AA, Lamhasni T, Abouarnadasse S, Chung IM. Phenol adsorption mechanism on the zinc oxide surface: Experimental, cluster DFT calculations, and molecular dynamics simulations. Journal of Molecular Liquids. 2021;**324**:114993

[19] Yu S, Wang X, Ai Y, Tan X, Hayat T, Hu W, et al. Experimental and theoretical studies on competitive adsorption of aromatic compounds on reduced graphene oxides. Journal of Materials Chemistry A. 2016;**4**: 5654-5662

[20] El Kassimi A, Boutouil A, El Himri M, Laamari R, El Haddad M. Selective and competitive removal of three basic dyes from single, binary and ternary systems in aqueous solutions: A combined experimental and theoretical study. Journal of Saudi Chemical Society. 2020;**24**(7):527-544

[21] Regti A, El Ayouchia HB, Laamari MR, Stiriba SE, Anane H, El Haddad M. Experimental and theoretical study using DFT method for the competitive adsorption of two cationic dyes from wastewaters. Applied Surface Science. 2016;**390**:311-319

[22] Achour Y, Khouili M, Abderrafia H, Melliani S, Laamari MR, El Haddad M. DFT investigations and experimental studies for competitive and adsorptive removal of two cationic dyes onto an eco-friendly material from aqueous media. International Journal of Environmental Research. 2018;**12**:789-802

[23] Liu Y, Ma X, Li HA, Hou J. Competitive adsorption behavior of hydrocarbon (s)/CO2 mixtures in a double-nanopore system using molecular simulations. Fuel. 2019;**252**: 612-621

[24] Kohn W, Sham LJ. Self consistent equations including exchange and correlation effects. Physics Review. 1965; **140**:A1133

[25] Koopmans T. Ordering of wave functions and eigenenergies to the individual electrons of an atom. Physica. 1933;**1**:104-113

[26] Schlegel HB. Geometry optimization on potential energy surfaces. In: Modern Electronic Structure Theory: Part I. 1995. pp. 459-500

[27] Pearson RG. Hard and soft acids and bases. Journal of the American Chemical Society. 1963;**85**:3533-3539

[28] Lesar A, Milošev I. Density functional study of the corrosion

inhibition properties of 1,2,4-triazole and its amino derivatives. Chemical Physics Letters. 2009;**483**:198-203

[29] Parr RG, Szentpaly LV, Liu S. Electrophilicity index. Journal of the American Chemical Society. 1999;**121**: 1922-1924

[30] Domingo LR, Chamorro E, Pérez P. Understanding the reactivity of captodative ethylenes in polar cycloaddition reactions. A theoretical study. The Journal of Organic Chemistry. 2008;**73**:4615-4624

[31] Guo T, Wang K, Zhang G, Wu X. A novel α-Fe2O3@ g-C3N4 catalyst: Synthesis derived from Fe-based MOF and its superior photo-fenton performance. Applied Surface Science. 2019;**469**:331-339

[32] El Gaidoumi A, Benabdallah AC, Lahrichi A, Kherbeche A. Adsorption du phénol en milieu aqueux par une pyrophyllite marocaine brute et traitée (Adsorption of phenol in aqueous medium by a raw and treated moroccan pyrophyllite). Journal of Materials and Environmental Science. 2015;**6**:2247-2259

[33] Bellamy LJ. Amides, proteins and polypeptides. In: The Infra-red Spectra of Complex Molecules. Dordrecht: Springer; 1975. pp. 231-262

[34] Saikia BJ, Parthasarathy G, Borah RR, Borthakur R. Raman and FTIR spectroscopic evaluation of clay minerals and estimation of metal contaminations in natural deposition of surface sediments from Brahmaputra river. International Journal of Geosciences. 2016;7:873-883

[35] Miyah Y, Lahrichi A, Idrissi M, Anis K, Kachkoul R, Idrissi N, et al. Removal of cationic dye "crystal violet" in aqueous solution by the local clay. JMES. 2017;**8**(10):3570-3582

[36] Aroke UO, Abdulkarim A, Ogubunka RO. Fourier-transform infrared characterization of kaolin, granite, bentonite and barite. ATBU Journal of Environmental Technology. 2013;**6**:42-53

[37] Kousha M, Daneshvar E, Salar-Sohrabi M, Jokar M, Bhatnagar A. Adsorption of acid orange II dye by raw and chemically modified brown macroalga *Stoechospermum marginatum*. Chemical Engineering Journal. 2012;**192**: 67-76

[38] Singh H, Choden S. Comparison of adsorption behaviour and kinetic modeling of bio-waste materials using basic dye as adsorbate. Indian Journal of Chemical Technology. 2015;**21**:359-367

[39] Ghasemian Lemraski E, Sharafinia S, Alimohammadi M. New activated carbon from Persian mesquite grain as an excellent adsorbent. Physical Chemistry Research. 2017;**5**:81-98

[40] Zhang L, Hu P, Wang J, Huang R. Crosslinked quaternized chitosan/ bentonite composite for the removal of Amino black 10B from aqueous solutions. International Journal of Biological Macromolecules. 2016;**93**:217-225

[41] Kausar A, Shahzad R, Iqbal J, Muhammad N, Ibrahim SM, Iqbal M. Development of new organic-inorganic, hybrid bionanocomposite from cellulose and clay for enhanced removal of Drimarine Yellow HF-3GL dye. International Journal of Biological Macromolecules. 2020;**149**:1059-1071

[42] Ngwabebhoh FA, Erdem A, Yildiz U. Synergistic removal of Cu (II) and nitrazine yellow dye using an eco-friendly chitosan-montmorillonite

hydrogel: Optimization by response surface methodology. Journal of Applied Polymer Science. 2016;**133**(29)

[43] Wang Y, Wang W, Wang A. Efficient adsorption of methylene blue on an alginate-based nanocomposite hydrogel enhanced by organo-illite/smectite clay. Chemical Engineering Journal. 2013;**228**:132-139

[44] El Mouzdahir Y, Elmchaouri A, Mahboub R, Gil A, Korili SA. Equilibrium modeling for the adsorption of methylene blue from aqueous solutions on activated clay minerals. Desalination. 2010;**250**:335-338

[45] Rahmat M, Rehman A, Rahmat S, Bhatti HN, Iqbal M, Khan WS, et al. Highly efficient removal of crystal violet dye from water by MnO2 based nanofibrous mesh/photocatalytic process. Journal of Materials Research and Technology. 2019;**8**(6):5149-5159

[46] Auta M, Hameed B. Chitosan–clay composite as highly effective and low-cost adsorbent for batch and fixed-bed adsorption of methylene blue. Chemical Engineering Journal. 2014;**237**:352-361

[47] Gao W, Zhao S, Wu H, Deligeer W, Asuha S. Direct acid activation of kaolinite and its effects on the adsorption of methylene blue. Applied Clay Science. 2016;**126**:98-106

[48] Gao Z, Li X, Wu H, Zhao S, Deligeer W, Asuha S. Magnetic modification of acid-activated kaolin: Synthesis, characterization, and adsorptive properties. Microporous and Mesoporous Materials. 2015;**100**(202):1-7

[49] Kausar A, Iqbal M, Javed A, Aftab K, Bhatti HN, Nouren S. Dyes adsorption using clay and modified clay: A review. Journal of Molecular Liquids. 2018;**256**: 395-407

[50] Şahin Ö, Kaya M, Saka C. Plasma-surface modification on bentonite clay to improve the performance of adsorption of methylene blue. Applied Clay Science. 2015;**116**:46-53

[51] Ai L, Zhou Y, Jiang J. Removal of methylene blue from aqueous solution by montmorillonite/CoFe2O4 composite with magnetic separation performance. Desalination. 2011;**266**:72-77

[52] Bolster CH, Hornberger GM. On the use of linearized Langmuir equations. Soil Science Society of America Journal. 2007;**71**(6):1796-1806

[53] Şişmanoğlu T. Corrigendum to "Removal of some fungicides from aqueous solution by the biopolymer chitin" [Colloid Surf. A: Physicochem. Eng. Aspects 297 (2007) 38–45]. Colloids and Surfaces A: Physicochemical and Engineering Aspects. 2007;**3**:257

[54] Freundlich HMF. Over the adsorption in solution. The Journal of Physical Chemistry. 1906;**57**(385471):1100-1107

[55] Ergene A, Ada K, Tan S, Katırcıoğlu H. Removal of Remazol Brilliant Blue R dye from aqueous solutions by adsorption onto immobilized *Scenedesmus quadricauda*: Equilibrium and kinetic modeling studies. Desalination. 2009;**249**:1308-1314

[56] Aziz EK, Abdelmajid R, Rachid LM, Mohammadine EH. Adsorptive removal of anionic dye from aqueous solutions using powdered and calcined vegetables wastes as low-cost adsorbent. Arab Journal of Basic and Applied Sciences. 2018;**25**(3):93-102

[57] Ho YS, McKay G. Sorption of dye from aqueous solution by peat. Chemical Engineering Journal. 1998; **70**(2):115-124

[58] Weber WJ, Morris JC, Sanit J. Kinetics of adsorption on carbon from solution. Journal of the Sanitary Engineering Division. 1963;**89**(2):31-59

[59] Varjani S, Rakholiya P, Ng HY, You S, Teixeira JA. Microbial degradation of dyes: an overview. Bioresource Technology. 2020;**314**:123728

[60] Almeida CAP, Debacher NA, Downs AJ, Cottet L, Mello CAD. Removal of methylene blue from colored effluents by adsorption on montmorillonite clay. Journal of Colloid and Interface Science. 2009;**332**(1):46-53

[61] Chamorro E, Pérez P, Domingo LR. On the nature of Parr functions to predict the most reactive sites along organic polar reactions. Chemical Physics Letters. 2013;**582**:141-143

[62] Sintakindi A, Ankamwar B. Fungal biosorption as an alternative for the treatment of dyes in waste waters: A review. Environmental Technology Reviews. 2021;**10**(1):26-43

Chapter 12

Classification of Clay Minerals

Praise Akisanmi

Abstract

The versatility of natural clay and their ability to adsorb a variety of environmental contaminants present in the water effluents has attracted esthetic concern among environmentalist. These practical applications rely primarily on the diversity of natural clay structure to retain harmful and undesirable substances from the immediate environment. The adsorptive capability of natural clays is related to the fundamental units of the clay-sized crystalline minerals which present in different combinations.

Keywords: clay mineral, classification, application, bentonite, kaolinite

1. Introduction

The significance of solid mineral resources has been of profound value to man since time immemorial. A mineral is (most of the time) an inorganic crystalline solid, natural, homogeneous, with a structure and a composition that give it defined macroscopic properties. Clays are minerals categorized under the clastic sedimentary rocks. Clay is a naturally occurring material composed of layered structures of fine-grained minerals which exhibit the property of plasticity at appropriate water content but becomes permanently hard when fired [1]. The clay material is formed from chemical alteration processes on the earth's surface and accounts for approximately 40% of the first-class grained sedimentary rocks (mudrocks) which incorporates dust stones, clay stones and shales. Clayey minerals are usually composed of aluminum silicates that are made up of tetrahedral and octahedral leaves that are bound together collectively via sharing of apical oxygen atoms [2]. The formation of clayey minerals depends on the physical-chemical conditions of the environment of the immediate altering environment, the nature of the raw materials and other related external environmental factors [3]. As such resulting in different types of clay materials. Hence, the potential for application of any clay mineral type in nature will depend on its will depend on its chemistry, structure and other intrinsic properties [1]. Natural clay minerals are widely recognized and acquainted to mankind since the first days of civilization. Owing to their low cost, plenty in most continents of the world, high sorption capability for ion exchange, clay substances are solid candidates as adsorbents [4]. Clay minerals share a fundamental set of structural and chemical traits but yet has its very own precise set of properties that determine its interplay with other chemical species. The variability of chemistry and structure between clays leads to their application in a wide variety of fields.

2. Clay mineral

Clay is usually fine-grained materials, with particle size lower than 0.002 mm with majorly clay minerals. Other minerals related to clay minerals in clays might embody quartz and feldspar, along with detritic materials that have been eroded off the earth's surface. Clay minerals do not seem to be the most precious among the minerals on the face of the earth, but they affect life on the earth to a great extent. Clay is one of the most ancient mineral substances used by mankind. Clay is a widely dispersed, plentiful mineral resource with substantial industrial significance for a wide range of applications. It is one of the most important minerals in the world, both in terms of value and annual production. The term "clay," like many geological concepts, is vague and has numerous meanings: clay minerals are a group of fine-grained minerals with a particle size (smaller than silt) and a kind of rock—a fine-grained sedimentary deposit dominated by clay particles. Clay also includes fine-grained non-aluminosilicate deposits like shale and some argillaceous soils under the later definition. It was immensely essential in ancient civilizations, with records kept in brick structures, monuments, and pottery, as well as writings on clay tablets. Clay continues to be a fundamental raw material in today's world. Clay and clay compounds are now used in far too many ways to mention them all. Clay has a variety of industrial applications, including the manufacture of refractories and drilling mud in the water, oil, and gas sectors. The success of these research hinges on the appropriate exploitation of local raw resources for use as drilling mud in order to achieve long-term economic growth and job creation. Regardless of how abundant clay is and how widely it is used in industry, certain property standards must be satisfied by either raw or refined clay. Clay refining improves the geotechnical qualities of natural clays, potentially increasing their industrial potential. The raw materials used to make mud are generally chosen clays with a significant quantity of montmorillonite, and their behavior in water is utilized to appraise them. Their suitability is decided through numerous criteria, among which can be the viscosity, the volumetric yield of given clay and filtration characteristics. The bentonite clay is the raw material that satisfies the majority of these drilling criteria. Drilling for oil, gas, and water resources often necessitates a large volume of drilling mud, which is frequently imported into the nation. Apart from eroding the country's hard currency reserves, such importation is also incompatible with the country's present local content strategy for the oil industry. As a result, the necessity for bentonitic clays or comparable alternatives to be sourced locally has become critical.

2.1 Classification of clay minerals

Clay minerals are classified into different groups as follows; Kaolinite, Smectite, Vermiculite, Illite and Chlorites.

Kaolinite is the purest of all clays, with little variation in composition. It also does not absorb water or expand when it comes into touch with it. As a result, kaolinite $Al_2Si_2O_5(OH)_2$ is the ceramic industry's chosen clay. Kaolinite clays have long been used in the ceramic industry, especially in fine porcelains, because they can be easily molded, have a fine texture, and are white when fired. These clays are also used as a filler in making paper. Clay minerals such as kaolinite, hallosite, nacrite, and dickite belong to the Kaolinite group, which is a 1:1 type clay mineral. It is made up of one layer of silica and one layer of alumina, which is created by advanced weathering processes or hydrothermal modifications of feldspars and other alluminosilicates in acidic circumstances [5]. The chemical formula of kaolinite is $Al_2O_3.2SiO_2.2H_2O$ (39% Al_2O_3, 46.5% SiO_2 and 14.0% H_2O) and its structure possesses strong binding forces between the layers which resists expansion when wetted [1]. Kaolin is mostly white in color, has a very tiny particle size, is nonabrasive (hardness 2–2.5 on the Mohs scale), and is chemically inert in most applications. The individual kaolin particle is a thin, flat, pseudohexagonal platelet, so small that if 10,000,000 of them were distributed on a postage stamp, the layer would be thinner than a human hair [1]. In many applications, the thin, flat particle form is advantageous. The particle size and color, or brightness, of commercially available kaolin is used to determine the grade. Kaolin delivers strength, dimensional stability, and smooth surfaces to completed whitewares and sanitary wares by providing a white body, easy molding qualities, and adding strength, dimensional stability, and smooth surfaces. Kaolins are ideal for particular refractories because of their refractoriness, dimensional stability, and chemical inertness. Kaolin's outstanding dielectric qualities, in addition to the foregoing, make it ideal for porcelain electric insulators. Industrial uses of kaoline includes production of paper, paint, rubber, ceramic, plastic, and medicinal products, as well as a catalyst for petroleum cracking and vehicle exhaust emission control systems, and as a cosmetics foundation and pigment [6]. Furthermore, kaolin is also used as an anti-cracking agent in the manufacturing of fertilizer prills, as a pesticide carrier, in the production of white cement (where it supplies alumina without iron), and in the production of glass fiber as a low-iron, low-alkali alumina source. Kaopectate and Rolaids, for example, are the main ingredients in the original formulation of anti-diarrhea medication in pharmaceutical applications. Through firm and selective binding of aflatoxins, plant secondary metabolites, pathogenic microorganisms, heavy metals, and other poisons in animal diets that could be harmful to the digestive system, kaolin can be used to decontaminate aflatoxins, plant secondary metabolites, pathogenic microorganisms, heavy metals, and other poisons that could be harmful to the digestive system [7]. Naturally, kaolin may be accompanied by other mineral impurities such as feldspar and mica, quartz, titanoferous, illite, montmorillonite, ilmenite, anastase, hematite, bauxite, zircon, rutile, silliminate, graphite, attapugite, halloysite and carbonaceous materials [8] thus reducing its industrial usefulness.

Smectite, which includes montmorillonite, beidellite, nantronite, saponite and hectorite, are 2:1 layer clay minerals formed from the weathering of soils, rocks (mainly bentonite) or volcanic ash and belongs to a group of hydroxyl aluminosilicate [1]. The most common smectite is montmorillonite, with a general chemical formula: $(1/2Ca,Na)(Al,Mg,Fe)\ 4(Si,Al)8O_{20}(OH)_4.nH_2O$ Smectites are a category of dioctahedral 2:1 expandable minerals having a charge of 0.2–0.6 per formula unit. The octahedral substitution of Mg^{2+} for Al^{3+} gives Montmorillonite, the most prevalent member of this group, its charge. Tetrahedral replacements provide much of the charge in beidellite and nontronite, which are less common in soils. The presence of iron in the octahedral sheet distinguishes nontronite from beidellite. Van der Waals connections and weak cation-to-oxygen links hold the 2:1 layers in smectites together. The presence of exchangeable cations in the interlayer between water molecules causes the crystal lattice to expand as the mineral hydrates. The basal spacing between layers can exceed 2 nm when the material is saturated with water, but it can be lowered to less than 1 ran when the mineral is dry. The major component of bentonite is montmorillinite, which is formed by weathering volcanic ash. When water comes into contact with montmorillinite, it expands by many times its original volume. It may be used as a drilling mud (to keep drill holes open) as well as to seal leaks in soil, rocks, and dams because of this. This expansion and contraction trait found in smectites, often referred to as shrink-swell potential, is problematic to engineers and farmers alike due to the propensity for crack formation and general instability of the soil surface. Montmorillinite, however, is a dangerous type of clay to encounter if it is found in tunnels or road cuts. Because of its expandable nature, it can lead to serious slope or wall failures. Differences in the degree of chemical substitution within the smectite structure, the nature of the exchangeable cations present, and the type and quantity of impurities present induce variation in the physical and chemical characteristics of bentonites within and across deposits [1]. Quartz, cristobalite, feldspars, zeolites, calcite, volcanic glass, and other clay minerals such as kaolinite are all minerals found in smectites [9]. Differences in chemical composition due to replacements of Al^{3+} or Fe^{3+} for Si^{4+} in the tetrahedral cation sites and Fe^{2+}, Mg^{2+}, or Mn^{2+} for Al^{3+} in the octahedral cation sites define the groups of smectite clays. Smectites contain very thin layers and microscopic particle sizes, resulting in a large surface area and hence a high degree of absorbency for a variety of compounds such as oil, water, and other chemicals [1]. Because of their high cation exchange capabilities, surface area, surface reactivity, adsorptive capacity, and catalytic activity, smectites are important minerals for industrial purposes. Bonding foundry sands, drilling fluids, iron ore pelletizing, agriculture (as a carrier material for pesticides, fertilizers, and seed coating), paper making, paints, pharmaceuticals, cosmetics, plastics, adhesives, decolorization, and ceramics are just a few of the applications for this group of clays [10]. The substance is also employed as a clarifier for oils and fats, as well as a chemical barrier, a liquid barrier, and a catalyst [11]. Purification and physicochemical changes of pure smectite are required for the preparation of several high-tech materials such as pillared clays, organoclays, and polymer/smectite-nano ccomposites [1]. Also they are used in many industries; the most important uses are as drilling muds and catalysts in the petroleum industry, as bonding clays in foundries, as bonding agents for taconite pellets, and as adsorbents in many industries. However, the commercial bentonites should contain not less than 60% smectite.

Vermiculite is a high-charge 2:1 phyllosilicate clay mineral. It is generally regarded as a weathering product of micas. Vermiculite is also hydrated and somewhat expansible though less so than smectite because of its relatively high charge. It has a layer charge of 0.9–0.6 per formula unit, and contains hydrated exchangeable cations primarily Ca, and Mg in the interlayer [12]. In soils, vermiculite exists as an Al^{3+} dominated dioctahedral and, to a lesser extent, Mg^{2+} dominated trioctahedral mineral. Water molecules and exchangeable cations—primarily Mg^{2+} and Ca^{2+}—are highly adsorbed within the interlayer region of vermiculites due to the tetrahedral charge origin. In vermiculites, unlike smectites, the strong bonding of the interlayer cations binds the 2:1 layers together, restricting basal spacing expansion to 1.5 nm. Vermiculite has a high cation exchange capacity due to its high charge per formula unit, and this clay type has a strong affinity for weakly hydrated cations including K^+, NH_4^+, and Cs^+. The water in raw flakes vermiculite flashes into steam and the flakes expand into accordion-like particles when heated rapidly to 900°C or higher [13], a phenomenon known as exfoliation [14]. Exfoliation, liberates bound water from between the mica-like layers of the mineral and literally expands the layers apart at right angles to the cleavage plane. The expanded or exfoliated material is low in density, chemically inert and adsorbent has excellent thermal and acoustic insulation properties, is fire resistant and odorless. Granular clay absorbents, such as vermiculite, have been used for over 75 years to clean up minor drips, spills and over sprays in factories and garages. Vermiculite is used to loosen and aerate soil mixes. Mixed with soil, it improves water retention and fertilizer release, making it ideal for starting seeds.

Also used as a medium for winter storage of bulbs and flower tubers. The common applications of exfoliated vermiculite include making of friction light weight aggregates, thermal insulator, brake linings, various construction products, animal feeds and in horticulture [1]. Vermiculite in fertilizers improves the efficiency with which nutrients are released, making fertilizers more cost-effective for customers [15]. Vermiculite's layered structure and surface qualities allow it to be employed in intumescent coatings and gaskets, as well as the treatment of hazardous waste and air freight. The internal pressure generated by the expansion of vermiculite when heated is sufficient to crush hard rock during tunneling activity [16]. Other minerals such as feldspars, pyroxenes, amphiboles, carbonates, and quartz, which develop alongside vermiculite in the rock and appear as major components, as well as minor components such as phosphates, iron oxides, titanium oxides, and zircon, can be found in vermiculite ores. Some impurities, such as asbestiform amphibole minerals present in vermiculite, have a negative impact on human health, as they can cause illnesses like malignant mesothelioma, asbestosis, or lung cancer; hence, clay characterization is necessary to detect such impurities [1].

Illite is the most frequent clay mineral, accounting for more than half of the claymineral suite in the deep sea. It is comparable to muscovite. They form in temperate climates or at high altitudes in the tropics, and they generally reach the ocean by rivers and wind transmission. Illite clays have a structure similar to muscovite, although they are generally low in alkalies and contain less Al substitution for Si. As a result, the typical formula for illites is KyAl4(Si8-y,Aly)O20(OH)4, with $1 < y < 1.5$, but always with $y < 2$. Ca and Mg can occasionally be used in place of K due to a charge imbalance. The interlayer cations of K, Ca, or Mg prohibit H_2O from entering the structure. As a result, the illite are non expanding clays. Clay micas are another name for the illite clay mineral group. Mica is a phyllosilicate mineral that may be split or delaminated into thin sheets that are platy, flexible, clean, elastic, transparent to opaque, robust, reflecting, refractive, dielectric, chemically inert, insulating, light weight, and hydrophilic [1]. Mica minerals' atoms are bound together into flat sheets, allowing for flawless cleavage of the minerals to generate durable sheets in a variety of colors, including brown, green, black, violet, and colorless, with a vitreous to pearly shine [17]. There are around 30 members of the mica group, but muscovite, biotite, phlogopite, lepidolite, fuchsite, and zinnwaldite are the six most frequent forms found in nature and employed in microscopy and other analytical applications [1]. Clay minerals are made up of three members (the illite group), which include illite, glauconite, and muscovite, and display clay-like characteristics, with illite being the most frequent. Illite is generated by alkaline weathering of potassium and aluminum-rich rocks such as muscovite and feldspar. Illite is a 2:1 layer silicate clay mineral that is non-expansive due to poorly hydrated potassium cations or calcium and magnesium ions filling the gap between the crystals of individual clay particles, preventing water molecules from entering the clay structure. Illite's cation exchange capacity ranges from 20 to 40 meq per 100 g. Minerals range in color from gray white to silvery white to greenish gray. Because of their high potassium concentration, illites are used in the structural clay industry and in agro minerals [18]. Quartz, feldspar, kaolin, and pyroxene are among the impurities found in mica clay ores [19]. The presence of these minerals in mica ores will affect the industrial value of the deposits as well as the processing complexity, lowering or boosting their value depending on the uses [20].

Chlorites are hydrous aluminosilicates with an interlayer organized in a 2:1 configuration. Chlorites are fundamental minerals found in soils that weather to generate vermiculite and smectite. Interlayered hydroxy-Al vermiculites or smectites, on the other hand, are regarded secondary minerals that develop as intermediate mineral weathering products or by the deposition of hydroxy-Al polymeric components inside the interlayer space of expanding minerals. There is no water adsorption within the interlayer space; thus, chlorites are considered nonexpansive minerals. These hydroxy-Al polycations balance a portion of the charge but they are not interchangeable. The CEC of the expandable 2:1 clays is lowered as a function of the quantity and valence of the hydroxy-Al polymer dwelling within the interlayer space because the level of hydroxy-Al occupancy within the interlayer space is varied. In the octahedral sheet inside the 2:1 layer and in the interlayer hydroxide sheet, they incorporate mostly Mg, Al, and Fe cations, with lesser amounts of Cr, Ni, Mn, V, Cu, and Li cations. In the tetrahedral sheet, they also show a substantial replacement of Si by Al cations [21]. Chlorites range in color from white to practically black or brown with a green tinge, and their optical qualities are linked to their chemical makeup [22]. In the study of phase interactions in low and intermediate grade metamorphic rock, understanding the chemical composition of chlorite is crucial [23]. The first mineralogical and chemical investigation of clay ores can be used to determine the appropriateness of the material for various uses, given the diversity of clay mineral groups in nature.

2.2 Other minerals found in clay

In addition to the phyllosilicates mentioned, the soil clay fraction may also contain minor amounts of oxides, hydroxides, and hydroxy-oxides (sesquioxides) of Si, Al, and Fe, as well as some weakly crystalline aluminosilicates. Quartz (owing to its excellent resistance to weathering) and opal (a weakly crystalline variant of quartz that precipitates from Si-supersaturated fluids or is of volcanic or biogenic origin) are two common Si-oxide minerals found in the clay fraction. Gibbsite ($Al(OH)_3$) is the most prevalent Al-hydroxide representative in severely worn soil clay fractions, whereas goethite (FeOOH) is the most common Fe-mineral in clay fractions. Allophane and imogolite, two weakly crystalline aluminosilicate clay minerals, are commonly found in clay fractions of volcanic soils. The primary varieties of commercial clays include ball clay, common clay and shale, and fire clay, in addition to bentonite (and Fuller's earth) and kaolin. Ball clay is mostly kaolinite, with traces of illite, chlorite, smectite minerals, quartz, and organic matter.

3. Applications of some essential clay minerals

3.1 Bentonite

The word "bentonite" is a bit of a misnomer. It is a rock consisting of extremely colloidal and plastic clays, mostly montmorillonite, a clay mineral of the smectite group, and created by in situ devitrification of volcanic ash, according to geologists [24]. Bentonite generated from ash falls is usually found in uniformly thick strata (ranging from a few millimeters to 15 meters) that cover enormous regions [24]. It is found in layers from ash falls and other sources all throughout the planet, although it is most plentiful in Cretaceous and newer rocks.

Bentonite is a rock made up of extremely colloidal and malleable clays, mostly montmorillonite, a smectite group clay mineral, that is created by devitrification of volcanic ash in situ. Bentonite may also contain feldspar, cristobalite, and crystalline quartz in addition to montmorillonite. The ability to produce thixotrophic gels with water, the ability to absorb huge amounts of water, and a high cation exchange capacity are all characteristics of bentonite.

In the foundry business, bentonites are used to bind sands into suitable forms in which metals may be cast. To hold the sand grains together, just 3–5% bentonite is required. Bentonites have a greater green, dry, and hot strength than any other form of clay due to their tiny particle size and the nature of their water adsorption. The taconite industry is quickly increasing the usage of bentonite as a bonding agent [25]. A variety of sodium bentonite's unique features are used in hazardous waste containment. Because of its swelling capacity, bentonite is an efficient soil sealant, since it fills the gaps in the soil by expanding inside the interstices of the soil with which it is combined, forming a very low permeability barrier. The capacity of bentonite to swell is facilitated by the potential of a very tiny average particle size, which allows it to fill even the tiniest of spaces. The strong cation exchange capacity improves waste retention, particularly for heavy metals. A combination of sodium bentonite and dirt also generates a stiff, flexible mastic that is exceptionally resistant and difficult to rupture. The capacity of bentonite to bind cationic metals and some pesticides has been used in experiments to detoxify paraquat poisoning patients and to reduce radiocaesium transfer to milk and other animal-derived foods.

3.2 Kaolinite

The word "kaolin" comes from the Chinese word Kau-Ling, which means "high ridge," and refers to a hill near Jau-chau Fu, where kaolin was first mined [26]. Kaolin, often known as china clay, is a clay that comprises 10–95% kaolinite and is generally composed primarily of kaolinite (85–95%).

Kaolin, sometimes known as china clay, is a mineral combination. Kaolinite is the primary component, however it also contains quartz, mica, feldspar, illite, and montmorillonite. Kaolinite is made up of triclinic crystal sheets with pseudohexagonal shape. Rock weathering is responsible for its formation. It has the ability to exchange cations. There are three methods for kaolinite to form:

- crumbling and transformation of rocks due to the effects of climatic factors (Zettlitz type);
- transformation of rocks due to hydrothermal effects (Cornwall type); and.
- formation by climatic and hydrothermal effects (mixed type).

The use of kaolin as a paper coating accounted for about half of total domestic consumption and roughly 80% of kaolin exports. Kaolin-coated papers are widely used in the production of cigarettes [24]. Smokers may be exposed to kaolinite particles by inhaling. Kaolin was also used as a filler in the manufacture of paint, paper, and rubber, as a component of fiber glass and mineral wool, as a landfill liner, and as a catalyst in the refinement of oil and gas. Historically, the use of kaolin in the making of porcelain and chinaware amounted for less than 1% of total domestic consumption in the United States. In 2003, ceramics accounted for 80–85% of overall manufacturing in China, with paper accounting for 5%, rubber for 3%, and paint accounting for 2%. Ceramics consumed 290,000 tonnes, paints 84,000 tonnes, paper/paperboard 68,000 tonnes, detergents 29,000 tonnes, and rubber 27,500 tonnes in India. Kaolinite has a variety of medically useful qualities. It is a good adsorbent that will adsorb not just lipids and proteins but also other substances [24] including viruses and bacteria [27]. It can be used to cause platelet aggregation and to start plasma coagulation by activating factor XII. [28] Also, non-specific haemaglutinin inhibitors should be removed from the serum. Kaolin is utilized as a local and gastrointestinal adsorbent in medicinal treatment (Kaopectate, bolus alba). Kaolin may be found in a variety of cosmetics, including eyeshadows, blushers, face powders, "powders," mascaras, foundations, and makeup bases. Kaolin was found in 509 different cosmetics in the United States in 1998, with concentrations ranging from 5 to 30%, but exceeding 84% in certain paste masks [29]. Medical, pharmaceutical, and cosmetic applications, on the other hand, accounted for around 0.01% of total kaolin usage in the United States.

4. Clay minerals AS potential application IN water purification

The assessment of water quality using the Water Quality Index (WOI) and every available means are being developed and used by researchers [30] and [31] applied the WOI assessment to the evaluation and validity of a river region with a single number that checked the multiple drinking water criteria The findings were able to certify whether the water was safe to drink or not, based on the set of criteria. Before being deemed safe for use, drinking water and water used for other purposes must be free of these toxins in order to encourage a healthy lifestyle free of illnesses linked to waterborne diseases [32]. Filtration, which is a typical procedure used in water purification, is one approach to remove harmful toxins. It entails the use of a medium, such as a membrane or aggregates, as well as the use of a membrane or aggregates. While adsorbing and absorbing pollutants on the media, filtration employs both physical and chemical processes. Ceramic (clay) membranes for point-of-use filtration, cloth or fiber membranes, compressed granular activated carbon (GAC), polymer membranes, sand, gravel, or crushed rock, and clay aggregates are only a few examples of filtration medium. Artifacts, medications, construction materials, electrochemical research, cosmetics, pharmaceuticals, earthen products, and agriculture are just a few of the economic benefits of clay [33], clay has played a vital role in water purification technology. Clay ceramic water filters have shown capabilities in removing water contaminants such as microbes [34], chemicals [18] and heavy metals. Ceramic water filters made from clay and clay minerals are very efficient for water filtration through adsorption/absorption, molecular sieving and ion exchange mechanisms [35]. Clay is also affordable, plentiful, ubiquitous, and easily accessible. User friendliness, cultural tolerance, and cheap maintenance costs are further factors to consider. Many scientific and technological advancements have been attributed to the characterization of materials in order to establish their important features for optimal use and application. Clay and clay minerals' suitability for use in water filtration media and other industrial applications is determined by their mineralogy, chemical composition, mechanical properties (such as plasticity), specific surface area, porosity, functionality, and structure, as well as their interactive behavior.

5. Conclusions

In this study, it has been established that clay minerals may appear not to be the most valuable among the minerals of the earth surface, yet they affect life on earth in far reaching ways. Little would be known to man in ages past that the uses of clay surpass molding bricks and all sort. Recent studies have shown that some of it classes are used for intumescent coatings and gaskets, treatment of toxic waste and air-freight. Due of its characteristic plastic property, clay has several industrial uses, which include manufacturing of refractories as well as in drilling mud in the water, oil and gas industries. Today, natural clay derivatives, a group of low-cost adsorbents, naturally with the hydrophilic characteristics have emerged to be a new solution to control the mobility and emission of water pollutants in the groundwater tables. Specifically, these unique adsorbents and its modified derivatives are found capable to remove anionic contaminants, hydrophobic or nonpolar organic pollutants through the interlayer (quaternary ammonium cations) exchange process.

Acknowledgements

My sincere gratitude goes to Almighty God for the gift of life, His grace, wisdom, knowledge and understanding, provisions, sound health and mind in the course of this write-up. Also, my profound gratitude goes to my parents; Pst. and Pst. (Mrs.) A.A. Akisanmi for their immerse support and inestimable love shown from the start of this write-up, it is unquantifiable. To my siblings; Peace and Precious, you are highly appreciated.

Author details

Praise Akisanmi
University of Ilorin, Ilorin, Nigeria

*Address all correspondence to: whizzprof07@gmail.com

References

[1] Ombaka O. Characterization and classification of clay minerals for potential applications in Rugi Ward, Kenya. African Journal of Environmental Science and Technology. 2016;**10**(11):415-431

[2] Nesbitt HW, Young GM, McLennan SM, Keays RR. Effects of chemical weathering and sorting on the petrogenesis of siliciclastic sediments, with implications for provenance studies. The Journal of Geology. 1996;**104**(5):525-542

[3] Wilson MJ. The origin and formation of clay minerals in soils; past, present and future perspectives. Clay Minerals. 1999;**34**(1):7-25

[4] Nayak PS, Singh BK. Instrumental characterization of clay by XRF, XRD and FTIR. Bulletin of Materials Science. 2007;**30**(3):235-238

[5] Ihekweme GO, Shondo JN, Orisekeh KI, Kalu-Uka GM, Nwuzor IC, Onwualu AP. Characterization of certain Nigerian clay minerals for water purification and other industrial applications. Heliyon. 2020;**6**(4)

[6] Olaremu AG. Physico-chemical characterization of Akoko mined kaolin clay. Journal of Minerals and Materials Characterization Engineering. 2015;**03**(05):353-361

[7] Trckova M et al. Kaolin, bentonite, and zeolites as feed supplements for animals: Health advantages and risks. Veterinarni Medicina. 2004;**49**(10):389

[8] Ramaswamy S, Raghavan P. Significance of impurity mineral identification in the value addition of kaolin–A case study with reference to an acidic kaolin from India. Journal of Minerals and Materials Characterization and Engineering. 2011;**10**(11):007-1025

[9] Abdou MI, Al-Sabagh AM, Dardir MM. Evalution of Egyptian bentonite and nano-bentonite as drilling mud. Journal of Petroleum. 2013;**25**:53-59

[10] Christidis GE. Physical and chemical properties of some bentonite deposits of kimolos island, Greece. Applied Clay Science. 1998;**13**(98):79-98

[11] Abubakar UAB, Yauri UAB, Faruz UZ, Noma SS, Sharif N. Characterization of Dabagi clay deposit for its ceramics potential. Journal of Environmental Science and Technology. 2014;**8**(8):455-459

[12] Schulze. Encyclopedia of soils in the environment. Clay Minerals. 2005;**1**:246-254

[13] Hillier S. On the mechanism of exfoliation of "vermiculite" clay minerals. 2013. African Journal of Environmental Science and Technology. 2016;17(9):819-830

[14] Belhouideg S, Lagache M. Prediction of the effective permeability coefficient in random porous media using the finite element method. Journal of Porous Media. 2014;**17**(9):819-830

[15] Abdel-Fattah MK, Merwad AMA. Effect of different sources of nitrogen fertilizers combined with vermiculite on productivity of wheat and availability of nitrogen in sandy soil in Egypt. American Journal of Plant Nutrition and Fertilization Technology 2015;**5**(2):50-60

[16] Ahn CH, Jong WH. Investigation of key parameters of rock cracking using the expansion of vermiculite materials.

Gyeonggi-do, Korea. Journals Materials. 2015;**8**(10):6950-6961

[17] Read HH. Certain physical properties of minerals. In: Rutley's Elements of Mineralogy. Dordrecht: Springer; 1970. pp. 48-72

[18] Ihekweme GO et al. Characterization of certain Nigerian clay minerals for water purification and other industrial applications. Heliyon. 2020;**6**(4):e03783

[19] Capedri S, Venturelli G, Photiades A. Assessory minerals. Journal of Culture. 2004:27-47

[20] Gaafar I, Cuney M, Gawad AA. Minerals chemistry of two mica granite rare metals: Impact of geophysics on the distribution of uranium mineralization at El Siela shear zone, Egypt. Journal of Geology. 2014;**4**:137-160

[21] Ako TA, Vishiti A, Ateh K, Kendia AC, Suh CE. Mineral alteration and chlorite geothermomentry in Platinum group elements(PGE)-Beearing meta-ultramafic Rocks from South East Cameroon. Journal of Geoscience Geomatics. 2015;**3**(4):96-108

[22] Saggerson EP, Tumer L. General comments on the identification of chlorites in tin sections. Mineral Magazine. 1982;**46**:468-473

[23] Albee AL. Relationship between the mineral association, physical properties of the chlorites series. American Mineralogist. 1962;**47**:851-870

[24] Adamis Z, Williams RB, Fodor J. Bentonite, Kaolin, and Selected Clay Minerals. Budapest, Hungary: World Health Organization; 2005;**9**(1):56-63

[25] Manukaji JU. An investigation into the use of local clays as a high temperature insulator for electric cookers. PhD Thesis mech. Eng. Dept. Federal University of Technology, Minna. International Journal of Engineering Research and Applications (IJERA). 2004;**3**(2):001-005

[26] Majeed MW. Effect of adding kaolin with natural and recycled coarse aggregates on asphalt mixture. Journal of Engineering and Sustainable Development. 2017;**21**(02):819-831

[27] Steel RF, Rf S. The interaction between kaolinite and *Staphylococcus aureus*. Journal of Pharmacy and Pharmacology. 1972;**24**:1-129

[28] Walsh PN. The effects of collagen and kaolin on the intrinsic coagulant activity of platelets. Evidence for an alternative pathway in intrinsic coagulation not requiring factor XII. British Journal of Haematology. 1972;**22**(4):393-405

[29] CIREP. Bentonite, Kaolin, and selected clay minerals. Identity. 2003;**2**:1

[30] Zhang L. Big data, knowledge mapping for sustainable development; a water quality index case study. Emergency Medicine. 2019;**3**(4):249-254

[31] Ismail S. Developing water quality index to access the quality of the drinking water. Civil Engineering. 2018;**4**:2345-2355

[32] WHO & UNICEF. 2001. Global Water Supply and Sanitation Assessment 2000 Report

[33] Lim SC, Gomes C, Kadir MZA. Characterizing of bentonite with chemical, physical and electrical prospectives for improvement of electrical grounding systems. International Journal of Electrochemical Science. 2013;**8**(9):11429-11447

[34] Sengco MR, Anderson DM. Controlling harmful algal blooms

through clay. Flocculation1. Journal of Eukaryotic Microbiology. 2004;**51**(2):169-172

[35] Wang Y, Guo L, Qi P, Liu X, Wei G. Synthesis of three-dimensional graphene-based hybrid materials for water purification: A review. Nanomaterials. 2019;**9**(8):1123

Chapter 13

Sedimentary Characteristics of the Member 1 of the Middle Devonian Suotoushan Formation and Its Ore-Bearing Potential in Northeastern Yunnan

Zizheng Wang, Xin-Sheng Jiang, Bang-Guo Zhou and Sheng-Lin Lu

Abstract

Super-large quartz sandstone deposit has been discovered in Northeastern Yunnan and the major ore-bearing bed is occurred in the Member 1 of the Middle Devonian Suotoushan Formation. Systematic study on the sedimentary characteristics of the Suotoushan Formation in Yiliang region, northeastern Yunnan Province has been conducted, Profile surveying, drilling and trenching show that, microscopic identification, the thickness of the Suotoushan Formation is 62.8 m ~ 222.7 m and principally composed of gray-white thick-bedded medium-grained quartz sandstone with well developed parallel bedding, swash bedding, oscillation cross-bedding and wave ripple, showing remarkable wave action features. Thin section observation shows that quartz sandstone in the Member 1 of the Suotoushan Formation is grain supported, siliceous cemented, concave-convex and line contact and the pressure solution and secondary overgrowth of quartz sand grains are obvious. Grain size analysis shows that the grain size is concentrated between 1.5Φ and 3.5Φ with saltation population up to 99% which is composed by tow sub populations and lacking of traction and suspension populations, showing a typical foreshore grain size distribution. Chemical analysis shows SiO_2 chemical purity of the quartz sandstone is 99.05 ~ 99.39 with weighted average 98.99 (N = 11) which has exceeded the industry standard for first-grade raw material of plate glass. Based our studies above, it can be concluded that the quartz sandstone in Member 1 of Suotoushan Formation is high energy coastal foreshore sediments with high texture and mineral maturity, and has been further purified by diagenesis, and both the grain-size characteristic and chemical purity have reached the standard of high-purity quartz sandstone, which can be expected raw material for high-purity quartz. The results in this article can provide accurate prospecting target to enlarge the resources potential of high-purity quartz deposits in northeastern Yunnan.

Keywords: Northeastern Yunnan, high-purity quartz sandstone, Middle Devonian, Suotoushan Formation, foreshore

1. Introduction

High purity quartz is a high-tech strategic mineral resource which is scarce in the world and in China ,and the quartz sandstone of littoral facies is a potential high-purity deposit [1]. It is the raw material of high-grade quartz products and the material basis of high-end products in silicon industry. It can be widely used in photovoltaic electronic information electric light source and optical communication industry. It plays an important role in strategic emerging industries such as new materials and new energy [2]. High purity quartz sandstone is a characteristic mineral in Wumeng mountain area, mainly distributed in northeast Yunnan province, which is the pillar mineral for zhaotong Government to build "Yunnan Silicon Valley" [3]. The current domestic production of high purity quartz sandstone strata in fluvial facies (Xiaodian in Suqian city, Sidaoliangzi in Inner Mongolia, Hewan in Lanzhou city), lacustrine facies (Dengjiatun in Shuangliao city, Yamenying and Zhangwutai in Tongliao city, Songfeng in Jiangxi), less littoral facies is found and always mainly loose quartz sand (Dongfang city, Beihai city, Dongshan city), it's low purity and impurity. The large-scale production of high-purity quartz sandstone produced in northeast Yunnan is restricted by the lag of mineral resources assessment and sedimentary facies research. Therefore, it's the strategic demand to carry out the potential evaluation of high-purity quartz ore in northeast Yunnan and expand its resource stock to support "Yunnan Silicon Valley". The Middle Devonian Suotoushan Formation distribute extensive in Yiliang area, northeast Yunnan, and it's an important potential area of high purity quartz sandstone deposit [4]. Previous studies based on small scale and limited outcrop sequence stratigraphy considered that the sedimentary facies of middle Devonian in northeast Yunnan is barrier coast facies [5], and more detailed sedimentary facies analysis has not been carried out systematically. Therefore, the author carried out a systematic study on the sedimentary characteristics of the Member 1 of Suotoushan Formation, collected 7 samples for grain size analysis and 22 samples for chemical analysis (**Figure 1**), and discussed the sedimentary facies characteristics of the first member of Suotoushan Formation and the genetic mechanism of the high purity quartz sandstone in northeast Yunnan.

The study area is located in the Kang-Dian fault depression belt at the western edge of the South China block, with kang-dian uplift in the west and Upper Yangtze uplift in the north, forming a southward opening bay, known as Zhaotong-Weining Depression (called Zhaotong Bay in this paper) [6]. In the late Middle Devonian, the study area is a littoral shallow sea clastic sedimentary area, mainly composed of quartz sand with high maturity in the littoral zone, distributed in a ring band on the east side of the Kang-Dian uplift and the southwest side of the Upper Yangtze uplift. Due to the influence of the Middle Devonian Haikou movement, small uplift is seen in the interior of the bay (**Figure 1**) [6]. Created by Xian Si-yuan in 1974 its mountain group name in Suotoushan of Qingmen in zhaotong city, the Middle Devonian strata group of Qingmen which is the shelf sea sedimentary contact the Late-Middle Devonian on Hongyapo group of tidal flat shelf sea sedimentary contact, main rock is fine-grained quartz sandstone, pure texture, well sorting and roundness. The lower and upper as a pure quartz sandstone, the middle part is sandstone with siltstone and mudstone [7]. The Suotoushan Formation in the area can be roughly divided into three members. The Member 1 and Member 3 of Suotoushan Formation are mainly quartz sandstone. The Member 2 of Suotoushan Formation contains more siltstone and mudstone. The quartz sandstone in the first member of Yutoushan Formation has the purest texture and is the host layer of high purity quartz sandstone in the study area.

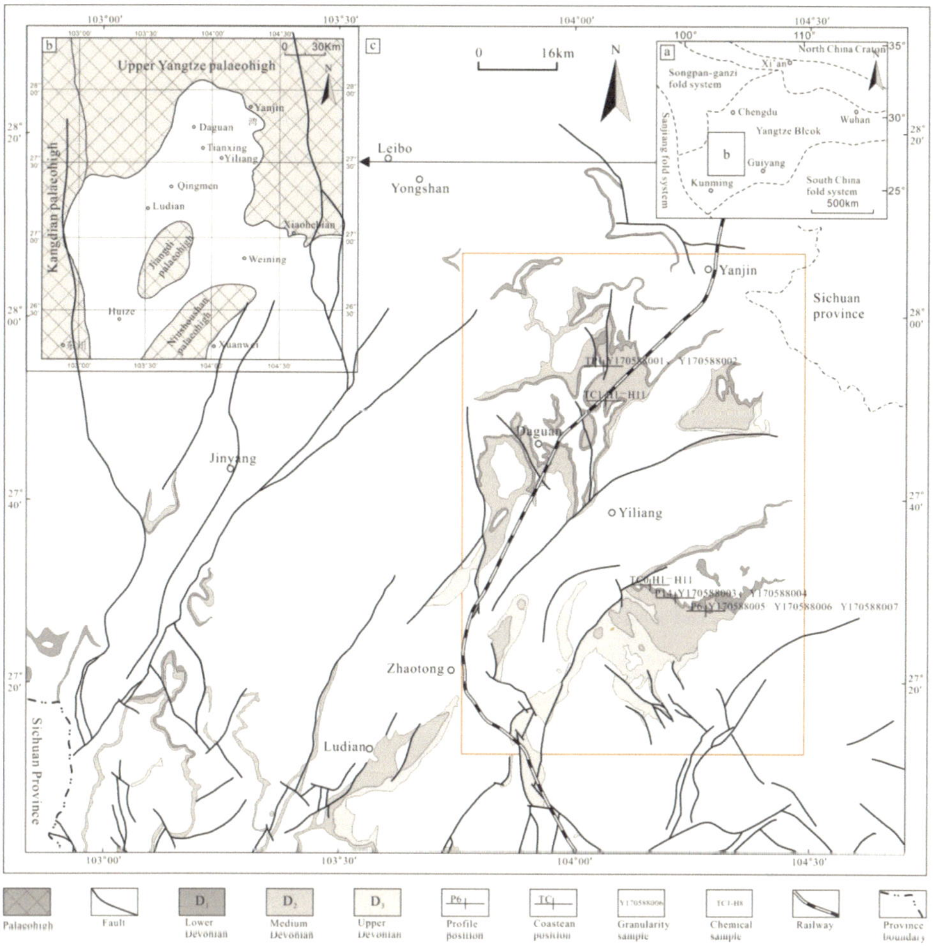

Figure 1.
Devonian tectonic setting, paleogeographic framework, brief geological map of Northeastern Yunnan and sampling position (after reference [5]).

2. Type section

In order to reveal the sedimentary characteristics and lithofacies paleogeographic evolution of quartz sandstone in the Member 1 of Suotoushan Formation in the study area, the author measured several sections, The type sections are listed as follows.

(1) The measured stratigraphic profile of Suotoushan Formation of Middle Devonian (D_2st) in Caijiawuji, Yiliang County (P14)	
Overlying strata: Hongyapo Formation of Middle Devonian (D_2hy)	
30. Dark gray shale, the upper part is gray white thin bedded quartz sandstone, containing mud and gravel, the grain size becomes coarser upward	0.9m
——————————conformable contact————————— —————	
Member 3 of Suotoushan Formation(D_2st^3)	
29. Dark gray shale, the upper part is gray white thin bedded quartz sandstone, containing mud gravel, upward grain size coarsening	9.8m

28. The grayish white thin bedded fine-grained siltstone sandstone intercalated with gray mud shale and wave-forming ripple developed	3.7m
27. The lower part is grayish white medium grained quartz sandstone with 8cm grayish yellow thin bedded argillaceous siltstone. the upper part is gray-white thin bedded fine grained quartz sandstone	8.8m
26. The gray-white thin bedded fine-grained quartz sandstone interbedded with dark gray mud shale in a ratio of approximately 3:1–5:1	30.7m
25. Gray-white middle-grained fine-grained quartz sandstone, with bluish silty mudstone at the bottom	8.8m
24. Gray-white thin bedded fine grained quartz sandstone	5.9m
23. Gray-white fine-grained quartz sandstone mixed with gray-green and purple mudstone, mud shale	12.6m
22. Gray thick bedded fine-grained quartz sandstone with cross bedding	14.4m
21. Gray thin-medium grained quartz sandstone, with argillaceous masses, oblique bedding and scour surface, gray thin-medium lithic quartz sandstone, silty mudstone intercalated mudstone can be seen locally	11m
20. Greyish-white thick bed – massive fine grained quartz sandstone	4.5m
——————————conformable contact———————— —————	
Member 2 of Suotoushan Formation(D_2st^2)	
19. Interbedded gray mudstone and gray-yellow medium siltstone with unequal thickness	20.1m
18. Gray – white meso-grained quartz sandstone developed in parallel bedding	36.5m
17. The lower part is gray and black-dark, yellow-gray, green mud shale, the upper part is dark gray mudstone	39.2m
16. Greyish-white mesomorphic powder-fine quartz sandstone	2.5m
15. Thin bedded mudstone and silty mudstone interspersed with gray-white medium – grained quartz sandstone	57.3m
14. The lower part is dark-gray thin-medium quartz siltstone and dark-gray argillaceous siltstone, and the middle and upper part is gray-yellow medium quartz siltstone. Pyrite clumps are seen in the quartz siltstone at the top	20.7m
13. The lower part is grayish yellow medium lithic siltstone, and the middle and upper part is grayish white medium fine siltstone sandstone	14.7m
12. Dark gray mudstone	14.9m
——————————conformable contact———————— —————	
Member 1 of Suotoushan Formation(D_2st^1)	
11. Gray-white thick bedded fine grained quartz sandstone	6.8m
10. The lower is yellow-gray medium siltstone and gray-white thick bedded fine grained quartz sandstone, the middle is covered by mantle rock, the upper is gray thin bedded argillaceous siltstone and gray-white thin bedded fine-grained quartz sandstone	45.6m
9. The gray-white thick bedded medium-grained quartz sandstone.the lower part contains 1–3cm purplish thin mudstone and 30cm argillaceous siltstone	31.1m
8. Gray-white thick bedded fine grained quartz sandstone	43.5m
7. Gray-white thick bedded fine-grained quartz sandstone with gray-white thin bedded fine-grained quartz sandstone in the middle, grayish yellow silty mudstone and mid-bedded fine-grained quartz sandstone can be seen at the top	4.4m
6. Gray-white medium-thick bedded fine-grained silty quartz sandstone	7m
5. Gray-white thick bedded fine grained quartz sandstone	28.8m

4. Gra-white thick bedded fine-grained quartz sandstone Light flesh-red, grayish white thick bedded medium-grained quartz sandstone, occasionally parallel bedded. The upper quartz sandstone is obliquely bedded. Gray argillaceous siltstone about 60cm in the lower part, gray-green mudstone 30–40cm in the middle, and purple red quartz sandstone 50cm thick in the top	48.2m
3. Gray-white thick bedded medium-fine grained quartz sandstone, with 15cm iron bearing fine grained quartz sandstone at the bottom. The upper medium-grained quartz sandstone reaches the industrial standard of silica ore	6.1m
——————————conformable contact——————————	
Lower strata: Middle Devonian Qingmen Formation (D2*qm*)	
2. Gray-green thin bedded siltstone intermixed with dark gray shale, 3–6cm purplish red argillaceous siltstone at the top, and 1.2m dark gray argillaceous mass at the cross section	1.2m
(2) Measured stratigraphic profile of Yutoushan Formation of Middle Devonian (D_2*st*) in Songlinpo, Yiliang County(P6)	
Overlying strata: Hongyapo Formation of Middle Devonian (D_2*hy*)	
24. Purplish-red and gray-green silty mudstone interbedded with grey thin bedded quartz silt-fine sandstone. The ratio of the two lithologies is about 1:1	9.77m
——————————conformable contact——————————	
Member 3 of Suotoushan Formation(D_2st^3)	
23. Gray-white thin-medium grained quartz sandstone	1.26m
22. Gray, thin-medium quartz sandstone. The quartz sand layers are interspersed with gray-green and 1–8cm purplish red thin-layer siltstone. From bottom to top, quartz sandstone becomes thinner and siltstone thickens. In the middle and upper part, the ratio of quartz sandstone to gray-green and purple siltstone approaches 1:1	2.84m
21. Light gray thick bed to massive quartz sandstone	5.77m
20. The bottom is gray-white thin bedded fine-grained quartz sandstone(gravels at the bottom), the middle is gray-yellow thin bedded siltstone, the top is gray-green thin bedded mudstone	1.29m
——————————conformable contact——————————	
Member 2 of Suotoushan Formation(D_2st^2)	
19. Gray white thin to medium grained fine-grained quartz sandstone interbedded with mud shale and very thin argillaceous siltstone	1.03m
18. Thick layers of gray-white massive fine grained quartz sandstone. 40cm thick gray-green silty mudstone in the middle and upper part, and it is separated from the next layer by 1–3cm thick gray-green silty mudstone at the top	24.63m
17. Light grey mudstone interbedded with grey white quartz siltstone	0.58m
16. Greyish medium grained quartz sandstone	2.02m
15. Thick gray-white massive fine and medium-grained quartz sandstone	20.73m
14. It is a set of siltstone, quartz sandstone, argillaceous siltstone and silty mudstone assemblage, which can be divided into five parts. The bottom is purplish red siltstone, the middle and lower part is gray white thin bedded quartz sandstone, the middle part is sandwiched with 50cm thick gray white thin bedded argillaceous siltstone, the middle and upper part is purple red thin bedded argillaceous siltstone, the top is gray green thin bedded argillaceous mudstone	5.37m
——————————conformable contact——————————	
Member 1 of Suotoushan Formation(D_2st^1)	
13. Thick layer of gray-white massive fine grained quartz sandstone	12.4m
12. Light grey thin – mesomorphic fine-grained quartz sandstone bounded by grey-green extremely thin mudstone at the bottom	5.72m

11. Grayish-white medium-thick fine-grained quartz sandstone	0.71m
10. Gray, thin – medium quartz sandstone	3.57m
9. This layer is a combination of mud shale and siltstone, true thickness of this layer is about 70cm. It can be divided into three parts. The bottom (25cm) is gray green thin layered mud shale; The middle part(25cm) is gray-yellow medium-lithic quartz sandstone. The upper part (25cm) is purplish red thin bedded siltstone	0.67m
8. Greyish medium – thick fine grained quartz sandstone	1.8m
7. Gray-yellow thin – intermediate lithic quartz sandstone	3.14m
6. Thick greyish – massive fine grained quartz sandstone	0.9m
5. Grayish white thin – medium grained quartz sandstone interbedded with very thin grayish green mudstone	10.53m
4. Gray thin bedded fine grained quartz sandstone	4.66m
——————————conformable contact———————— —————	
Lower strata: Qingmen Formation of Middle Devonian (D_2*qm*)	
3. Gray-green silty mudstone	5.5m
(3) Measured stratigraphic profile of Suotoushan Formation (D_2*st*) in Haibatou, Daguan County(TP1)	
Overlying strata: Hongyapo Formation of Middle Devonian (D_2*hy*)	
13. Gray-green mudstone and silty mudstone interspersed with a small amount of gray-white thin bedded fine-grained quartz sandstone	28.40m
——————————conformable contact———————— —————	
Member 3 of Suotoushan Formation(D_2st^3)	
12. Gray – white thick bedded fine grained quartz sandstone	14.0m
——————————conformable contact———————— —————	
Member 2 of Suotoushan Formation(D_2st^2)	
11. Gray mudstone with little iron	1.30m
10. Gray thin bedded fine grained quartz sandstone	0.80m
9. Gray mudstone with little iron	0.90m
——————————conformable contact———————— —————	
Member 1 of Suotoushan Formation(D_2st^1)	
8. Gray medium-thick bedded fine-grained quartz sandstone, little pyrite nodules included	8.10m
7. Gray thick and blocky fine-grained quartz sandstone. fossil plant fragment included, large oblique bedding developed	16.50m
6. Gray medium-thick fine grained quartz sandstone	5.30m
5. Thin-bedded fine-grained quartz sandstone interbedded with iron argillaceous quartz siltstone of unequal thickness	8.40m
4. Thick gray blocky fine-grained quartz sandstone, horizontal bedding developed	19.60m
3. Gray thin bedded fine-grained quartz sandstone	2.10m
2. Gray medium-thick fine-grained quartz sandstone	2.80m
——————————conformable contact———————— —————	
Lower strata: Qingmen Formation of Middle Devonian (D_2*qm*)	
1. Gray silty mudstone	14.60m

From the above three representative measured sections, the boundary between top and bottom of Suotoushan Formation is clearly marked and the internal tripartite is extremely obvious, which can be compared regionally (**Figure 2**). The Member 1 of Suotoushan Formation is mainly composed of gray-white medium-thick bedded quartz sandstone with relatively few muddy interbeds, which is a concentrated production layer of high purity quartz sandstone. Many muddy interlayers in Member 2 of Suotoushan Formation, and the quartz sandstone is dark in color and mainly thin bedded, which is not the enrichment layer of high purity quartz sandstone. Although Member 3 of Suotoushan Formation contains more quartz sandstone, there are more mud interlayers, which is not the ideal formation for high purity quartz sandstone production. Therefore, only the sedimentary structure, tectonic characteristics and sedimentary environment of Member 1 of Suotoushan Formation are discussed detailedly in this paper.

3. Sediment structure

Member 1 of Suotoushan Formation in the study area is mainly gray-white medium thick bedded quartz sandstone. Microscopic identification results show that the minerals are mainly medium-fine quartz, with a content of 99% the roundness is subedge, grain support, Moderate sorted, concave-line contact, and the cementation mode is press-inlay – pore type. Field profile shows the Member 1 of Suotoushan Formation is dominated by gray-white medium-thick bedded quartz sandstone with sheet sand body. Parallel bedding (**Figure 3a**) and wash bedding (**Figure 3b**) are extremely well developed. Straight ridge wave-forming marks (**Figure 3c**), small flowing wave marks and exfoliated lineation can be seen on the plane. Skolithos in the vertical plane can also be seen in the Luohanba section (**Figure 3e**). Sedimentological studies show [8, 9] that parallel bedding, wash bedding and wave-forming ripple marks are typical structures of the foreshore wash zone, while small flow ripple marks are products of wave backflow deposition. Therefore, it can be considered that the sedimentary environment of Member 1 of Suotoushan formation is the foreshore zone. The existence of Skolithos [10] also indicates that the sedimentary environment of Member 1 of Suotoushan Formation is dominated by the foreshore zone.

4. Grain size analysis

The grain size distribution characteristics of sedimentary rocks can reflect the hydrodynamic conditions of sedimentation, and the results of grain size analysis are an important means to distinguish the sedimentary environment [11]. Sediment can be transported in three ways in normal traction flow: suspension, saltation, bottom passage or traction. Different sedimentary environments have different transport modes, among which the eolian dune sand and beach sand have narrow grain size interval, saltation is specially developed in general, while the passage is not developed in general and there is a lack of suspension. The biggest characteristic of beach sand is that there is a cut-off point inside the saltation whole, which represents the slight change of wave round-trip energy [11]. The quartz sandstone in the Member 1 of Suotoushan Formation is mostly medium and fine grained quartz sandstone

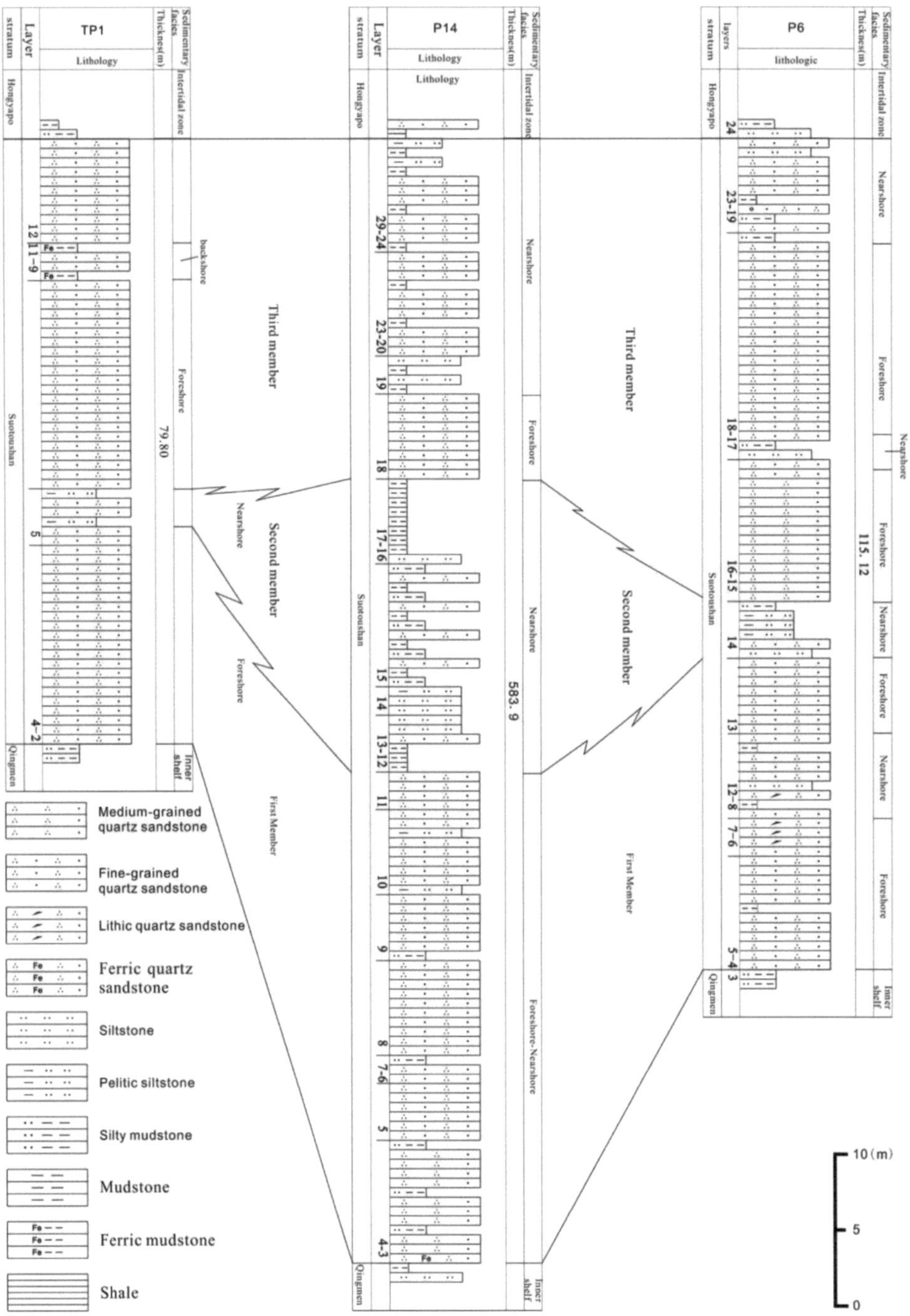

Figure 2.
Histogram correlation diagram for the Middle Devonian Suotoushan Formation in Northeastern Yunnan.

(**Figure 3f**), with excellent sorting, grain support, voidaceous cementation, and void filled by secondary enlarged quartz (**Figure 3g**). Its mineral maturity and structural maturity are extremely high. In order to quantitatively reveal the grain size distribution characteristics of quartz sandstone in the Member 1 of Suotoushan Formation,

DOI: http://dx.doi.org/10.5772/intechopen.103826

Figure 3.
Typical sedimentary texture and structure of the Member 1of the Suotoushan Formation.

Sample	Y170588001	Y170588002	Y170588003	Y170588004	Y170588005	Y170588006	Y170588007
separation coefficient(S_0)	0.78	0.90	0.89	0.80	0.83	0.81	0.92
Standard deviation (σ)	0.41	0.35	0.38	0.34	0.32	0.64	0.81
Skewness (SK)	0.37	0.23	0.03	0.23	0.14	−0.01	4.76
Kurtosis (K)	2.92	3.22	3.52	2.90	3.11	3.07	29.92
Average (Mz)	2.31	2.27	2.13	2.25	2.40	1.99	3.01
Coarse sand($0 < \Phi < 1$)	0.00per	0.00per	0.00per	0.00per	0.00per	6.06per	0.00per
Medium sand($1 < \Phi < 2$)	23.24per	21.61per	34.76per	25.39per	9.77per	43.27per	0.00per
Fine sand($2 < \Phi < 3$)	71.63per	76.63per	63.61per	73.38per	86.83per	45.05per	52.15per
Very fine sand($3 < \Phi < 4$)	5.12	1.76per	1.63per	1.23per	3.39per	5.59	45.03per
Coarse silt($4 < \Phi < 5$)	0.01per	0.00per	0.00per	0.00per	0.00per	0.00per	0.82per
Fine powder sand($5 < \Phi < 8$)	0.00per	0.00per	0.00per	0.00per	0.00per	0.00per	0.00per
Clay($\geq 8\Phi$)	0.00per	0.00per	0.00per	0.00per	0.00per	0.00per	0.00per

Table 1.
Grain size parameters of the quartz sandstone from the Member 1 of the Suotoushan Formation.

seven samples were randomly selected for grain size analysis in this paper (**Table 1**, **Figure 4a–c**).

The results showed that the samples were dominated by fine sand, followed by medium sand, containing a certain proportion of very fine sand, most of which did not contain coarse silt, only one sample detected coarse sand (Y170588006, 6.06per), completely missing gravel particles, silt and clay particles. The separation coefficient (S_0) is between 0.78 and 0.92. Comparison sorting coefficient grading standard (Fine: 1–1.23; Good: 1.23–1.41; Medium: 1.41 to 1.74; Poor: 1.74–2.0; Bad: > 2.0) [11], and the

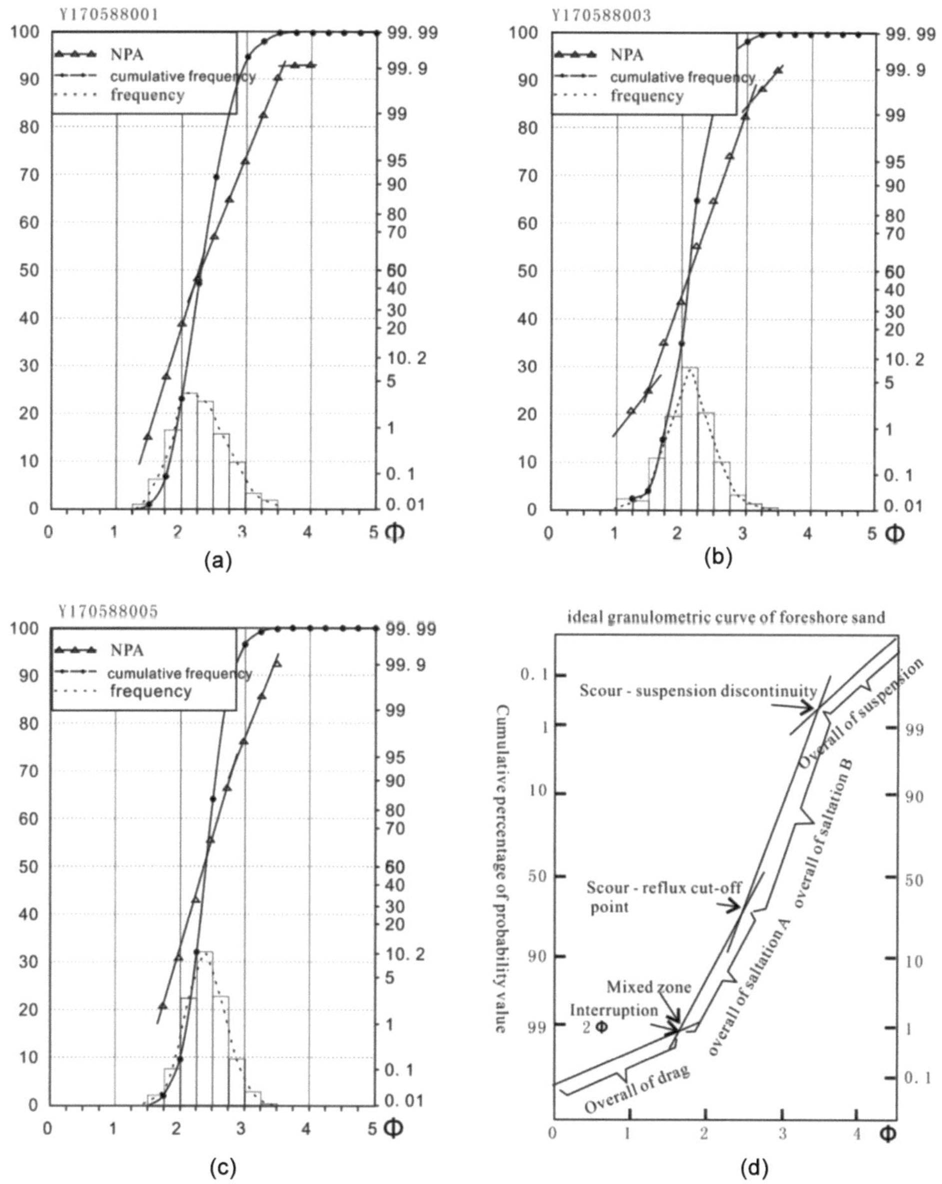

Figure 4.
Grain size analysis diagram of the quartz sandstone from the Member 1 of the Suotoushan Formation (diagram (d) is after reference [11]) (NPA-Normal probability accumulation).

sorting degree of the analyzed samples was better than "very good" grade. Standard deviation (σ) can also describe the sorting degree, with most samples ranging from good (0.35–0.50) to excellent (< 0.35). The skewness (SK) values of the samples were 2 near symmetrical ($-0.10 \sim +0.10$), 3 positive skewness ($+0.10 \sim +0.30$) and 2 polar positive skewness ($+3 \sim +1.0$), and the grain size peak value was inclined to the fine end. It is generally believed that beach sand is mostly negatively skewed and dune sand is mostly positively skewed [11]. The characteristics of sample skewness of the Member 1 of Suotoushan Formation may be related to the dissolution and fining of quartz particles during diagenesis. Kurtosis mainly describes the convexity of grain size curve, so as to reflect the concentration degree of grain size, and is also an indicator of the degree of sediment sorting. The classification levels are: very wide (< 0.67 =, wide ($0.67 \sim 0.90$), narrow ($1.11 \sim 1.50$), very narrow ($1.50 \sim 3.00$), and very narrow (> 3.0) [11]. The kurtosis of the quartz sand in the Member 1 of Suotoushan Formation is very narrow and the grain size distribution is unusually concentrated, indicating that the quartz sand has undergone very sufficient sorting and transformation, thus laying the foundation of high purity quartz sandstone. Appears above a mountain group quartz sandstone grain size analysis results show that the majority of samples to a single modal nearly normal distribution, the lack of time overall, overall and suspended saltation overall content above 99per, and can distinguish two general, show wave flow and reflow sedimentary characteristics, and the ideal of foreshore with particle size distribution (**Figure 4d**). It should be noted that the internal intercept points of the saltation of quartz sandstone in the Member 1 of Suotoushan Formation are not clear, indicating that the sorting ability of quartz sandstone is close to eolian dune sand and superior to ordinary beach sand. This is also an obvious feature of quartz sand in the Member 1 of Suotoushan Formation, which is different from other quartz sandstone along the shore.

Member 3 of Suotoushan Fomation	**Sample**	**SiO_2**	**Al_2O_3**	**Fe_2O_3**	**Member 1 of Suotoushan Fomation**	**Sample**	**SiO_2**	**Al_2O_3**	**Fe_2O_3**
	TC1-H1	90.21	4.20	2.35		TC0-H1	99.07	0.22	0.25
	TC1-H2	94.64	1.88	2.29		TC0-H2	99.05	0.21	0.23
	TC1-H3	97.08	0.41	2.25		TC0-H3	98.83	0.16	0.51
	TC1-H4	98.28	0.45	1.07		TC0-H4	99.14	0.26	0.26
	TC1-H5	98.18	0.33	1.32		TC0-H5	99.35	0.21	0.10
	TC1-H6	97.31	0.37	2.11		TC0-H6	99.26	0.19	0.14
	TC1-H7	97.84	0.40	1.48		TC0-H7	99.39	0.14	0.14
	TC1-H8	97.02	0.63	2.05		TC0-H8	99.20	0.14	0.32
	TC1-H9	98.50	0.56	0.60		TC0-H9	98.97	0.22	0.42
	TC1-H10	90.20	4.65	2.07		TC0-H10	98.28	0.43	0.33
	TC1-H11	98.06	0.29	1.51		TC0-H11	98.34	0.48	0.32
Industry standard of Flat glass	Superior	≥98.5	≤1.0	≤0.05	Standard according to the quality requirements of silicon materials for flat glass **(DZ/T027-2002)**				
	Level 1	≥98.0	≤1.0	≤0.10					

Table 2.
Chemical analysis data of the Member 1 and Member 3 quartz sandstone of the Soutoushan Formation (%).

5. Chemical purity

Chemical purity is one of the evaluation factors of high purity quartz sand [1], and also a standard to measure whether it meets the plate glass industry index [12]. In view of this, the chemical composition of quartz sandstone in the first and Member 3 of Suotoushan Formation in the study area was analyzed (**Table 2**). The analysis results show that, except for a small amount of Fe_2O_3, the SiO_2 index of the Member 1 of Suotoushan Formation quartz sandstone exceeds the standard of the first grade industrial index for flat glass, while the Fe_2O_3 and SiO_2 quartz sandstone of the Member 3 of Suotoushan Formation basically cannot reach the standard of the first grade. Therefore, the Member 1 of Suotoushan Formation is a favorable horizon for high purity quartz sandstone deposit.

6. Discussion

6.1 Correlation with grain size characteristics of high-purity quartz sand

The grain size and its distribution are two of the four basic characteristics of high purity quartz sand (chemical purity, grain size, grain size distribution and particle shape) [1] and an important factor which directly affects the equilibrium melting of high purity quartz sand and a material control index to reduce the defects of quartz glass [13]. Therefore, Shi Bin et al. established a set of grain size parameter evaluation system for high purity quartz sand [14]. The grain size parameters of typical high-purity quartz sand are as follows: average grain size (Mz) = 3.125Φ; Standard deviation (σ) = 0.274; Partial degrees (SK) = 0.000. The grain size parameters of samples (ZOZ) are as follows: average grain size (Mz) = 2.64Φ; Standard deviation (σ) = 0.33; Partial degrees (SK) = 0.36. The comparison results show that the grain size, grain size distribution and grain size curve of quartz sand in the Member 1 of Suotoushan Formation reach or approach the index of high quality and high purity quartz sandstone. Considering the common grain size characteristics of foreshore deposits, the quartz sandstone of the Member 1 of Suotoushan Formation should be the most ideal high purity quartz sandstone.

6.2 Relationship between sea level rise and high purity quartz sandstone

The Qingmen Formation in the lower strata of Suotoushan Formation is dominated by gray-green, yellow-green and yellow-gray mudstone, intermixed with sandy mudstone, siltstone, argillaceous limestone and limestone. The fossils of normal shallow sea creatures are abundant, mainly brachiopods and corals. Trilobites, ammonites and bambusitic stones are found in the shallow sea shelf environment [7]. The Member 1 of Suotoushan Formation shoreline quartz sandstone is obviously lower than Qingmen Formation shallow sea shelf deposition, which records an event of sea level decline and shallow sea shelf uplift and exposure. Extensive shelf exposure can lead to quartz sand sorting, transformation and redeposition, which may be one of the reasons for the high structural and compositional maturity of quartz sandstone in the Member 1 of Suotoushan Formation. The Member 1 of Suotoushan Formation in the study area is gray medium fine quartz sandstone, with almost no argillaceous interlayer, indicating that the sedimentation of sedimentary basement and sedimentary filling were in a long-term equilibrium state, and the sedimentary environment was always in the foreshore zone, which was conducive to the full sorting and enrichment of quartz sand. The Member 2 of Suotoushan Formation is mainly composed of thin bedded sandstone and mudstone, which is estimated to be littoral zone-inner shelf deposition, indicating that the sea level begins to rise and the weakening of hydrodynamic conditions is not conducive to quartz sand sorting and enrichment. Although the sea level of the Member 3 of Suotoushan Formation fell again during the sedimentary period, the sedimentary environment was not stable and the medium thickness bedded quartz sandstone was not developed, which was not conducive to the formation of high purity quartz sandstone deposits. Chemical analysis results (**Table 2**) also prove that the purity of quartz sandstone in the Member 1 of Suotoushan Formation is higher than that in the Member 3 of Suotoushan Formation.

6.3 Diagenetic purification process of quartz sandstone in the Member 1 of Suotoushan Formation

The microscopic observation shows that the contact relationship between quartz particles in the Member 1 of Suotoushan Formation has developed from simple grain support to concave and convex intercalation, and the intergranular pores are filled by secondary expansion edge of quartz. It is generally believed that pressure dissolution first occurs at the contact position of quartz particles. Simple pressure can locally improve the solubility of minerals or rocks, and pressure on a certain point on the surface of dry solid material can improve the solubility of the material at that point, which is the basic theoretical basis of pressure dissolution [15]. In the process of pressure dissolution, elements such as silicon, aluminum, sodium, potassium and calcium enter pore water with mineral dissolution, leading to material redistribution [15]. SiO_2 can be precipitated around quartz sand particles again, forming quartz enlarged edge, while other materials may be taken away with pore water, further purification of quartz sandstone. Pressure solution phenomenon not only exists between particles, but also generally exists between layers and cracks. Sutures and pressure solution pits are relatively developed (**Figure 3h**). It can be seen that the quartz sandstone of the Member 1 of Suotoushan Formation has undergone the diagenetic process of pressure dissolution and SiO_2 reprecipitation, which undoubtedly further purified the quartz sandstone of the Member 1 of Suotoushan Formation. It can be considered that the high purity quartz sandstone of the Member 1 of Suotoushan Formation not only has the deposition of high purity quartz sand but also has the diagenetic purification process. The specific diagenetic purification process and mechanism need to be further studied.

7. Conclusion

1. The Member 1 of Suotoushan Formation is a quartz sandstone deposit with high mineral maturity and structural maturity in the foreshore zone, which is a favorable horizon for the exploration of high-purity quartz sandstone deposit.

2. The high-purity quartz sandstone of the Member 1 of Suotoushan Formation is recycled quartz sand, which is the result of uplifting, exposure and redeposition of the underlying shallow sea shelf.

DOI: http://dx.doi.org/10.5772/intechopen.103826

3. The diagenesis (pressure dissolution and siliceous cementation) of the Member 1 of Suotoushan Formation improves the purity of SiO_2 further.

4. The grain size characteristics of the Member 1 of Suotoushan Formation are up to or close to the index of high quality and high purity quartz sandstone, and the chemical purity exceeds the standard of the industrial index Grade I for flat glass, so it can be used as the raw material of high-purity quartz sand.

Author details

Zizheng Wang*, Xin-Sheng Jiang, Bang-Guo Zhou and Sheng-Lin Lu
Chengdu Center, China Geological Survey, Chengdu, China

*Address all correspondence to: 15523327@qq.com

References

[1] Zhang P-C, Liu X-F, Li J-F, et al. Study on high-purity quartz mineral resource engineering. Mineralogy and Petrology. 2012;**32**(2):38-44 (in Chinese with English abstract)

[2] Wang L, Dang C-P, Li C-X, et al. Technology of high-puiry quartz in China: Status quo and prospect. Earthscience Frontiers. 2014;**21**(5):267-273 (in Chinese with English abstract)

[3] Jiang W-T. Study on geological characteristics and beneficiability of quartz sandstone in Yiliang County, Yunnan Province. China Non-metallic Minerals Industry. 2016;**12102**:10-13 (in Chinese)

[4] Qiu Z-J, Jiang W-T. Analysis of quartz sandstone prospect in Yiliang County, Yunnan Province. China Non-metallic Minerals Industry. 2016; **124**(04):9-21 (in Chinese)

[5] Zeng Y-F, Zhang J-Q, Liu W-J, et al. Devonian lithofacies paleogeography and mineralization in south China. Beijing: Geological Publishing House; 1993. pp. 1-123 (in Chinese)

[6] Zheng R-C, Zhang J-Q. The tectonic framework and the evolution of lithofacies and paleogeography of Devonian in eastern Yunnan and southwestern Guizhou. Journal of Chengdu College of Geology. 1989;**16**(4): 51-60 (in Chinese with English abstract)

[7] Yunnan Bureau of Geology and Mineral Resources. Petrostratigraphy of Yunnan Province. Wuhan: China University of Geosciences Press; 1980. pp. 1-366 (in Chinese)

[8] Liu B-J. Sedimentary Petrology. Beijing: Geological Publishing House; 1980. pp. 1-497 (in Chinese)

[9] Liu B-J, Zeng Y-F. Basic and Working Methods of Lithofacies Paleogeography. Beijing: Geological Publishing House; 1985. pp. 1-442 (in Chinese)

[10] Yang S-B, Zhang J-P, Yang M-F. Trace Fossils of China. Beijing: Science Press; 2004. pp. 1-353 (in Chinese)

[11] Team of ShanBei, Chengdu College of Geology. Grain-size Analysis of Sedimentary Rocks and Its Application. Beijing: Geological Publishing House; 1978. pp. 1-147 (in Chinese)

[12] Ministry of Land and Resources, PRC. Glassy siliceous raw materials—General industrial requirements—Geological and Mineral Industry standards of the People's Republic of China (DZ/T 0207-2002). 2002:569–570 (in Chinese)

[13] Hu X-Q, Li G-L, Sun M, et al. Crystal and grain size distribution control of silica powder for quartz glass. China Non-metallic Minerals Industry. 2007; **26**(4):41-43 (in Chinese)

[14] Shi B, Liu Y-F, Zhang P-C, et al. Grain-size analysis and grain evaluation system of the super high quartz sand. Mineralogy and Petrology. 2013;**33**(01): 16-21 (in Chinese with English abstract)

[15] Liu B-J, Zhang J-Q. Sedimentary Digenesis. Beijing: Science Press; 1993. pp. 1-271 (in Chinese)